Method and appraisal in the physical sciences

Method and appraisal in the physical sciences

The critical background to modern science, 1800-1905

EDITED BY

COLIN HOWSON

DEPARTMENT OF PHILOSOPHY, LOGIC AND SCIENTIFIC METHOD, LONDON SCHOOL OF ECONOMICS

CAMBRIDGE UNIVERSITY PRESS

CAMBRIDGE

LONDON · NEW YORK · MELBOURNE

Published by the Syndics of the Cambridge University Press
The Pitt Building, Trumpington Street, Cambridge CB2 1RP
Bentley House, 200 Euston Road, London NW1 2DB
32 East 57th Street, New York, NY 10022, USA
296 Beaconsfield Parade, Middle Park, Melbourne 3206, Australia

First published 1976

Printed in Great Britain at the
University Printing House, Cambridge
(Euan Phillips, University Printer)

Library of Congress Cataloguing in Publication Data

Main entry under title:
Method and appraisal in the physical sciences.

Bibliography: p.
Includes index.
CONTENTS: Lakatos, I. History of science and its rational reconstructions.
–Clark, P. Atomism versus thermodynamics.–Worrall, J.
Thomas Young and the 'refutation' of Newtonian optics. [etc.]
1. Science–Methodology. 2. Science–History. I. Howson, Colin.
Q 175.M5417 500.2'01 75–44580
ISBN 0 521 21110 7

Contents

Editorial preface

This volume constitutes the first collected edition of work so far done in illustrating an important new development in the philosophy of science, 'the methodology of scientific research programmes', with case studies drawn from the history of the physical sciences. This material, no doubt the forerunner of more complete accounts of the fit between Lakatos's ideas and scientific practice, is prefaced with an exposition of the methodology of scientific research programmes by its author, Imre Lakatos, who, sadly, died before the material was assembled in this volume; and there is a concluding critical appraisal both of Lakatos's theory and of the illustrations of it, by Paul Feyerabend.

Briefer versions of Clark's, Musgrave's and Worrall's papers were delivered at a conference on *Research Programmes in Physics and Economics* in Nafplion, Greece in September 1974. Lakatos's paper was published originally in *Boston Studies in the Philosophy of Science*, VII; Zahar's paper was first published in the *British Journal for the Philosophy of Science*, **24**, 1973.

I should like to thank Miss Gillian Page and D. Reidel Publishing Company for permission to reprint Lakatos's paper, and the author and editor of the *British Journal for the Philosophy of Science* for permission to reprint Zahar's paper. Finally, I should like to express my profound gratitude to the secretaries in the Philosophy Department at the London School of Economics and Political Science, and particularly Anne Smith, for their help in preparing these papers for publication, and to thank Mr Alex Bellamy for compiling the index to this volume.

L.S.E., June 1975 C. HOWSON

History of science and its rational reconstructions*

IMRE LAKATOS

Introduction

Introduction

'Philosophy of science without history of science is empty; history of science without philosophy of science is blind.' Taking its cue from this paraphrase of Kant's famous dictum, this paper intends to explain *how* the historiography of science should learn from the philosophy of science and *vice versa*. It will be argued that (*a*) philosophy of science provides normative methodologies in terms of which the historian reconstructs 'internal history' and thereby provides a rational explanation of the growth of objective knowledge; (*b*) two competing methodologies can be evaluated with the help of (normatively interpreted) history; (*c*) any rational reconstruction of history needs to be supplemented by an empirical (socio-psychological) 'external history'.

The vital demarcation between normative–internal and empirical–external is different for each methodology. Jointly, internal and external historiographical theories determine to a very large extent the choice of problems for the historian. But some of external history's most crucial

* Earlier versions of this paper were read and criticised by Colin Howson, Alan Musgrave, John Watkins, Elie Zahar, and especially John Worrall. The present paper further develops some of the theses proposed in my [1970]. I have tried, at the cost of some repetition, to make it self-contained.

problems can be formulated only in terms of one's methodology; thus internal history, so defined, is primary, and external history only secondary. Indeed, in view of the autonomy of internal (but not of external) history, external history is irrelevant for the understanding of science.[1]

1 *Rival methodologies of science; rational reconstructions as guides to history*

There are several methodologies afloat in contemporary philosophy of science; but they are all very different from what used to be understood by 'methodology' in the seventeenth or even eighteenth century. Then it was hoped that methodology would provide scientists with a mechanical book of rules for solving problems. This hope has now been given up: modern methodologies or 'logics of discovery' consist merely of a set of (possibly not even tightly knit, let alone mechanical) rules for the *appraisal* of ready, articulated theories.[2] Often these rules, or systems of appraisal, also serve as 'theories of scientific rationality', 'demarcation criteria' or 'definitions of science'.[3] Outside the legislative domain of these normative rules there is, of course, an empirical psychology and sociology of discovery.

I shall now sketch four different 'logics of discovery'. Each will be characterised by rules governing the (scientific) *acceptance* and *rejection* of theories or research programmes.[4] These rules have a double function. Firstly, they function as *a code of scientific honesty* whose violation is intolerable; secondly, as hard cores of (*normative*) *historiographical research programmes*. It is their second function on which I should like to concentrate.

A. *Inductivism*

One of the most influential methodologies of science has been inductivism. According to inductivism only those propositions can be accepted into the body of science which either describe hard facts or are infallible inductive generalisations from them.[5] When the inductivist *accepts* a scientific proposition, he accepts it as provenly true; he *rejects*

[1] 'Internal history' is usually defined as intellectual history; 'external history' as social history (cf. e.g. Kuhn [1968]). My unorthodox, new demarcation between 'internal' and 'external' history constitutes a considerable problemshift and may sound dogmatic. But my definitions form the hard core of a historiographical research programme; their evaluation is part and parcel of the evaluation of the fertility of the whole programme.

[2] This is an all-important shift in the problem of normative philosophy of science. The term 'normative' no longer means rules for arriving at solutions, but merely directions for the appraisal of solutions already there. Thus *methodology* is separated from *heuristics*, rather as value judgments are from 'ought' statements. (I owe this analogy to John Watkins.)

[3] This profusion of synonyms has proved to be rather confusing.

[4] The epistemological significance of scientific 'acceptance' and 'rejection' is, as we shall see, far from being the same in the four methodologies to be discussed.

[5] '*Neo*-inductivism' demands only (provably) highly probable generalisations. In what follows I shall only discuss classical inductivism; but the watered down neo-inductivist variant can be similarly dealt with.

it if it is not. His scientific rigour is strict: a proposition must be either proven from facts, or – deductively of inductively – derived from other propositions already proven.

Each methodology has its specific epistemological and logical problems. For example, inductivism has to establish with certainty the truth of 'factual' ('basic') propositions and the validity of inductive inferences. Some philosophers get so preoccupied with their epistomological and logical problems that they never get to the point of becoming interested in actual history; if actual history does not fit their standards they may even have the temerity to propose that we start the whole business of science anew. Some others take some crude solution of these logical and epistemological problems for granted and devote themselves to a rational reconstruction of history without being aware of the logico-epistomological weakness (or, even, untenability) of their methodology.[6]

Inductivist criticism is primarily sceptical: it consists in showing that a proposition is unproven, that is, pseudoscientific, rather than in showing that it is false.[7] When the inductivist historian writes the *prehistory* of a scientific discipline, he may draw heavily upon such criticisms. And he often explains the early dark age – when people were engrossed by 'unproven ideas' – with the help of some 'external' explanation, like the socio-psychological theory of the retarding influence of the Catholic Church.

The inductivist historian recognises only two sorts of *genuine scientific discoveries: hard factual propositions* and inductive *generalisations*. These and only these constitute the backbone of his *internal history*. When writing history, he looks out for them – finding them is quite a problem. Only when he finds them, can he start the construction of his beautiful pyramids. Revolutions consist in unmasking [irrational] errors which then are exiled from the history of science into the history of pseudoscience, into the history of mere beliefs: genuine scientific progress starts with the latest scientific revolution in any given field.

Each internal historiography has its characteristic victorious paradigms.[8] The main paradigms of inductivist historiography were Kepler's generalisations from Tycho Brahe's careful observations; Newton's discovery of his law of gravitation by, in turn, inductively generalising Kepler's 'phenomena' of planetary motion; and Ampère's discovery of his law of electrodynamics by inductively generalising his observations of electric currents. Modern chemistry too is taken by some inductivists as having really started with Lavoisier's experiments and his 'true explanations' of them.

But the inductivist historian cannot offer a *rational* 'internal' explana-

[6] Cf. p. 19.

[7] For a detailed discussion of inductivist (and, in general, justificationist) criticism cf. my [1966].

[8] I am now using the term 'paradigm' in its pre-Kuhnian sense.

tion for *why* certain facts rather than others were selected in the first instance. For him this is a *non-rational, empirical, external* problem. Inductivism as an 'internal' theory of rationality is compatible with many different supplementary empirical or external theories of problem-choice. It is, for instance, compatible with the vulgar-Marxist view that problem-choice is determined by social needs;[9] indeed, some vulgar-Marxists identify major phases in history of science with the major phases of economic development.[10] But choice of facts need not be determined by social factors; it may be determined by extra-scientific intellectual influences. And inductivism is equally compatible with the 'external' theory that the choice of problems is primarily determined by inborn, or by arbitrarily chosen (or traditional) theoretical (or 'metaphysical') frameworks.

There is a radical brand of inductivism which condemns all external influences, whether intellectual, psychological or sociological, as creating impermissible bias: radical inductivists allow only a [random] selection by the empty mind. Radical inductivism is, in turn, a special kind of *radical internalism*. According to the latter once one establishes the existence of some external influence on the acceptance of a scientific theory (or factual proposition) one must withdraw one's acceptance: proof of external influence means invalidation:[11] but since external influences always exist, radical internalism is utopian, and, as a theory of rationality, self-destructive.[12]

When the radical inductivist historian faces the problem of why some great scientists thought highly of metaphysics and, indeed, why they thought that their discoveries were great for reasons which, in the light of inductivism, look very odd, he will refer these problems of 'false consciousness' to psychopathology, that is, to external history.

B. *Conventionalism*

Conventionalism allows for the building of any system of pigeon holes which organises facts into some coherent whole. The conventionalist decides to keep the centre of such a pigeonhole system intact as long as possible: when difficulties arise through an invasion of anomalies, he only changes and complicates the peripheral arrangements. But the conventionalist does not regard any pigeonhole system as provenly true, but only as 'true by convention' (or possibly even as neither true nor false). In

[9] This compatibility was pointed out by Agassi on pp. 23–7 of his [1963]. But did he not point out the analogous compatibility within his own falsificationist historiography; cf. below, pp. 8, 9.

[10] Cf. e.g. Bernal [1965], p. 377.

[11] Some logical positivists belonged to this set: one recalls Hempel's horror at Popper's casual praise of certain external metaphysical influences upon science (Hempel [1937]).

[12] When German obscurantists scoff at 'positivism', they frequently mean radical internalism, and in particular, radical inductivism.

revolutionary brands of conventionalism one does not have to adhere forever to a given pigeonhole system: one may abandon it if it becomes unbearably clumsy and if a simpler one is offered to replace it.[13] This version of conventionalism is epistomologically, and especially logically, much simpler than inductivism: it is in no need of valid inductive inferences. Genuine *progress* of science is cumulative and takes place on the ground level of 'proven' facts;[14] the *changes* on the theoretical level are merely instrumental. Theoretical 'progress' is only in convenience ('simplicity'), and not in truth-content.[15] One may, of course, introduce revolutionary conventionalism also at the level of 'factual' propositions, in which case one would accept 'factual' propositions by decision rather than by experimental 'proofs'. But then, if the conventionalist is to retain the idea that the growth of 'factual' science has anything to do with objective, factual truth, he must devise some metaphysical principle which he then has to superimpose on his rules for the game of science.[16] If he does not, he cannot escape scepticism or, at least, some radical form of instrumentalism.

(It is important to clarify the *relation between conventionalism and instrumentalism.* Conventionalism rests on the recognition that false assumptions may have true consequences; therefore false theories may have great predictive power. Conventionalists had to face the problem of comparing rival false theories. Most of them conflated truth with its signs and found themselves holding some version of the pragmatic theory of truth. It was Popper's theory of truth-content, verisimilitude and corroboration which finally laid down the basis of a philosophically flawless version of conventionalism. On the other hand some conventionalists did not have sufficient logical eduction to realise that some propositions may be true whilst being unproven; and others false whilst having true consequences, and also some which are both false and approximately true. These people opted for 'instrumentalism': they came to regard theories as neither true nor false but merely as 'instruments' for prediction. Conventionalism, as here defined, is a philosophically sound position; instrumentalism is a degenerate

[13] For what I here call *revolutionary conventionalism*, see my [1970], pp. 105–6 and 187–9.

[14] I mainly discuss here only one version of revolutionary conventionalism, the one which Agassi, in his [1966], called 'unsophisticated': the one which assumes that factual propositions – unlike pigeonhole systems – can be 'proven'. (Duhem, for instance, draws no clear distinction between facts and factual propositions.)

[15] It is important to note that most conventionalists are reluctant to give up inductive generalisations. They distinguish between the '*floor of facts*', the '*floor of laws*' (i.e. inductive generalisations from 'facts') and the '*floor of theories*' (or of pigeonhole systems) which classify, conveniently, both facts and inductive laws. (Whewell, the conservative conventionalist, and Duhem, the revolutionary conventionalist, differ less than most people imagine.)

[16] One may call such metaphysical principles 'inductive principles'. For an 'inductive principle' which – roughly speaking – makes Popper's 'degree of corroboration' (a conventionalist appraisal) the measure of Popper's verisimilitude (truth-content minus falsity-content) see my [1968*a*], pp. 390–408 and my [1974], §2. (Another widely spread 'inductive principle' may be formulated like this: 'What the group of trained – or up-to-date, or suitably purged – scientists decide to *accept* as "true", is true.')

version of it, based on a mere philosophical muddle caused by lack of elementary logical competence.)

Revolutionary conventionalism was born as the Bergsonians' philosophy of science: free will and creativity were the slogans. The code of scientific honour of the conventionalist is less rigorous than that of the inductivist: it puts no ban on unproven speculation, and allows a pigeonhole system to be built around *any* fancy idea. Moreover, conventionalism does not brand discarded systems as unscientific: the conventionalist sees much more of the actual history of science as rational ('internal') than does the inductivist.

For the conventionalist historian, major discoveries are primarily inventions of new and simpler pigeonhole systems. Therefore he constantly compares for simplicity: the complications of pigeonhole systems and their revolutionary replacement by simpler ones constitute the backbone of his internal history.

The paradigmatic case of a scientific revolution for the conventionalist has been the Copernican revolution.[17] Efforts have been made to show that Lavoisier's and Einstein's revolutions too were replacements of clumsy theories by simple ones.

Conventionalist historiography cannot offer a *rational* explanation of why certain facts were selected in the first instance or of why certain particular pigeonhole systems were tried rather than others at a stage when their relative merits were yet unclear. Thus conventionalism, like inductivism, is compatible with various supplementary empirical–'externalist' programmes.

Finally, the conventionalist historian, like his inductivist colleague, frequently encounters the problem of 'false consciousness'. According to conventionalism for example, it is a 'matter of fact' that great scientists arrive at their theories by flights of their imaginations. Why then do they often claim that they derived their theories from facts? The conventionalist's rational reconstruction often differs from the great scientists' own reconstruction – the conventionalist historian relegates these problems of false consciousness to the externalist.[18]

[17] Most historical accounts of the Copernican revolution are written from the conventionalist point of view. Few claimed that Copernicus' theory was an 'inductive generalisation' from some 'factual discovery'; or that it was proposed as a bold theory to replace the Ptolemaic theory which had been 'refuted' by some celebrated 'crucial' experiment.

For a further discussion of the historiography of the Copernican revolution, cf. Lakatos and Zahar [1976].

[18] For example, for non-inductivist historians Newton's '*Hypotheses non fingo*' represents a major problem. Duhem, who unlike most historians did not over-indulge in Newton-worship, dismissed Newton's inductivist methodology as logical nonsense; but Koyré, whose many strong points did not include logic, devoted long chapters to the 'hidden depths' of Newton's muddle.

C. *Methodological falsificationism*

Contemporary falsificationism arose as a logico-epistemological criticism of inductivism and of Duhemian conventionalism. Inductivism was criticised on the grounds that its two basic assumptions, namely, that factual propositions can be 'derived' from facts and that there can be valid inductive (content-increasing) inferences, are themselves unproven and even demonstrably false. Duhem was criticised on the grounds that comparison of intuitive simplicity can only be a matter for subjective taste and that it is so ambiguous that no hard-hitting criticism can be based on it. Popper, in his *Logik der Forschung*, proposed a new 'falsificationist' methodology.[19] This methodology is another brand of revolutionary conventionalism: the main difference is that it allows factual, spatio-temporally singular 'basic statements', rather than spatio-temporally universal theories, to be accepted by convention. In the code of honour of the falsificationist a theory is scientific only if it can be *made* to conflict with a basic statement; and a theory must be eliminated if it conflicts with an accepted basic statement. Popper also indicated a further condition that a theory must satisfy in order to qualify as scientific: it must predict facts which are *novel*, that is, unexpected in the light of previous knowledge. Thus, it is against Popper's code of scientific honour to propose unfalsifiable theories or '*ad hoc*' hypotheses (which imply no *novel* empirical predictions) – just as it is against the [classical] inductivist code of scientific honour to propose unproven ones.

The great attraction of Popperian methodology lies in its clarity and force. Popper's deductive model of scientific criticism contains empirically falsifiable spatio-temporally universal propositions, initial conditions and their consequences. The weapon of criticism is the *modus tollens*: neither inductive logic nor intuitive simplicity complicate the picture.[20]

(Falsificationism, though logically impeccable, has epistemological difficulties of its own. In its 'dogmatic' proto-version it assumes the provability of propositions from facts and thus the disprovability of theories – a false assumption.[21] In its Popperian 'conventionalist' version it needs some [extra-methodological] 'inductive principle' to lend epistemological weight to its decisions to accept 'basic' statements, and in general to connect its rules of the scientific game with verisimilitude.[22])

The Popperian historian looks for great, 'bold', falsifiable theories and for great negative crucial experiments. These form the skeleton of his

[19] *In this paper I use this term to stand exclusively for one version of falsificationism, namely for 'naive methodological falsificationism', as defined in my* [1970], pp. 93–116.

[20] Since in his methodology the *concept* of intuitive simplicity has no place, Popper was able to use the term 'simplicity' for 'degree of falsifiability'. But there is more to simplicity than this: cf. my [1970], pp. 131ff.

[21] For a discussion cf. my [1970], especially pp. 99–100.

[22] For further discussion cf. pp. 20–1.

rational reconstruction. The Popperians' favourite paradigms of great falsifiable theories are Newton's and Maxwell's theories, the radiation formulas of Rayleigh, Jeans and Wien, and the Einsteinian revolution; their favourite paradigms for crucial experiments are the Michelson–Morley experiment, Eddington's eclipse experiment, and the experiments of Lummer and Pringsheim. It was Agassi who tried to turn this naive falsificationism into a systematic historiographical research programme.[23] In particular he predicted (or 'postdicted', if you wish) that behind each great experimental discovery lies a theory which the discovery contradicted; the importance of a factual discovery is to be measured by the importance of the theory refuted by it. Agassi seems to accept at face value the value judgments of the scientific community concerning the importance of factual discoveries like Galvani's, Oersted's, Priestley's, Roentgen's and Hertz's; but he denies the 'myth' that they were chance discoveries (as the first four were said to be) or confirming instances (as Hertz first thought his discovery was).[24] Thus Agassi arrives at a bold prediction: all these five experiments were successful refutations – in some cases even *planned* refutations – of theories which he proposes to unearth, and, indeed, in most cases, claims to have unearthed.[25]

Popperian internal history, in turn, is readily supplemented by external theories of history. Thus Popper himself explained that [on the positive side] (1) the main *external* stimulus of scientific theories comes from unscientific 'metaphysics', and even from myths (this was later beautifully illustrated mainly by Koyré); and that [on the negative side] (2) facts do *not* constitute such external stimulus – factual discoveries belong completely to internal history, emerging as refutations of some scientific theory, so that facts are only noticed if they conflict with some previous expectation. Both theses are cornerstones of Popper's *psychology* of discovery.[26] Feyerabend developed another interesting *psychological* thesis of Popper, namely, that proliferation of rival theories may – *externally* – speed up *internal* Popperian falsification.[27]

[23] Agassi [1963].

[24] An experimental discovery is *a chance discovery in the objective sense* if it is neither a confirming nor a refuting instance of some theory in the objective body of knowledge of the time; it is *a chance discovery in the subjective sense* if it is made (or recognised) by the discoverer neither as a confirming nor as a refuting instance of some theory he personally had entertained at the time.

[25] Agassi [1963], pp. 64–74.

[26] Within the Popperian circle, it was Agassi and Watkins who particularly emphasized the importance of unfalsifiable or barely testable '*empirical*' theories in providing an *external* stimulus to later properly *scientific* developments. (Cf. Agassi [1964] and Watkins [1958].) This idea, of course, is already there in Popper's [1935] and [1960]. Cf. my [1970], p. 184; but the new formulation of the difference between their approach and mine which I am going to give in this paper will, I hope, be much clearer.

[27] Popper occasionally – and Feyerabend systematically – stressed the catalytic (*external*) role of alternative theories in devising so-called 'crucial experiments'. But alternatives are not merely catalysts, which can be later removed in the rational reconstruction, they are *necessary* parts of the falsifying process. Cf. Popper [1940] and Feyerabend [1965]; but cf. also Lakatos [1970], especially p. 121, footnote 4.

But the external supplementary theories of falsificationism need not be restricted to purely intellectual influences. It has to be emphasised (*pace* Agassi) that falsificationism is no less compatible with a vulgar-Marxist view of what makes science progress than is inductivism. The only difference is that while for the latter Marxism might be invoked to explain the discovery of *facts*, for the former it might be invoked to explain the invention of *scientific theories*; while the choice of facts (that is, for the falsificationist, the choice of 'potential falsifiers') is primarily determined internally by the theories.

'False awareness' – 'false' from the point of view of *his* rationality theory – creates a problem for the falsificationist historian. For instance, why do some scientists believe that crucial experiments are positive and verifying rather than negative and falsifying? It was the falsificationist Popper who, in order to solve these problems, elaborated better than anybody else before him the cleavage between objective knowledge (in his 'third world') and its distorted reflections in individual minds.[28] Thus he opened up the way for my demarcation between internal and external history.

D. *Methodology of scientific research programmes*

According to my methodology the great scientific achievements are research programmes which can be evaluated in terms of progressive and degenerating problemshifts; and scientific revolutions consist of one research programme superseding (overtaking in progress) another.[29] This methodology offers a new rational reconstruction of science. It is best presented by contrasting it with falsificationism and conventionalism, from both of which it borrows essential elements.

From conventionalism, this methodology borrows the licence rationally to accept by convention not only spatio-temporally singular 'factual statements' but also spatio-temporally universal theories: indeed, this becomes the most important clue to the continuity of scientific growth.[30] The basic unit of appraisal must be not an isolated theory or conjunction of theories but rather a '*research programme*', with a conventionally accepted (and thus by provisional decision 'irrefutable') '*hard core*' and with a '*positive heuristic*' which defines problems, outlines the construction of a belt of auxiliary hypotheses, foresees anomalies and turns them victoriously into examples, all according to a preconceived plan. The scientist lists anomalies, but as long as his research programme sustains its momentum,

[28] Cf. Popper [1968*a*] and [1968*b*].

[29] The terms 'progressive' and 'degenerating problemshifts', 'research programmes' 'superseding' will be crudely defined in which follows – for more elaborate definitions see my [1968*b*] and especially my [1970].

[30] Popper does not permit this: 'There is a vast difference between my views and conventionalism. I hold that what characterises the empirical method is just this: our conventions determine the acceptance of the *singular*, not of the *universal* statements' (Popper [1935], section 30).

he may freely put them aside. *It is primarily the positive heuristic of his programme, not the anomalies, which dictate the choice of his problems.*[31] Only when the driving force of the positive heuristic weakens, may more attention be given to anomalies. The methodology of research programmes can explain in this way *the high degree of autonomy of theoretical science*; the naive falsificationist's disconnected chains of conjectures and refutations cannot. What for Popper, Watkins and Agassi is *external*, influential metaphysics, here turns into the *internal* 'hard core' of a programme.[32]

The methodology of research programmes presents a very different picture of the game of science from the picture of the methodological falsificationist. The best opening gambit is not a falsifiable (and therefore consistent) hypothesis, but a research programme. Mere 'falsification' (in Popper's sense) must not imply rejection.[33] Mere 'falsifications' (that is, anomalies) are to be recorded but need not be acted upon. Popper's great negative crucial experiments disappear; 'crucial experiment' is an honorific title, which may, of course, be conferred on certain anomalies, but only *long after the event*, only when one programme has been defeated by another one. According to Popper, a crucial experiment is described by an accepted basic statement which is inconsistent with a theory – according to the methodology of scientific research programmes, no accepted basic statement *alone* entitles the scientist to reject a theory. Such a clash may present a problem (major or minor), but in no circumstance a 'victory'. Nature may shout *no*, but human ingenuity – contrary to Weyl and Popper[34] – may always be able to shout louder. With sufficient resourcefulness and some luck, any theory can be defended 'progressively' for a long time, even if it is false. The Popperian pattern of 'conjectures and refutations', that is the pattern of trial-by-hypothesis followed by error-shown-by-experiment, is to be abandoned: no experiment is crucial at the time – let alone before – it is performed (except, possibly, psychologically).

It should be pointed out, however, that the methodology of scientific research programmes has more teeth than Duhem's conventionalism: instead of leaving it to Duhem's unarticulated common sense[35] to judge

[31] The falsificationist hotly denies this: 'Learning from experience is learning from a refuting instance. The refuting instance then becomes a problematic instance' (Agassi [1964], p. 201). In his [1969] Agassi attributed to Popper the statement that 'we learn from experience by refutations' (p. 169), and adds that according to Popper one can learn *only* from refutation but not from corroboration (p. 167). Feyerabend, even in his [1969], says that '*negative instances suffice in science*'. But these remarks indicate a very one-sided theory of learning from experience. (Cf. my [1970], p. 121, footnote 1, and p. 123.)

[32] Duhem, as a staunch positivist within philosophy of science, would, no doubt, exclude most 'metaphysics' as unscientific and would not allow it to have any influence on science proper.

[33] Cf. my [1968*a*], pp. 383–6, my [1968*b*], pp. 162–7, and my [1970], pp. 116 ff. and pp. 155 ff.

[34] Cf. Popper ([1935], section 85.

[35] Cf. Duhem [1906], part II, chapter VI, §10.

when a 'framework' is to be abandoned, I inject some hard Popperian elements into the appraisal of whether a programme progresses or degenerates or of whether one is overtaking another. That is, I give criteria of progress and stagnation within a programme and also rules for the 'elimination' of whole research programmes. A research programme is said to be *progressing* as long as its theoretical growth anticipates its empirical growth, that is, as long as it keeps predicting novel facts with some success (*'progressive problemshift'*); it is *stagnating* if its theoretical growth lags behind its empirical growth, that is, as long as it gives only *post-hoc* explanations either of chance discoveries or of facts anticipated by, and discovered in, a rival programme (*'degenerating problemshift'*).[36] If a research programme progressively explains more than a rival, it 'supersedes' it, and the rival can be eliminated (or, if you wish, 'shelved').[37]

(*Within* a research programme a theory can only be eliminated by a better theory, that is, by one which has excess empirical content over its predecessors, some of which is subsequently confirmed. And for this replacement of one theory by a better one, the first theory does not even have to be 'falsified' in Popper's sense of the term. Thus, progress is marked by instances verifying excess content rather than by falsifying instances;[38] empirical 'falsification' and actual 'rejection' become independent.[39] Before a theory has been modified we can never know in what way it had been 'refuted', and some of the most interesting modifications are motivated by the 'positive heuristic' of the research programme

[36] In fact, I define a research programme as degenerating even if it anticipates novel facts but does so in a patched-up development rather than by a coherent, pre-planned positive heuristic. I distinguish three types of *ad hoc* auxiliary hypotheses: those which have no excess empirical content over their predecessor (*'ad hoc$_1$'*), those which do have such excess content but none of it is corroborated (*'ad hoc$_2$'*) and finally those which are not *ad hoc* in these two senses but do not form an integral part of the positive heuristic (*'ad hoc$_3$'*). Examples for an *ad hoc$_1$* hypothesis are provided by the linguistic prevarications of pseudosciences, or by the conventionalist stratagems discussed in my [1963–4], like 'monsterbarring', 'exceptionbarring', 'monsteradjustment', etc. A famous example of an *ad hoc$_2$* hypothesis is provided by the Lorentz–Fitzgerald contraction hypothesis; an example of an *ad hoc$_3$* hypothesis is Planck's first correction of the Lummer–Pringsheim formula (also cf. p. 103). Some of the cancerous growth in contemporary social 'sciences' consists of a cobweb of such *ad hoc$_3$* hypotheses, as shown by Meehl and Lykken. (For references, cf. my [1970], p. 175, footnotes 2 and 3.)

[37] The rivalry of two research programmes is, of course, a protracted process during which it is rational to work in either (*or, if one can, in both*). The latter pattern becomes important, for instance, when one of the rival programmes is vague and its opponents wish to develop it in a sharper form in order to show up its weakness. Newton elaborated Cartesian vortex theory in order to show that it is inconsistent with Kepler's laws. (Simultaneous work on rival programmes, of course, undermines Kuhn's thesis of the psychological incommensurability of rival paradigms.)

The progress of one programme is a vital factor in the degeneration of its rival. If programme P_1 constantly produces 'novel facts' these, by definition, will be anomalies for the rival programme P_2. If P_2 accounts for these novel facts only in an *ad hoc* way, it is degenerating by definition. Thus the more P_1 progresses, the more difficult it is for P_2 to progress.

[38] Cf. especially my [1970], pp. 120–1.

[39] Cf. especially my [1968*a*], p. 385 and [1970], p. 121.

rather than by anomalies. This difference alone has important consequences and leads to a rational reconstruction of scientific change very different from that of Popper's.[40])

It is very difficult to decide, especially since one must not demand progress at each single step, when a research programme has degenerated hopelessly or when one of two rival programmes has achieved a decisive advantage over the other. In this methodology, as in Duhem's conventionalism, there can be no instant – let alone mechanical – rationality. *Neither the logician's proof of inconsistency nor the experimental scientist's verdict of anomaly can defeat a research programme in one blow.* One can be 'wise' only after the event.[41]

In this code of scientific honour modesty plays a greater role than in other codes. One *must* realise that one's opponent, even if lagging badly behind, may still stage a comeback. No advantage for one side can ever be regarded as absolutely conclusive. There is never anything inevitable about the triumph of a programme. Also, there is never anything inevitable about its defeat. Thus pigheadedness, like modesty, has more 'rational' scope. *The scores of the rival sides, however, must be recorded*[42] *and publicly displayed at all times.*

(We should here at least refer to the main epistemological problem of the methodology of scientific research programmes. As it stands, like Popper's methodological falsificationism, it represents a very radical version of conventionalism. One needs to posit some extra-methodological inductive principle to relate – even if tenuously – the scientific gambit of pragmatic acceptances and rejections to verisimilitude.[43] Only such an 'inductive principle' can turn science from a mere game into an epistemologically rational exercise; from a set of lighthearted sceptical gambits pursued for intellectual fun into a – more serious – fallibilist venture of approximating the Truth about the Universe.[44])

The methodology of scientific research programmes constitutes, like any other methodology, a historiographical research programme. The historian who accepts this methodology as a guide will look in history for

[40] For instance, a rival theory, which acts as an *external* catalyst for the Popperian falsification of a theory, here becomes an *internal* factor. In Popper's (and Feyerabend's) reconstruction such a theory, after the falsification of the theory under test, can be removed from the rational reconstruction; in my reconstruction it has to stay within the internal history lest the falsification be undone. (Cf. note 27.)

Another important consequence is the difference between Popper's discussion of the Duhem–Quine argument and mine; cf. on the one hand Popper [1935], last paragraph of section 18 and section 19, footnote 1; Popper [1957*b*], pp. 131–3; Popper [1963*a*], p. 112, footnote 26, pp. 238–9 and p. 243; and on the other hand, my [1970], pp. 184–9.

[41] For the falsificationist this is a repulsive idea; cf. e.g. Agassi [1963], pp. 48ff.

[42] Feyerabend seems now to deny that even this is a possibility; cf. his [1970*a*] and especially [1970*b*] and [1971].

[43] I use 'verisimilitude' here in Popper's technical sense, as the difference between the truth content and falsity content of a theory. Cf. his [1963*a*], chapter 10.

[44] For a more general discussion of this problem, cf. pp. 20–1.

rival research programmes, for progressive and degenerating problem shifts. Where the Duhemian historian sees a revolution merely in simplicity (like that of Copernicus), he will look for a large scale progressive programme overtaking a degenerating one. When the falsificationist sees a crucial negative experiment, he will 'predict' that there was none, that behind any alleged crucial experiment, behind any alleged single battle between theory and experiment, there is a hidden war of attrition between two research programmes. The outcome of the war is only later linked in the falsificationist reconstruction with some alleged single 'crucial experiment'.

The methodology of research programmes – like any other theory of scientific rationality – must be supplemented by empirical–external history. No rationality theory will ever solve problems like why Mendelian genetics disppeared in Soviet Russia in the 1950's, or why certain schools of research into genetic racial differences or into the economics of foreign aid came into disrepute in the Anglo-Saxon countries in the 1960s. Moreover, to explain different speeds of development of different research programmes we may need to invoke external history. Rational reconstruction of science (in the sense in which I use the term) cannot be comprehensive since human beings are not *completely* rational animals; and even when they act rationally they may have a false theory of their own rational actions.[45]

But the methodology of research programmes draws a demarcation between internal and external history which is markedly different from that drawn by other rationality theories. For instance, what for the falsificationist looks like the (regrettably frequent) phenomenon of irrational adherence to a 'refuted' or to an inconsistent theory and which he therefore relegates to *external* history, may well be explained in terms of my methodology *internally* as a rational defence of a promising research programme. Or, the successful *pre*dictions of novel facts which constitute serious evidence for a research programme and therefore vital parts of internal history, are irrelevant both for the inductivist and for the falsificationist.[46] For the inductivist and the falsificationist it does not really matter whether the discovery of a fact preceded or followed a theory: only their logical relation is decisive. The 'irrational' impact of the historical coincidence that a theory happened to have *anticipated* a factual discovery, has no internal significance. Such anticipations constitute 'not proof but [mere] propaganda'.[47] Or again, take Planck's discontent with his own

[45] Also cf. pp. 4, 7, 9, 17, 20.

[46] The reader should remember that in this paper I discuss only naive falsificationism; cf. note 19.

[47] This is Kuhn's comment on Galileo's successful *pre*diction of the phases of Venus (Kuhn [1957], p. 224). Like Mill and Keynes before him, Kuhn cannot understand why the historic order of theory and evidence should count, and he cannot see the importance of the fact that Copernicans *pre*dicted the phases of Venus, while Tychonians only explained them by *post hoc* adjustments. Indeed, since he does not see the importance of the fact, he does not even care to mention it.

1900 radiation formula, which he regarded as 'arbitrary'. For the falsificationist the formula was a bold, falsifiable hypothesis and Planck's dislike of it a non-rational mood, explicable only in terms of psychology. However, in my view, Planck's discontent can be explained internally: it was a rational condemnation of an '*ad hoc*$_3$' theory.[48] To mention yet another example: for falsificationism irrefutable 'metaphysics' is an external intellectual influence, in my approach it is a vital part of the rational reconstruction of science.

Most historians have hitherto tended to regard the solution of some problems as being the monopoly of externalists. One of these is the problem of the high frequency of *simultaneous discoveries*. For this problem vulgar-Marxists have an easy solution: a discovery is made by many people at the same time, once a social need for it arises.[49] Now what constitutes a 'discovery', and especially a major discovery, depends on one's methodology. For the inductivist, the most important discoveries are factual, and, indeed, such discoveries are frequently made simultaneously. For the falsificationist a *major* discovery consists in the discovery of a theory rather than of a fact. Once a theory is discovered (or rather invented), it becomes public property; and nothing is more obvious than that several people will test it simultaneously and make, simultaneously, (minor) factual discoveries. Also, a published theory is a challenge to devise higher-level, independently testable explanations. For example, given Kepler's ellipses and Galileo's rudimentary dynamics, simultaneous 'discovery' of an inverse square law is not so very surprising: a problem-situation being public, simultaneous solutions can be explained on *purely internal* grounds.[50] The discovery of a new problem, however, may not be so readily explicable. If one thinks of the history of science as of one of rival research programmes, then most simultaneous discoveries, theoretical or factual, are explained by the fact that research programmes being public property, many people work on them in different corners of the world, possibly not knowing of each other. However, really *novel*, *major*, *revolutionary* developments are rarely invented simultaneously. Some alleged simultaneous discoveries of novel programmes are seen as having been simultaneous discoveries only with false hindsight: in fact they are *different* discoveries, merged only later into a single one.[51]

A favourite hunting ground of externalists has been the related problem of why so much importance is attached to – and energy spent on – *priority disputes*. This can be explained only *externally* by the inductivist, the naive falsificationist, or the conventionalist; but in the light of the methodology

[48] Cf. note 36.

[49] For a statement of this position and an interesting critical discussion cf. Polanyi [1951], pp. 4ff and pp. 78ff.

[50] Cf. Popper [1963*b*] and Musgrave [1969].

[51] This was illustrated convincingly, by Elkana, for the case of the co-called simultaneous discovery of the conservation of energy; cf. his [1971].

of research programmes some priority disputes are vital *internal* problems, since in this methodology *it becomes all-important for rational appraisal which programme was first in anticipating a novel fact and which fitted in the by now old fact only later*. Some priority disputes can be explained by rational interest and not simply by vanity and greed for fame. It then becomes important that Tychonian theory, for instance, succeeded in explaining – only *post hoc* – the observed phases of, and the distance to, Venus which were orginally precisely anticipated by Copernicans;[52] or that Cartesians managed to explain everything that the Newtonians *pre*dicted – but only *post hoc*. Newtonian optical theory explained *post hoc* many phenomena which were anticipated and first observed by Huyghensians.[53]

All these examples show how the methodology of scientific research programmes turns many problems which had been *external* problems for other historiographies into internal ones. But occasionally the borderline is moved in the opposite direction. For instance there may have been an experiment which was accepted *instantly* – in the absence of a better theory – as a negative crucial experiment. For the falsificationist such acceptance is part of internal history; for me it is not rational and has to be explained in terms of external history.

Note. The methodology of research programmes was criticised both by Feyerabend and by Kuhn. According to Kuhn: '[Lakatos] must specify criteria which can be used *at the time* to distinguish a degenerative from a progressive research programme; and so on. Otherwise, *he has told us nothing at all*'.[54] Actually, I *do* specify such criteria. But Kuhn probably meant that '[my] standards have practical force only if they are combined with a *time limit* (what looks like a degenerating problemshift may be the the beginning of a much longer period of advance)'.[55] Since I specify no such time limit, Feyerabend concludes that my standards are no more than '*verbal ornaments*'.[56] A related point was made by Musgrave in a letter containing some major constructive criticisms of an earlier draft, in which he demanded that I specify, for instance, at what point dogmatic adherence to a programme ought to be explained 'externally' rather than 'internally'.

Let me try to explain why such objections are beside the point. One may rationally stick to a degenerating programme until it is overtaken by a rival *and even after*. What one must *not* do is to deny its poor public record. Both Feyerabend and Kuhn conflate

[52] Also cf. note 47.

[53] For the Mertonian brand of functionalism – as Alan Musgrave pointed out to me – priority disputes constitute a *prima facie* disfunction and therefore an anomaly for which Merton has been labouring to give a general socio-psychological explanation. (Cf. e.g. Merton [1957], [1963] and [1969].) According to Merton 'scientific *knowledge* is not the richer or the poorer for having credit given where credit is due: it is the social *institution* of science and individual men of science that would suffer from repeated failures to allocate credit justly' (Merton [1957], p. 648). But Merton overdoes his point: in important cases (like in some of Galileo's priority fights) there was more at stake than institutional interests: the problem was whether the Copernican research programme was progressive or not. (Of course, not all priority disputes have scientific relevance. For instance, the priority dispute between Adams and Leverrier about who was first to discover Neptune had no such relevance: whoever discovered it, the discovery strengthened the same (Newtonian) programme. In such cases Merton's external explanation may well be true.)

[54] Kuhn [1970], p. 239; my italics.

[55] Feyerabend [1970*a*], p. 215.

[56] *Ibid.*

methodological appraisal of a programme with firm *heuristic* advice about what to do.[57] It is perfectly rational to play a risky game: what is irrational is to deceive oneself about the risk.

This does not mean as much licence as might appear for those who stick to a degenerating programme. For they can do this mostly only in private. Editors of scientific journals should refuse to publish their papers which will, in general, contain either solemn reassertions of their position or absorption of counterevidence (or even of rival programmes) by *ad hoc*, linguistic adjustments. Research foundations, too, should refuse money.[58]

These observations also answer Musgrave's objection by separating rational and irrational (or honest and dishonest) adherence to a degenerating programme. They also throw further light on the demarcation between internal and external history. They show that internal history is self-sufficient for the presentation of the history of disembodied science, including degenerating problemshifts. External history explains why some people have false beliefs about scientific progress, and how their scientific activity may be influenced by such beliefs.

E. *Internal and external history*

Four theories of the rationality of scientific progress – or logics of scientific discovery – have been briefly discussed. It was shown how each of them provides a theoretical framework for the rational reconstruction of the history of science.

Thus the internal history of *inductivists* consists of alleged discoveries of hard facts and of so-called inductive generalisations. The internal history of *conventionalists* consists of factual discoveries and of the erection of pigeonhole systems and their replacement by allegedly simpler ones.[59] The internal history of *falsificationists* dramatises bold conjectures, improvements which are said to be *always* content-increasing and, above all, triumphant 'negative crucial experiments'. The *methodology of research programmes*, finally, emphasises long-extended theoretical and empirical

[57] Cf. note 2.

[58] I do, of course, *not* claim that such decisions are necessarily uncontroversial. In such decisions one has to use also one's *common sense*. Common sense (that is, judgment in *particular* cases which is not made according to mechanical rules but only follows general principles which leave some *Spielraum*) plays a role in all brands of non-mechanical methodologies. The Duhemian conventionalist needs common sense to decide when a theoretical framework has become sufficiently cumbersome to be replaced by a 'simpler' one. The Popperian falsificationist needs common sense to decide when a basic statement is to be 'accepted', or to which premise the *modus tollens* is to be directed. (Cf. my [1970], pp. 106ff.) But neither Duhem nor Popper gives a blank cheque to 'common sense'. They give very definite guidance. The Duhemian judge directs the jury of common sense to agree on comparative simplicity; the Popperian judge directs the jury to look out primarily for, and agree upon, accepted basic statements which clash with accepted theories. My judge directs the jury to agree on appraisals of progressive and degenerating research programmes. But, for example, there may be conficting views about whether an accepted basic statement expresses a *novel* fact or not. Cf. my [1970], p. 156.

Although it is important to reach agreement on such verdicts, there must also be the possibility of appeal. In such appeals inarticulated common sense is questioned, articulated and criticised. (The criticism may even turn from a criticism of law interpretation into a criticism of the law itself.)

[59] Most conventionalists have also an intermediate inductive layer of 'laws' between facts and theories; cf. note 15.

rivalry of major research programmes, progressive and degenerating problemshifts, and the slowly emerging victory of one programme over the other.

Each rational reconstruction produces some characteristic pattern of rational growth of scientific knowledge. But all of these *normative* reconstructions may have to be supplemented by *empirical* external theories to explain the residual non-rational factors. The history of science is always richer than its rational reconstruction. *But rational reconstruction or internal history is primary, external history only secondary, since the most important problems of external history are defined by internal history.* External history either provides non-rational explanation of the speed, locality, selectiveness etc. of historic events *as interpreted* in terms of internal history; or, when history differs from its rational reconstruction, it provides an empirical explanation of why it differs. But the *rational* aspect of scientific growth is fully accounted for by one's logic of scientific discovery.

Whatever problem the historian of science wishes to solve, he has first to reconstruct the relevant section of the growth of objective scientific knowledge, that is, the relevant section of 'internal history'. As it has been shown, what constitutes for him internal history, depends on his philosophy, whether he is aware of this fact or not. Most theories of the growth of knowledge are theories of the growth of disembodied knowledge: whether an experiment is crucial or not, whether a hypothesis is highly probable in the light of the available evidence or not, whether a problemshift is progressive or not, is not dependent in the slightest on the scientists' beliefs, personalities or authority. These subjective factors are of no interest for any internal history. For instance, the 'internal historian' records the Proutian programme with its hard core (that atomic weights of pure chemical elements are whole numbers) and its positive heuristic (to overthrow, and replace, the contemporary false observational theories applied in measuring atomic weights). This programme was later carried through.[60] The internal historian will waste little time on Prout's *belief* that if the 'experimental techniques' *of his time* were 'carefully' applied, and the experimental findings properly interpreted, the anomalies would *immediately* be seen as mere illusions. The internal historian will regard this historical fact as a fact in the second world which is only a caricature of

[60] The proposition 'the Proutian programme was carried through' looks like a 'factual' proposition. But there are no 'factual' propositions: the phrase only came into ordinary language from dogmatic empiricism. *Scientific 'factual' propositions* are theory-laden: the theories involved are 'observational theories'. *Historiographical 'factual' propositions* are also theory-laden: the the theories involved are methodological theories. In the decision about the truth-value of the 'factual' proposition, 'the Proutian programme was carried through', two methodological theories are involved. First, the theory that the units of scientific appraisal are research programmes; secondly, some *specific* theory of how to judge whether a programme was 'in fact' carried through. For all these considerations a Popperian internal historian will not need to take any interest whatsoever in the *persons* involved, or in their beliefs about their own activities.

its counterpart in the third world.[61] *Why* such caricatures come about is none of his business; he might – in a footnote – pass on to the externalist the problem of why certain scientists had 'false beliefs' about what they were doing.[62]

Thus, in constructing internal history the historian will be highly selective: he will omit everything that is irrational in the light of his rationality theory. But this normative selection still does not add up to a fully fledged rational reconstruction. For instance, Prout never articulated the 'Proutian programme': the Proutian programme is not Prout's programme. *It is not only the ('internal') success or the ('internal') defeat of a programme which can only be judged with hindsight: it is frequently also its content.* Internal history is not just a *selection* of methodologically interpreted facts: it may be, on occasions, their *radically improved version.* One may illustrate this using the Bohrian programme. Bohr, in 1913, may not have even thought of the possibility of electron spin. He had more than enough on his hands without the spin. Nevertheless, the historian, describing with hindsight the Bohrian programme, should include electron spin in it, since electron spin fits naturally in the original outline of the programme. Bohr might have referred to it in 1913. Why Bohr did not do so, is an interesting problem which deserves to be indicated in a footnote.[63] (Such problems might then be solved either internally by pointing to rational reasons in the growth of objective, impersonal knowledge; or externally by pointing to psychological causes in the development of Bohr's personal beliefs.)

One way to indicate discrepancies between history and its rational reconstruction is to relate the internal history *in the text*, and indicate *in the footnotes* how actual history 'misbehaved' in the light of its rational reconstruction.[64]

[61] The 'first world' is that of matter, the 'second' the world of feelings, beliefs, consciousness, the 'third' the world of objective knowledge, articulated in propositions. This is an age-old and vitally important trichotomy; its leading contemporary proponent is Popper. Cf. Popper [1968*a*], [1968*b*] and Musgrave [1969] and [1974].

[62] Of course what, in this context, constitutes 'false belief' (or 'false consciousness'), depends on the rationality theory of the critic: cf. pp. 4, 6 and 8. But no rationality theory can ever succeed in leading to 'true consciousness'.

[63] If the publication of Bohr's programme had been delayed by a few years, further speculation might even have led to the spin problem without the previous observation of the anomalous Zeeman effect. Indeed, Compton raised the problem in the context of the Bohrian programme in his [1919].

[64] I first applied this expositional device in my [1963–4]; I used it again in giving a detailed account of the Proutian and the Bohrian programmes; cf. my [1970], pp. 138, 140, 146. This practice was criticised at the 1969 Minneapolis conference by some historians. McMullin, for instance, claimed that this presentation may illuminate a *methodology*, but certainly not real *history*: the text tells the reader what ought to have happened and the footnotes what in fact happened (cf. McMullin [1970]). Kuhn's criticism of my exposition ran essentially on the same lines: he thought that it was a specifically *philosophical* exposition: 'a *historian* would not include *in his narrative* a factual report which he knows to be false. If he had done so, he would be so sensitive to the offence that he could not conceivably compose a footnote calling attention to it.' (Cf. Kuhn [1970], p. 256.)

Many historians will abhor the idea of *any* rational reconstruction. They will quote Lord Bolingbroke: 'History is philosophy teaching by example.' They will say that before philosophising 'we need a lot more examples'.[65] But such an inductivist theory of historiography is utopian.[66] *History without some theoretical 'bias' is impossible.*[67] Some historians look for the discovery of hard facts, inductive generalisations, others for bold theories and crucial negative experiments, yet others for great simplifications, or for progressive and degenerating problemshifts; all of them have *some* theoretical 'bias'. This bias, of course, may be obscured by an eclectic variation of theories or by theoretical confusion: but neither eclecticism nor confusion amounts to an atheoretical outlook. What a historian regards as an external problem is often an excellent guide to his implicit methodology: some will ask why a 'hard fact' or a 'bold theory' was discovered exactly when and where it actually was discovered; others will ask why a 'degenerating problemshift' could have wide popular acclaim over an incredibly long period or why a 'progressive problemshift' was left 'unreasonably' unacknowledged.[68] Long texts have been devoted to the problem of whether, and if so, why, the emergence of science was a purely European affair; but such an investigation is bound to remain a piece of confused rambling until one clearly defines 'science' according to some normative philosophy of science. One of the most interesting problems of external history is to specify the psychological, and indeed, social conditions which are necessary (but, of course, never sufficient) to make scientific progress possible; but in the very formulation of this 'external' problem *some* methodological theory, *some* definition of science is bound to enter. History of *science* is a history of events which are selected and interpreted in a normative way.[69] This being so, the hitherto neglected problem of appraising rival logics of scientific discovery and, hence, rival reconstructions of history, acquires paramount importance. I shall now turn to this problem.

[65] Cf. L. P. Williams [1970].

[66] Perhaps I should emphasise the difference between on the one hand, *inductivist historiography of science*, according to which *science* proceeds through discovery of hard facts (in nature) and (possibly) inductive generalisations, and, on the other hand, the *inductivist theory of historiography of science* according to which *historiography of science* proceeds through discovery of hard facts (in history of science) and (possibly) inductive generalisations. 'Bold conjectures', 'crucial negative experiments', and even 'progressive and degenerating research programmes' may be regarded as 'hard historical facts' by some inductivist historiographers. One of the weaknesses of Agassi's [1963] is that he omitted to emphasise this distinction between scientific and historiographical inductivism.

[67] Cf. Popper [1957*b*], section 31.

[68] This thesis implies that the work of those 'externalists' (mostly trendy 'sociologists of science') who claim to do social history of some scientific discipline without having mastered the discipline itself, and its internal history, is worthless. Also cf. Musgrave [1974].

[69] Unfortunately there is only one single word in most languages to denote history_1 (the set of historical events) and history_2 (a set of historical propositions). Any history_2 is a theory- and value-laden reconstruction of history_1.

2 *Critical comparison of methodologies: history as a test of its rational reconstructions*

Theories of scientific rationality can be classified under two main heads.

(1) *Justificationist methodologies* set very high epistemological standards: for classical justificationists a proposition is 'scientific' only if it is *proven*, for neojustificationists, if it is *probable* (in the sense of the probability calculus) or *corroborated* (in the sense of Popper's third note on corroboration) to a proven degree.[70] Some philosophers of science gave up the idea of proving or of (provably) probabilifying scientific theories but remained dogmatic empiricists: whether inductivists, probabilists, conventionalists or falsificationists, they still stick to the provability of 'factual' propositions. By now, of course, all these different forms of justificationism have crumbled under the weight of *epistemological and logical criticism.*

(2) The only alternatives with which we are left are *pragmatic–conventionalist methodologies*, crowned by some global principle of induction. Conventionalist methodologies first lay down rules about 'acceptance' and 'rejection' of factual and theoretical propositions – without yet laying down rules about proof and disproof, truth and falsehood. We then get *different systems of rules of the scientific game*. The inductivist game would consist of collecting 'acceptable' (not proven) data and drawing from then 'acceptable' (not proven) inductive generalisations. The conventionalist game would consist of collecting 'acceptable' data and ordering them into the simplest possible pigeonhole systems (or devising the simplest possible pigeonhole systems and filling them with acceptable data). Popper specified yet another game as 'scientific'.[71] Even methodologies which have been epistemologically and logically discredited, may go on functioning, in these emasculated versions, as guides for the rational reconstruction of history. But these *scientific games* are without any genuine epistemological relevance *unless* we superimpose on them some sort of metaphysical (or, if you wish, 'inductive') principle which will say that the game, as specified by the methodology, gives us the best chance of approaching the Truth. Such a principle then turns the pure conventions

[70] That is, a hypothesis h is scientific only if there is a number q such that $p(h, e) = q$ where e is the available evidence and $p(h, e) = q$ can be *proved*. It is irrelevant whether p is a Carnapian confirmation function or a Popperian corroboration function as long as $p(h, e) = q$ is allegedly proved. (Popper's third note on corroboration, of course, is only a curious slip which is out of tune with his philosophy: cf. my [1968*a*], pp. 411–7.)

Probablism has never generated a programme of historiographical reconstruction; it has never emerged from grappling – unsuccessfully – with the very problems it created. As an epistemological programme it has been degenerating for a long time; as a historiographical programme it never even started.

[71] Popper [1935], sections 11 and 85. Also cf. the comment in my [1974], footnote 13. The methodology of research programmes too is, in the first instance, defined as a game; cf. especially pp. 9–12.

of the game into fallible conjectures; but without such a principle the scientific game is just like any other game.[72]

It is very difficult to criticise conventionalist methodologies like Duhem's and Popper's. There is no obvious way to criticise either a game or a metaphysical principle of induction. In order to overcome these difficulties I am going to propose a new theory of how to appraise such methodologies of science (the ones, which – at least in the first stage, before the introduction of an inductive principle – are conventionalist). I shall show that methodologies may be criticised without any direct reference to any epistemological (or even logical) theory, and without using directly any logico-epistemological criticism. The basic idea of this criticism is that *all methodologies function as historiographical (or meta-historical) theories (or research programmes) and can be criticised by criticising the rational historical reconstructions to which they lead.*

I shall try to develop this historiographical method of criticism in a dialectical way. I start with a special case: I first 'refute' falsificationism by 'applying' falsificationism (on a normative historiographical meta-level) to itself. Then I shall apply falsificationism also to inductivism and conventionalism, and, indeed, argue that all methodologies are bound to end up 'falsified' with the help of this Pyrrhonian *machine de guerre*. Finally, I shall 'apply' not falsificationism but the methodology of scientific research programmes (again on a normative–historiographical meta-level) to inductivism, conventionalism, falsificationism and to itself, and show that – on this meta-criterion – methodologies can be constructively criticised and compared. This normative–historiographical version of the methodology of scientific research programmes supplies a general theory of how to compare rival logics of discovery in which (in a sense carefully to be specified) *history may be seen as a 'test' of its rational reconstructions.*

A. *Falsificationism as a meta-criterion: history 'falsifies' falsificationism (and any other methodology)*

In their purely 'methodological' versions scientific appraisals, as has already been said, are *conventions* and can always be formulated as a definition of science.[73] How can one criticise such a definition? If one interprets it nominalistically,[74] a definition is a mere abbreviation, a terminological suggestion, a tautology. How can one criticise a tautology? Popper, for one, claims that his definition of science is 'fruitful' because 'a great many points can be clarified and explained with its help'. He

[72] This whole problem area is the subject of my [1968*a*], pp. 390ff, but especially of my [1974].

[73] Cf. Popper [1935], sections 4 and 11. Popper's definition of science is, of course, his celebrated 'demarcation criterion'.

[74] For an excellent discussion of the distinction between nominalism and realism (or, as Popper prefers to call it, 'essentialism') in the theory of definitions, cf. Popper [1945], vol. 2, chapter 11, and [1963*a*], p. 20.

quotes Menger: 'Definitions are dogmas; only the conclusions drawn from them can afford us any new insight'.[75] But how can a definition have explanatory power or afford new insights? Popper's answer is this: 'It is only from the consequences of my definition of empirical science, and from the methodological decisions which depend upon this definition, that the scientist will be able to see how far it conforms to his intuitive idea of the goal of his endeavours'.[76]

The answer complies with Popper's general position that conventions can be criticised by discussing their 'suitability' relative to some purpose: 'As to the suitability of any convention opinions may differ; and a reasonable discussion of these questions is only possible between parties having some purpose in common. The choice of that purpose...goes beyond rational argument'.[77] Indeed, Popper never offered a theory of rational criticism of consistent conventions. He does not raise, let alone answer, the question: '*Under what conditions would you give up your demarcation criterion?*'[78]

But the question can be answered. I give my answer in two stages: I propose first a naive and then a more sophisticated answer. I start by recalling how Popper, according to his own account,[79] arrived at his criterion. He thought, like the best scientists of his time, that Newton's theory, although refuted, was a wonderful scientific achievement; that Einstein's theory was still better; and that astrology, Freudianism and twentieth century Marxism were pseudoscientific. His problem was to find a definition of science which yielded these '*basic judgments*' concerning particular theories; and he offered a novel solution. Now let us consider the proposal that *a rationality theory – or demarcation criterion – is to be rejected if it is inconsistent with an accepted 'basic value judgment' of the scientific élite.* Indeed, this meta-methodological rule (*meta-falsificationism*) would seem to correspond to Popper's methodological rule (falsificationism) that a scientific theory is to be rejected if it is inconsistent with an ('empirical') basic statement unanimously accepted by the scientific community. Popper's whole methodology rests on the contention that there exist (relatively) singular statements on whose truth-value scientists can reach unanimous agreement; without such agreement there would be a new

[75] Popper [1935], section 11.

[76] *Ibid.*

[77] Popper [1935], section 4. But Popper, in his *Logik der Forschung*, never specifies a *purpose* of the game of science that would go beyond what is contained in its rules. The thesis that the *aim* of science is *truth*, occurs only in his writings since 1957. All that he says in his *Logik der Forschung* is that the quest for truth may be a psychological *motive* of scientists. For a detailed discussion cf. my [1974].

[78] This flaw is the more serious since Popper himself has expressed qualifications about his criterion. For instance in his [1963*a*] he describes 'dogmatism', that is, treating anomalies as a kind of 'background noise', as something that is 'to some extent necessary' (p. 49). But on the next page he identifies this 'dogmatism' with 'pseudoscience'. Is then pseudoscience 'to some extent necessary'? Also, cf. my [1970], p. 177, footnote 3.

[79] Cf. Popper [1963*a*], pp. 33–7.

Babel and 'the soaring edifice of science would soon lie in ruins'.[80] But even if there were an agreement about 'basic' statements, if there were no agreement about how to appraise scientific achievement relative to this 'empirical basis', would not the soaring edifice of science equally soon lie in ruins? No doubt it would. While there has been little agreement concerning a *universal* criterion of the scientific character of theories, there has been considerable agreement over the last two centuries concerning *single* achievements. While there has been no *general* agreement concerning a theory of scientific rationality, there has been considerable agreement concerning whether a particular single step in the game was scientific or crankish, or whether a particular gambit was played correctly or not. A general definition of science thus must reconstruct the acknowledgedly best gambits as 'scientific': if it fails to do so, it has to be rejected.[81]

Then let us propose tentatively that *if a demarcation criterion is inconsistent with the 'basic' appraisals of the scientific élite, it should be rejected.*

Now *if* we apply this quasi-empirical meta-criterion (which I am going to reject later), Popper's demarcation criterion – that is, Popper's rules of the game of science – has to be rejected.[82]

Popper's basic rule is that the scientist must specify in advance under what experimental conditions he will give up even his most basic assumptions. For instance, he writes, when criticising psychoanalysis: '*Criteria of refutation* have to be laid down beforehand: it must be agreed which observable situations, if actually observed, mean that the theory is refuted. But what kind of clinical responses would refute to the satisfaction of the analyst *not merely a particular analytic diagnosis but psychoanalysis itself*? And have such criteria ever been discussed or agreed upon by analysts?'[83] In

[80] Popper [1935], section 29.

[81] This approach, of course, does not imply that we *believe* that the scientists 'basic judgments' are unfailingly rational; it only means that we *accept* them in order to criticise universal definitions of science. (If we were to add that no such *universal* definition has been found and no such *universal* definition will ever be found, the stage would be set for Polanyi's conception of the lawless closed autocracy of science.)

My meta-criterion may be seen as a 'quasi-empirical' self-application of Popperian falsificationism. I introduced this 'quasi-empiricalness' earlier in the context of mathematical philosophy. We may abstract from *what* flows in the logical channels of a deductive system, whether it is something certain or something fallible, whether it is truth and falsehood or probability and improbability, or even moral or scientific desirability and undesirability: it is the *how* of the flow which decides whether the system is negativist, 'quasi-empirical', dominated by *modus tollens* or whether it is justificationist, 'quasi-Euclidean', dominated by *modus ponens*. (Cf. my [1967].) This 'quasi-empirical' approach may be applied to *any* kind of normative knowledge: Watkins has already applied it to ethics in his [1963] and [1967]. But now I prefer another approach: cf. note 123.

[82] It may be noted that this meta-criterion does not have to be construed as psychological, or 'naturalistic' in Popper's sense. (Cf. his [1935], section 10.) The definition of the 'scientific *élite*' is not simply an empirical matter.

[83] Popper [1963*a*], p. 38, footnote 3; my italics. This, of course, is equivalent to his celebrated 'demarcation criterion' between [internal, rationally reconstructed] science and non-science (or 'metaphysics'). The latter may be [externally] 'influential' and has to be branded as pseudoscience only if it declares itself to be science.

the case of psychoanalysis Popper was right: no answer has been forthcoming. Freudians have been nonplussed by Popper's basic challenge concerning scientific honesty. Indeed, they have refused to specify experimental conditions under which they would give up their basic assumptions. For Popper this was the hallmark of their intellectual dishonesty. But what if we put Popper's question to the Newtonian scientist: 'What kind of observation would refute to the satisfaction of the Newtonian not merely a particular Newtonian explanation but Newtonian dynamics and gravitational theory itself? And have such criteria ever been discussed or agreed upon by Newtonians?' The Newtonian will, alas, scarcely be able to give a positive answer.[84] But then if analysts are to be condemned as dishonest by Popper's standards, Newtonians must also be condemned. Newtonian science, however, in spite of this sort of 'dogmatism', is highly regarded by the greatest scientists, and, indeed, by Popper himself. Newtonian 'dogmatism' then is a 'falsification' of Popper's definition: it defies Popper's rational reconstruction.

Popper may certainly withdraw his celebrated challenge and demand falsifiability – and rejection on falsification – only for systems of theories, including initial conditions and all sorts of auxiliary and observational theories.[85] This is a considerable withdrawal, for it allows the imaginative scientist to save his pet theory by suitable lucky alterations in some odd, obscure corner on the periphery of his theoretical maze. But even Popper's mitigated rule will show up even the most brilliant scientists as irrational dogmatists. For in large research programmes there are always known anomalies: normally the researcher puts them aside and follows the positive heuristic of the programme.[86] In general he rivets his attention on the positive heuristic rather than on the distracting anomalies, and hopes that the 'recalcitrant instances' will be turned into confirming instances as the programme progresses. On Popper's terms the greatest scientists in these situations used forbidden gambits, *ad hoc* stratagems: instead of regarding Mercury's anomalous perihelion as a falsification of the Newtonian theory of our planetary system and thus as a reason for its rejection, most physicists shelved it as a problematic instance to be solved at some later stage – or offered *ad hoc* solutions. This methodological attitude of treating as (mere) *anomalies* what Popper would regard as (dramatic) counterexamples is commonly accepted by the best scientists. Some of the research programmes now held in highest esteem by the scientific community progressed in an ocean of anomalies.[87] That in their choice of problems the greatest scientists 'uncritically' ignore anomalies (and that they isolate them with the help of *ad hoc* stratagems) offers, at least on our metacriterion, a further falsification of Popper's methodology. He cannot interpret as rational some most important patterns in the growth of science.

[84] Cf. my [1970], pp. 100–1.
[85] Cf. e.g. his [1935], section 18.
[86] Cf. my [1970], especially pp. 135ff.
[87] *Ibid.*, pp. 138ff.

Furthermore, for Popper, working on *an inconsistent system* must invariably be regarded as irrational: 'a self-contradictory system must be rejected... [because it] is uninformative...No statement is singled out...since all are derivable'.[88] But some of the greatest scientific research programmes progressed on inconsistent foundations.[89] Indeed in such cases the best scientists' rule is frequently: '*Allez en avant et la foi vous viendra*'. This anti-Popperian methodology secured a breathing space both for the infinitesimal calculus and for naive set theory when they were bedevilled by logical paradoxes.

Indeed, if the game of science had been played according to Popper's rule book, Bohr's 1913 paper would never have been published because it was inconsistently grafted on to Maxwell's theory, and Dirac's delta functions would have been suppressed until Schwartz. All these examples of research based on inconsistent foundations constitute further 'falsifications' of falsificationist methodology.[90]

Thus several of the 'basic' appraisals of the scientific *élite* 'falsify' Popper's definition of science and scientific ethics. The problem then arises, to what extent, given these considerations, can falsificationism function as a guide for the historian of science. The simple answer is, to a very small extent. Popper, the leading falsificationist, never wrote any history of science; possibly because he was too sensitive to the judgment of great scientists to pervert history in a falsificationist vein. One should remember that while in his autobiographical recollections he mentions Newtonian science as the paradigm of scientificness, that is, of falsifiability, in his classical *Logik der Forschung* the falsifiability of Newton's theory is nowhere discussed. The *Logik der Forschung*, on the whole, is dryly abstract and highly ahistorical.[91] Where Popper does venture to remark casually on the falsifiability of major scientific theories, he either plunges into some logical blunder,[92] or distorts history to fit his rationality theory. If a historian's methodology provides a poor rational reconstruction, he may either misread history in such a way that it coincides with his rational reconstruction, or he will find that the history of science is highly irrational.

[88] Cf. Popper [1935], section 24. [89] Cf. my [1970], especially pp. 140ff.

[90] In general Popper stubbornly overestimates the immediate striking force of purely negative criticism. 'Once a mistake, or a contradiction, is pinpointed, there can be no verbal evasion: it can be proved, and that is that' (Popper [1959], p. 394). He adds: 'Frege did not try evasive manoeuvres when he received Russell's criticism.' But of course he did. (Cf. Frege's *Postscript* to the second edition of his *Grundgesetze*.)

[91] Interestingly, as Kuhn points out, 'a consistent interest in historical problems and a willingness to engage in original historical research distinguishes the men [Popper] has trained from the members of any other current school in the philosophy of science' (Kuhn [1970], p. 236). For a hint at a possible explanation of the apparent discrepancy cf. note 130.

[92] For instance, he claims that a perpetual motion machine would 'refute' (on his terms) the first law of thermodynamics ([1935], section 15). But how can one interpret, on Popper's own terms, the statement that '*K* is a perpetual motion machine' as a 'basic', that is, as a spatio-*temporally* singular statement?

Popper's respect for great science made him choose the first option, while the disrespectful Feyerabend chose the second.[93] Thus Popper, in his historical asides, tends to turn anomalies into 'crucial experiments' and to exaggerate their immediate impact on the history of science. Through his spectacles, great scientists accept refutations readily and this is the primary source of their problems. For instance, in one place he claims that the Michelson–Morley experiment decisively overthrew classical ether theory; he also exaggerates the role of this experiment in the emergence of Einstein's relativity theory.[94] It takes a naive falsificationist's simplifying spectacles to see, with Popper, Lavoisier's classical experiments as refuting (or as 'tending to refute') the phlogiston theory; or to see the Bohr–Kramers–Slater theory as being knocked out with a single blow from Compton; or to see the parity principle 'rejected' by 'counterexample'.[95]

Furthermore, if Popper wants to reconstruct the provisional acceptance of theories as rational on *his* terms, he is bound to ignore the historical fact that most important theories are born refuted and that some laws are further explained, rather than rejected, in spite of the known counterexamples. He tends to turn a blind eye on all anomalies known before the one which later was enthroned as 'crucial counterevidence'. For instance, he mistakenly thinks that 'neither Galileo's nor Kepler's theories were refuted before Newton'.[96] The context is significant. Popper holds that the most important pattern of scientific progress is when a crucial experi-

[93] I am referring to Feyerabend's [1970*b*] and [1971].

[94] Cf. Popper [1935], section 30 and Popper [1945], vol. 2, pp. 220–1. He stressed that Einstein's problem was how to account for experiments 'refuting' classical physics and he 'did not...set out to criticise our conceptions of space and time'. But Einstein certainly did. His Machian criticism of our concepts of space and time, and, in particular his operationalist criticism of the concept of simultaneity played an important role in his thinking.

I discussed the role of the Michelson–Morley experiments at some length in my [1970].

Popper's competence in physics would never, of course, have allowed him to distort the history of relativity theory as much as Beveridge, who wanted to persuade economists to an empirical approach by setting them Einstein as an example. According to Beveridge's falsificationist reconstruction, Einstein 'started [in his work on gravitation] from facts [which refuted Newton's theory, that is,] from the movements of the planet Mercury, the unexplained aberrancies of the moon' (Beveridge [1937]). Of course, Einstein's work on gravitation grew out from a 'creative shift' in the positive heuristic of his special relativity programme, and certainly not from pondering over Mercury's anomalous perihelion or the moon's devious, unexplained aberrancies.

[95] Popper [1963*a*], pp. 220, 239, 242–3 and [1963*b*], p. 965. Popper, of course, is left with the problem why 'counterexamples' (that is, anomalies) are not recognised immediately as causes for rejection. For instance, he points out that in the case of the breakdown of parity 'there had been many observations – that is, photographs of particle tracks – from which we might have read off the result, but the observations had been either ignored or misinterpreted' ([1963*b*], p. 965). Popper's – external – explanation seems to be that scientists have not yet learned to be sufficiently critical and revolutionary. But is it not a better – and internal – explanation that the anomalies *had* to be ignored until some progressive alternative theory was offered which turned the counterexamples into examples?

[96] *Op. cit.*, p. 246.

ment leaves one theory *unrefuted* while it refutes a rival one. But, as a matter of fact, in most, if not in all, cases where there are two rival theories, both are known to be simultaneously infected by anomalies. In such situations Popper succumbs to the temptation to simplify the situation into one to which his methodology is applicable.[97]

Falsificationist historiography is then 'falsified'. But if we apply the same meta-falsificationist method to inductivist and conventionalist historiographies, we shall 'falsify' them too.

The best logico-epistemological demolition of inductivism is, of course, Popper's; but even if we assumed that inductivism were philosophically (that is, epistemologically and logically) sound, Duhem's historiographical criticism falsifies it. Duhem took the most celebrated '*successes*' *of inductivist historiography*: Newton's law of gravitation and Ampère's electromagnetic theory. These were said to be two most victorious applications of inductive method. But Duhem (and, following him, Popper and Agassi) showed that they were not. Their analyses illustrate how the inductivist, if he wants to show that the growth of actual science is rational, must falsify actual history out of all recognition.[98] Therefore, if the rationality of science is inductive, actual science is not rational; if it is rational, it is not inductive.[99]

Conventionalism – which, unlike inductivism, is no easy prey to logical or epistemological criticism[100] – can also be historiographically falsified. One can show that the clue to scientific revolutions is not the replacement of cumbersome frameworks by simpler ones.

The Copernican revolution was generally taken to be the *paradigm of conventionalist historiography*, and it is still so regarded in many quarters. For instance Polanyi tells us that Copernicus's 'simpler picture' had 'striking beauty' and '[justly] carried great powers of conviction'.[101] But modern study of primary sources, particularly by Kuhn,[102] has dispelled

[97] As I mentioned, one Popperian, Agassi, did write a book on the historiography of science (Agassi [1963]). The book has some incisive critical sections flogging inductivist historiography, but he ends up by replacing inductivist mythology by falsificationist mythology. For Agassi *only* those facts have scientific (internal) significance which can be expressed in propositions which conflict with some extant theory: only their discovery deserves the honorific title 'factual discovery'; factual propositions which *follow from* rather than *conflict with* known theories are irrelevant; so are factual propositions which are *independent of* them. If some valued factual discovery in the history of science is known as a confirming instance or chance discovery, Agassi boldly predicts that on *close* investigation they will turn out to be refuting instances, and he offers five case studies to support his claim (pp. 60–74). Alas, on *closer* investigation it turns out that Agassi got wrong all the five examples which he adduced as confirming instances of his historiographical theory. In fact all the five examples (in our normative meta-falsificationist sense) 'falsify' his historiography.

[98] Cf. Duhem [1906], Popper [1948] and [1957*a*], Agassi [1963].

[99] Of course, an inductivist may have the temerity to claim that genuine science has not yet started and may write a history of extant science as a history of bias, superstition and false belief.

[100] Cf. Popper [1935], section 19.

[101] Cf. Polanyi [1951], p. 70.

[102] Kuhn [1957]. Also cf. Price [1959].

this myth and presented a clear-cut historiographical refutation of the conventionalist account. It is now agreed that the Copernican system was 'as least as complex as the Ptolemaic'.[103] But if this is so, then, if the acceptance of Copernican theory was rational, it was not for its superlative objective simplicity.[104]

Thus inductivism, falsificationism and conventionalism can be falsified as rational reconstructions of history with the help of the sort of historiographical criticism I have adduced.[105] Historiographical falsification of inductivism, as we have seen, was initiated already by Duhem and continued by Popper and Agassi. Historiographical criticisms of [naive] falsificationism have been offered by Polanyi, Kuhn, Feyerabend and Holton.[106] The most important historiographical criticism of conventionalism is to be found in Kuhn's – already quoted – masterpiece on the Copernican revolution.[107] The upshot of these criticisms is that all these rational reconstructions of history force history of science into the Procrustean bed of their hypocritical morality, thus creating fancy histories, which hinge on mythical 'inductive bases', 'valid inductive generalisations', 'crucial experiments', 'great revolutionary simplifications', etc. But critics of falsificationism and conventionalism drew very different conclusions from the falsification of these methodologies than Duhem, Popper and Agassi did from their own falsification of inductivism. Polanyi (and, seemingly, Holton) concluded that while proper, rational scientific appraisal can be made in *particular* cases, there can be no *general* theory of scientific rationality.[108] *All* methodologies, *all* rational reconstructions can

[103] Cohen [1960], p. 61. Bernal, in his [1954], says that '[Copernicus's] reasons for [his] revolutionary change were essentially philosophic and aesthetic [that is, in the light of conventionalism, scientific]; but in later editions he changes his mind: '[Copernicus's] reasons were mystical rather than scientific.'

[104] For a more detailed sketch cf. my [1971*b*].

[105] Other types of criticism of methodologies may, of course, be easily devised. We may, for instance, apply the standards of each methodology (not only falsificationism) to itself. The result, for most methodologies, will be equally destructive: inductivism cannot be proved inductively, simplicity will be seen as hopelessly complex. (For the latter cf. end of note 107.)

[106] Cf. Polanyi [1958], Kuhn [1962], Holton [1969], Feyerabend [1970*b*] and [1971]. I should also add Lakatos [1963–4], [1968*b*], and [1970].

[107] Kuhn [1957]. Such historiographical criticism can easily drive some rationalists into an irrational defence of their favourite falsified rationality theory. Kuhn's historiographical criticism of the simplicity theory of the Copernican revolution shocked the conventionalist historian Richard Hall so much that he published a polemic article in which he singled out and re-asserted those aspects of Copernican theory which Kuhn himself had mentioned as possibly having a claim to higher simplicity, and ignored the rest of Kuhn's – valid – argument (Hall [1970]). No doubt, simplicity can always be defined for *any* pair of theories T_1 and T_2 in such a way that the simplicity of T_1 is greater than that of T_2.

For further discussion of conventionalist historiography cf. Lakatos and Zahar [1976].

[108] Thus Polanyi is a conservative rationalist concerning science, and an 'irrationalist' concerning the philosophy of science. But, of course, this meta-'irrationalism' is a perfectly respectable brand of rationalism: to claim that the concept of 'scientifically acceptable' cannot be further defined, but only transmitted by the channels of 'personal knowledge',

be historiographically 'falsified': science *is* rational, but its rationality cannot be subsumed under the general laws of any methodology.[109] Feyerabend, on the other hand, concluded that not only can there be no general theory of scientific rationality but also that there is no such thing as scientific rationality.[110] Thus Polanyi swung towards conservative authoritarianism, while Feyerabend swung towards sceptical anarchism. Kuhn came up with a highly original vision of irrationally changing rational authority.[111]

Although, as it transpires from this section, I have high regard for Polanyi's, Feyerabend's and Kuhn's criticisms of extant ('internalist') theories of method, I drew a conclusion completely different from theirs. I decided to look for an improved methodology which offers a better *rational* reconstruction of science.

Feyerabend and Kuhn immediately tried to 'falsify' my improved methodology in turn.[112] I soon had to discover that, at least in the sense described in the present section, my methodology too – and any methodology whatsoever – *can* be 'falsified', for the simple reason that no set of human judgments is completely rational and thus no rational reconstruction can ever coincide with actual history.[113]

This recognition led me to propose a new *constructive* criterion by which methodologies *qua* rational reconstructions of history might be appraised.

does not make one an outright irrationalist, only an outright conservative. Polany's position in the philosophy of natural science corresponds closely to Oakeshott's ultra-conservative philosophy of political science. (For references and an excellent criticism of the latter cf. Watkins [1952]. Also cf. pp. 35–6).

[109] Of course, none of the critics were aware of the exact logical character of meta-methodological falsificationism as explained in this section and none of them applied it completely consistently. One of them writes: 'At this stage we have not yet developed a general theory of criticism even for scientific theories, let alone for theories of rationality: therefore if we want to falsify methodological falsificationism, we have to do it before having a theory of how to do it' (Lakatos [1970], p. 114).

[110] I use the critical machinery developed in this paper against Feyerabend's epistemological anarchism in Lakatos and Zahar [1976].

[111] Kuhn's vision was criticised from many quarters; cf. Shapere [1964] and [1967], Scheffler [1967] and especially the critical comments by Popper, Watkins, Toulmin, Feyerabend and Lakatos – and Kuhn's reply – in Lakatos and Musgrave [1970]. But none of these critics applied a systematic *historiographical* criticism to his work. One should also consult Kuhn's 1970 *Postscript* to the second edition of his [1962] and its review by Musgrave (Musgrave [1971]).

[112] Cf. Feyerabend [1970*a*], [1970*b*] and [1971]; and Kuhn [1970].

[113] For instance, one may refer to the actual immediate impact of at least *some* 'great' negative crucial experiments, like that of the falsification of the parity principle. Or one may quote the high respect for at least *some* long, pedestrian, trial-and-error procedures which occasionally precede the announcement of a major research programme, which in the light of my methodology is, at best, 'immature science'. (Cf. my [1970], p. 175; also cf. L. P. Williams's reference to the history of spectroscopy between 1870 and 1900 in his [1970].) Thus the judgment of the scientific élite, on occasions, goes also against *my* universal rules too.

B. *The methodology of historiographical research programmes. History – to varying degrees – corroborates its rational reconstructions*

I should like to present my proposal in two stages. First, I shall amend slightly the falsificationist historiographical meta-criterion just discussed, and then replace it altogether with a better one.

First, the slight amendment. If a universal rule clashes with a particular 'normative basic judgment', one should allow the scientific community time to ponder the clash: they may give up their particular judgment and submit to the general rule. 'Second-order' – historiographical – falsifications must not be rushed any more than 'first order' – scientific – ones.[114]

Secondly, since we have abandoned naive falsificationism in *method*, why should we stick to it in *meta-method*? We can easily replace it with a methodology of scientific research programmes of second order, or if you wish, a methodology of historiographical research programmes.

While maintaining that a theory of rationality has to try to organise basic value judgments in universal, coherent frameworks, we do not have to reject such a framework immediately merely because of some anomalies or other inconsistencies. We should, of course, insist that a good rationality theory must anticipate further basic value judgments unexpected in the light of its predecessors or that it must even lead to the revision of previously held basic value-judgments.[115] We then reject a rationality theory only for a better one, for one which, in this 'quasi-empirical' sense, represents a *progressive shift* in the sequence of research programmes of rational reconstructions. Thus this new – more lenient – meta-criterion enables us to compare rival logics of discovery and discern growth in 'meta-scientific' – methodological – knowledge.

For instance, Popper's theory of scientific rationality need not be rejected simply because it is 'falsified' by some actual 'basic judgments' of leading scientists. Moreover, on our new criterion, Popper's demarcation criterion clearly represents progress over its justificationist predecessors, and in particular, over inductivism. For, contrary to these predecessors, it rehabilitated the scientific status of falsified theories like phlogiston theory, thus reversing a value judgment which had expelled the latter from the history of science proper into the history of irrational beliefs.[116]

[114] There is a certain analogy between this pattern and the occasional appeal procedure of the theoretical scientist against the verdict of the experimental jury; cf. my [1970], pp. 127–31.

[115] This latter criterion is analogous to the exceptional 'depth' of a theory which clashes with some basic statements available at the time and, at the end, emerges from the clash victoriously. (Cf. Popper's [1957*a*].) Popper's example was the inconsistency between Kepler's laws and the Newtonian theory which set out to explain them.

[116] Conventionalism, of course, had performed this historic role to a great extent before Popper's version of falsificationism.

Also, it successfully rehabilitated the Bohr–Kramers–Slater theory.[117] In the light of most justificationist theories of rationality the history of science is, at its best, a history of *pre*scientific preludes to some *future* history of science.[118] Popper's methodology enabled the historian to interpret more of the *actual* basic value judgments in the history of science as rational: in *this* normative–historiographical sense Popper's theory constituted progress. In the light of better rational reconstructions of science one can always reconstruct more of actual great science as rational.[119]

I hope that my modification of Popper's logic of discovery will be seen, in turn – on the criterion I specified – as yet a further step forward. For it seems to offer a coherent account of *more* old, isolated basic value judgments; moreover, it has led to new and, at least for the justificationist or naive falsificationist, surprising basic value judgments. For instance, according to Popper's theory, it was irrational to retain and further elaborate Newton's gravitational theory after the discovery of Mercury's anomalous perihelion; or again, it was irrational to develop Bohr's old quantum theory based on inconsistent foundations. From my point of view these were perfectly rational developments: some rearguard actions in the defence of defeated programmes – even after the so-called 'crucial experiments' – are perfectly rational. Thus my methodology leads to the reversal of those historiographical judgments which deleted these rearguard actions both from inductivist and from falsificationist party histories.[120]

Indeed, this methodology confidently predicts that where the falsificationist sees the instant defeat of a theory through a simple battle with some fact, the historian will detect a complicated war of attrition, starting long before, and ending after, the alleged 'crucial experiment'; and where the falsificationist sees consistent and unrefuted theories, it predicts the existence of hordes of known anomalies in research programmes progressing on possibly inconsistent foundations.[121] Where the conventionalist sees the clue to the victory of a theory over its predecessor in the former's intuitive simplicity, this methodology predicts that it will be found that victory was due to empirical degeneration in the old and empirical progress in the new programme.[122] Where Kuhn and Feyerabend see irrational

[117] van der Waerden had thought that the Bohr–Kramers–Slater theory was bad: Popper's theory showed it to be good. Cf. van der Waerden [1967], p. 13 and Popper [1963*a*], pp. 242ff; for a critical discussion cf. my [1970], p. 168, footnote 4 and p. 169, footnote 1.

[118] The attitude of some modern logicians to the history of mathematics is a typical example; cf. my [1963–4], p. 3.

[119] This formulation was suggested to me by my friend Michael Sukale.

[120] Cf. my [1970], section 3(*c*).

[121] Cf. my [1970], pp. 138–73.

[122] Duhem himself gives only one explicit example: the victory of wave optics over Newtonian optics [1906], chapter VI, §10 (also see chapter IV, §4). But where Duhem relies on intuitive 'common sense', I rely on an analysis of rival problemshifts.

change, I predict that the historian will be able to show that there has been rational change. The methodology of research programmes thus predicts (or, if you wish, 'postdicts') novel historical facts, unexpected in the light of extant (internal and external) historiographies and these predictions will, I hope, be corroborated by historical research. If they are, then the methodology of scientific research programmes will itself constitute a progressive problemshift.

Thus progress in the theory of scientific rationality is marked by discoveries of novel historical facts, by the reconstruction of a growing bulk of value-impregnated history as rational.[123] In other words, the theory of scientific rationality progresses if it constitutes a 'progressive' historiographical research programme. I need not say that no such historiographical research programme can or should explain *all* history of science as rational: even the greatest scientists make false steps and fail in their judgment. Because of this *rational reconstructions remain for ever submerged in an ocean of anomalies. These anomalies will eventually have to be explained either by some better rational reconstruction or by some 'external' empirical theory.*

This approach does not advocate a cavalier attitude to the 'basic normative judgments' of the scientist. 'Anomalies' may be rightly ignored by the internalist *qua* internalist and relegated to external history only as long as the internalist historiographical research programme is *progressing*; or if a supplementary empirical externalist historiographical programme absorbs them *progressively*. But if in the light of a rational reconstruction the history of science is seen as increasingly irrational *without* a progressive externalist explanation (such as an explanation of the degeneration of science in terms of political or religious terror, or of an antiscientific ideological climate, or of the rise of a new parasitic class of pseudoscientists with vested interests in rapid 'university expansion'), then historiographical innovation, proliferation of historiographical theories, is vital. Just as scientific progress is possible even if one never gets rid of scientific anomalies, progress in rational historiography is also possible even if one never gets rid of historiographical anomalies. The rationalist historian need not be disturbed by the fact that actual history is more than, and, on occasions, even different from, internal history, and that he may have to relegate the explanation of such anomalies to external history. But this unfalsifiability of internal history does not render it immune to construc-

[123] One may introduce the notion of '*degree of correctness*' into the meta-theory of methodologies, which would be analogous to Popper's empirical content. Popper's empirical 'basic statements' would have to be replaced by quasi-empirical 'normative basic statements' (like the statement that 'Planck's radiation formula is arbitrary').

Let me point out here that the methodology of research programmes may be applied not only to norm-impregnated historical knowledge but to any normative knowledge, including even ethics and aesthetics. This would then supersede the naive falsificationist 'quasi-empirical' approach as outlined on note 81.

tive, but only to negative, criticism – just as the unfalsifiability of a scientific research programme does not render it immune to constructive, but only to negative, criticism.

Of course, one can criticise internal history only by making the historian's (usually latent) methodology explicit, showing how it functions as a historiographical research programme. Historiographical criticism frequently succeeds in destroying much of fashionable externalism. An 'impressive', 'sweeping', 'far-reaching' external explanation is usually the hallmark of a weak methodological substructure; and, in turn, the hallmark of a relatively weak internal history (in terms of which most actual history is either inexplicable or anomalous) is that it leaves too much to be explained by external history. When a better rationality theory is produced, internal history may expand and reclaim ground from external history. The competition, however, is not as open in such cases as when two rival scientific research programmes compete. Externalist historiographical programmes which supplement internal histories based on naive methodologies (whether aware or unaware of the fact) are likely either to degenerate quickly or never even to get off the ground, for the simple reason that they set out to offer psychological or sociological 'explanations' of methodologically induced fantasies rather than of (more rationally interpreted) historical facts. Once an externalist account uses, whether consciously or not, a naive methodology (which can so easily creep into its 'descriptive' language), it turns into a fairy tale which, for all its apparent scholarly sophistication, will collapse under historiographical scrutiny.

Agassi has already indicated how the poverty of inductivist history opened the door to the wild speculations of vulgar-Marxists.[124] His falsificationist historiography, in turn, flings the door wide open to those trendy 'sociologists of knowledge' who try to explain the further (possibly unsuccessful) development of a theory 'falsified' by a 'crucial experiment' as the manifestation of the irrational, wicked, reactionary resistance by established authority to enlightened revolutionary innovation.[125] But in the light of the methodology of scientific research programmes such rear-guard skirmishes are perfectly explicable *internally*: where some externalists

[124] Cf. text to note 9. (The terminology 'wild speculation' is, of course, inherited from inductivist methodology. It should now be reinterpreted as 'degenerating programme'.)

[125] The fact that even degenerating externalist theories have been able to achieve some respectability was to a considerable extent due to the weakness of their previous internalist rivals. Utopian Victorian morality either creates false, hypocritical accounts of bourgeois decency, or adds fuel to the view that mankind is totally depraved; utopian scientific standards either create false, hypocritical accounts of scientific perfection, or add fuel to the view that scientific theories are no more than mere beliefs bolstered by some vested interests. This explains the 'revolutionary' aura which surrounds some of the absurd ideas of contemporary sociology of knowledge: some of its practitioners claim to have unmasked the bogus rationality of science, while, at best, they exploit the weakness of outdated theories of scientific rationality.

see power struggle, sordid personal controversy, the rationalist historian will frequently find rational discussion.[126]

An interesting example of how a poor theory of rationality may impoverish history is the treatment of degenerating problemshifts by historiographical positivists.[127] Let us imagine for instance that in spite of the objectively progressing astronomical research programmes, the astronomers are suddenly all gripped by a feeling of Kuhnian 'crisis'; and then they all are converted, by an irresistible *Gestalt*-switch, to astrology. I would regard this catastrophe as a horrifying *problem*, to be accounted for by some empirical externalist explanation. But not a Kuhnian. All he sees is a 'crisis' followed by a mass conversion effect in the scientific community: an ordinary revolution. Nothing is left as problematic and unexplained.[128] The Kuhnian psychological epiphenomena of 'crisis' and 'conversion' can accompany either objectively progressive or objectively degenerating changes, either revolutions or counterrevolutions. But this fact falls outside Kuhn's framework. Such historiographical anomalies cannot be formulated, let alone be progressively absorbed, by his historiographical research programme, in which there is no way of distinguishing between, say, a 'crisis' and 'degenerating problemshift'. But such anomalies might even be predicted by an externalist historiographical theory based on the methodology of scientific research programmes that would specify social conditions under which degenerating research programmes may achieve socio-psychological victory.

[126] For examples cf. Cantor [1971] and the Forman–Ewald debate (Forman [1969] and Ewald [1969]).

[127] I call '*historiographical positivism*' the position that history can be written as a completely *external* history. For historiographical positivists history is a purely empirical discipline. They deny the existence of objective standards as opposed to mere beliefs about standards. (Of course, they too hold beliefs about standards which determine the choice and formulation of their historical problems.) This position is typically Hegelian. It is a special case of *normative positivism*, of the theory that sets up might as the criterion of right. (For a criticism of Hegel's ethical positivism cf. Popper [1945], vol. 1, pp. 71–2, vol. 2, pp. 305–6 and Popper [1961].) Reactionary Hegelian obscurantism pushed values back completely into the world of facts; thus reversing their separation by Kantian philosophical enlightenment.

[128] Kuhn seems to be in two minds about objective scientific progress. I have no doubt that, being a devoted scholar and scientist, he *personally* detests relativism. But his *theory* can either be interpreted as denying scientific progress and recognising only scientific change; or, as recognising scientific progress but as 'progress' marked solely by the march of actual history. Indeed, on his criterion, he would have to describe the catastrophe mentioned in the text as a proper 'revolution'. I am afraid this might be one clue to the unintended popularity of his theory among the New Left busily preparing the 1984 'revolution'.

C. *Against aprioristic and antitheoretical approaches to methodology*

Finally, let us contrast the theory of rationality here discussed with the strictly aprioristic (or, more precisely, 'Euclidean') and with the antitheoretical approaches.[129]

'Euclidean' methodologies lay down *a priori general rules* for scientific appraisal. This approach is most powerfully represented today by Popper. In Popper's view there must be the constitutional authority of an *immutable statute law* (laid down in his demarcation criterion) to distinguish between good and bad science.

Some eminent philosophers, however, ridicule the idea of statute law, the possibility of any valid demarcation. According to Oakeshott and Polanyi there must be – and can be – no statute law at all: only case law. They may also argue that even if one mistakenly allowed for statute law, statute law too would need authoritative interpreters. I think that Oakeshott's and Polanyi's position has a great deal of truth in it. After all, one must admit (*pace* Popper) that until now all the 'laws' proposed by the apriorist philosophers of science have turned out to be wrong in the light of the verdicts of the best scientists. Up to the present day it has been the scientific standards, as applied 'instinctively' by the scientific *élite* in *particular* cases, which have constituted the main – although not the exclusive – yardstick of the philosopher's *universal* laws. But if so, methodological progress, as least as far as the most advanced sciences are concerned, still lags behind common scientific wisdom. Is it not then *hubris* to try to impose some *a priori* philosophy of science on the most advanced sciences? Is it not *hubris* to demand that if, say, Newtonian or Einsteinean science turns out to have violated Bacon's, Carnap's or Popper's *a priori* rules of the game, the business of science should be started anew?

I think it is. And, indeed, the methodology of historiographical research programmes implies a pluralistic system of authority, partly because the wisdom of the scientific jury and its case law has not been, and cannot be, fully articulated by the philosopher's statute law, and partly because the philosopher's statute law may occasionally be right when the scientists' judgment fails. I disagree, therefore, both with those philosophers of science who have taken it for granted that general scientific standards are immutable and reason can recognise them *a priori*,[130] and with those who have thought that the light of reason illuminates only particular cases.

[129] The technical term 'Euclidean' (or rather 'quasi-Euclidean') means that one starts with universal, high level propositions ('axioms') rather than singular ones. I suggested in my [1967] and [1962] that the 'quasi-Euclidean' versus 'quasi-empirical' distinction is more useful than the '*a priori*' versus '*a posteriori*' distinction.

Some of the 'apriorists' are, of course, empiricists. But empiricists may well be apriorists (or, rather, 'Euclideans') on the meta-level here discussed.

[130] Some might claim that Popper does *not* fall into this category. After all, Popper defined 'science' in such a way that it should include the refuted Newtonian theory and exclude unrefuted astrology, Marxism and Freudianism.

The methodology of historiographical research programmes specifies ways both for the philosopher of science to learn from the historian of science and *vice versa*.

But this two-way traffic need not always be balanced. The statute law approach should become much more important when a tradition degenerates[131] or a new bad tradition is founded.[132] In such cases statute law may thwart the authority of the corrupted case law, and slow down or even reverse the process of degeneration.[133] When a scientific school degenerates into pseudoscience, it may be worthwhile to force a methodological debate in the hope that working scientists will learn more from it than philosophers (just as when ordinary language degenerates into, say, journalese, it may be worthwhile to invoke the rules of grammar).[134]

D. *Conclusion*

In this paper I have proposed a 'historical' method for the evaluation of rival methodologies. The arguments were primarily addressed to the philosopher of science and aimed at showing how he can – and should – learn from the history of science. But the same arguments also imply that the historian of science must, in turn, pay serious attention to the philosophy of science and decide upon which methodology he will base his internal history. I hope to have offered some strong arguments for the following theses. First, each methodology of science determines a characteristic (and sharp) demarcation between (primary) internal history and (secondary) external history and, secondly, both historians and philosophers of science must make the best of the critical interplay between internal and external factors.

Let me finally remind the reader of my favourite – and by now well-worn – joke that history of science is frequently a caricature of its rational reconstructions; that rational reconstructions are frequently caricatures of actual history; and that some histories of science are caricatures both of actual history and of its rational reconstructions.[135] This paper, I think, enables me to add: *Quod erat demonstrandum.*

[131] This seems to be the case in modern particle physics; or according to some philosophers and physicists even in the Copenhagen school of quantum physics.

[132] This is the case with some of the main schools of modern sociology, psychology and social psychology.

[133] This, of course, explains why a good methodology – 'distilled' from the mature sciences – may play an important role for immature and, indeed, dubious disciplines. While Polanyiite academic autonomy should be defended for departments of theoretical physics, it must not be tolerated, say, in institutes for computerised social astrology, science planning or social imagistics. (For an authoritative study of the latter, cf. Priestley [1968].)

[134] Of course, a critical discussion of scientific standards, possibily leading even to their improvement, is impossible without articulating them in general terms; just as if one wants to challenge a language, one has to articulate its grammar. Neither the conservative Polanyi nor the conservative Oakeshott seem to have grasped (or to have been inclined to grasp) the *critical* function of language – Popper has. (Cf. especially Popper [1963*a*], p. 135.)

[135] Cf. e.g. my [1962], p. 157 or my [1968*a*], p. 387, footnote 1.

References

Agassi, J. [1963]: *Towards an Historiography of Science.*

Agassi, J. [1964]: 'Scientific Problems and their Roots in Metaphysics', in M. Bunge (ed.): *The Critical Approach to Science and Philosophy*, pp. 189–211.

Agassi, J. [1966]: 'Sensationalism', *Mind*, **75**, pp. 1–24.

Agassi, J. [1969]: 'Popper on Learning from Experience', in N. Rescher (ed.): *Studies in the Philosophy of Science*, pp. 162–71.

Bernal, J. D. [1954]: *Science in History*, 1st edition.

Bernal, J. D. [1965]: *Science in History*, 3rd edition.

Beveridge, W. [1937]: 'The Place of the Social Sciences in Human Knowledge', *Politica*, **2**, pp. 459–79.

Cantor, G. [1971]: 'Henry Brougham and the Scottish Methodological Tradition', *Studies in the History and Philosophy of Science*, **2**, pp. 69–89.

Cohen, I. B. [1960]: *The Birth of a New Physics.*

Compton, A. H. [1919]: 'The Size and Shape of the Electron', *Physical Review*, **14**, pp. 20–43.

Duhem, P. [1906]: *La théorie physique, son objet et sa structure* (English translation of 2nd (1914) edition: *The Aim and Structure of Physical Theory*, 1954).

Elkana, Y. [1971]: 'The Conservation of Energy: a Case of Simultaneous Discovery?', *Archives Internationales d'Histoire des Sciences*, **24**, pp. 31–60.

Ewald, P. [1969]: 'The Myth of Myths', *Archive for History of Exact Sciences*, **6**, pp. 72–81.

Feyerabend, P. K. [1964]: 'Realism and Instrumentalism: Comments on the Logic of Factual Support', in M. Bunge (ed.): *The Critical Approach to Science and Philosophy*, pp. 280–308.

Feyerabend, P. K. [1965]: 'Reply to Criticism', in R. S. Cohen and M. Wartofsky (eds.): *Boston Studies in the Philosophy of Science* **2**, pp. 223–61.

Feyerabend, P. K. [1969]: 'A Note on Two "Problems" of Induction', *British Journal for the Philosophy of Science*, **19**, pp. 251–3.

Feyerabend, P. K. [1970*a*]: 'Consolations for the Specialist', in I. Lakatos and A. Musgrave (eds.): *Criticism and the Growth of Knowledge*, pp. 197–230.

Feyerabend, P. K. [1970*b*]: 'Against Method', in *Minnesota Studies for the Philosophy of Science* **4**.

Feyerabend, P. K. [1971]: *Against Method* [expanded version of Feyerabend [1970*b*]].

Forman, P. [1969]: The Discovery of the Diffraction of X-Rays by Crystals: A Critique of the Critique of the Myths, *Archive for History of Exact Sciences*, **6**, pp. 38–71.

Hall, R. J. [1970]: 'Kuhn and the Copernican Revolution', *British Journal for the Philosophy of Science*, **21**, pp. 196–7.

Hempel, C. G. [1937]: Review of Popper [1934], *Deutsche Literaturzeitung*, pp. 309–14.

Holton, G. [1969]: 'Einstein, Michelson, and the "Crucial" Experiment', *Isis*, **6**, pp. 133–97.

Kuhn, T. S. [1957]: *The Copernican Revolution.*

Kuhn, T. S. [1962]: *The Structure of Scientific Revolutions.*

Kuhn, T. S. [1968]: 'Science: The History of Science', in D. L. Sills (ed.): *International Encyclopedia of the Social Sciences*, vol. 14, pp. 74–83.

Kuhn, T. S. [1970]: 'Reflections on my Critics', in I. Lakatos and A. Musgrave (eds.): *Criticism and the Growth of Knowledge*, pp. 237–78.

Lakatos, I. [1962]: 'Infinite Regress and the Foundations of Mathematics', *Aristotelian Society Supplementary Volume*, **36**, pp. 155–84.

Lakatos, I. [1963–4]: 'Proofs and Refutations', *The British Journal for the Philosophy of Science*, **14**, pp. 1–25, 120–39, 221–43, 296–342.
Lakatos, I. [1966]: 'Popkin on Skepticism', in W. Yourgrau and A. D. Breck (eds.): *Logic, Physics and History*, 1970, pp. 220–3.
Lakatos, I. [1967]: 'A Renaissance of Empiricism in the Recent Philosophy of Mathematics', in I. Lakatos (ed.): *Problems in the Philosophy of Mathematics*, pp. 199–202.
Lakatos, I. [1968*a*]: 'Changes in the Problem of Inductive Logic', in I. Lakatos (ed.): *The Problem of Inductive Logic*, pp. 315–417.
Lakatos, I. [1968*b*]: 'Criticism and the Methodology of Scientific Research Programmes', *Proceedings of the Aristotelian Society*, **69**, pp. 149–86.
Lakatos, I. [1970]: 'Falsification and the Methodology of Scientific Research Programmes', in I. Lakatos and A. Musgrave (eds.): *Criticism and the Growth of Knowledge*, pp. 91–196.
Lakatos, I. [1974]: 'Popper on Demarcation and Induction' in P. A. Schilpp (ed.): *The Philosophy of Sir Karl Popper*. (Available in German in *Neue Aspekte der Wissenschaftstheorie*, ed. H. Lenk.)
Lakatos, I. and Musgrave, A. [1970]: *Criticism and the Growth of Knowledge*.
Lakatos, I. and Zahar, E. G. [1976]: 'Why did Copernicus's Programme Supersede Ptolemy's?', in R. Westman (ed.): *The Copernican Achievement*.
McMullin, E. [1970]: 'The History and Philosophy of Science: a Taxonomy', *Minnesota Studies in the Philosophy of Science*, **5**, pp. 12–67.
Merton, R. [1957]: 'Priorities in Scientific Discovery', *American Sociological Review*, **22**, pp. 635–59.
Merton, R. [1963]: 'Resistance to the Systematic Study of Multiple Discoveries in Science', *European Journal of Sociology*, **4**, pp. 237–82.
Merton, R. [1969]: 'Behaviour Patterns of Scientists', *American Scholar*, **38**, pp. 197–225.
Musgrave, A. [1969]: *Impersonal Knowledge: A Criticism of Subjectivism*, Ph.D. thesis, University of London.
Musgrave, A. [1971]: 'Kuhn's Second Thoughts', *British Journal for the Philosophy of Science*, **22**, pp. 287–97.
Musgrave, A. [1974]: 'The Objectivism of Popper's Epistemology', in P. A. Schilpp (ed.): *The Philosophy of Sir Karl Popper*.
Polanyi, M. [1951]: *The Logic of Liberty*.
Polanyi, M. [1958]: *Personal Knowledge, Towards a Post-Critical Philosophy*.
Popper, K. R. [1935]: *Logik der Forschung*.
Popper, K. R. [1940]: 'What is Dialectic?', *Mind*, **49**, pp. 403–26; reprinted in Popper [1963], pp. 312–35.
Popper, K. R. [1945]: *The Open Society and Its Enemies*, 2 vols.
Popper, K. R. [1948]: 'Naturgesetze und theoretische Systeme', in *Gesetz und Wirklichkeit* (ed. S. Moser), pp. 65–84.
Popper, K. R. [1957*a*]: 'The Aim of Science', *Ratio* **1**, pp. 24–35.
Popper, K. R. [1957*b*]. *The Poverty of Historicism*.
Popper, K. R. [1959]: *The Logic of Scientific Discovery*.
Popper, K. R. [1960]: 'Philosophy and Physics', *Atti del XII Congresso Internazionale di Filosofia*, **2**, pp. 363–74.
Popper, K. R. [1961]: 'Facts, Standards, and Truth: A Further Criticism of Relativism', *Addendum* to the Fourth Edition of Popper [1945].
Popper, K. R. [1963*a*]: *Conjectures and Refutations*.
Popper, K. R. [1963*b*]: 'Science: Problems, Aims, Responsibilities', *Federation Proceedings*, **22**, pp. 961–72.

Popper, K. R. [1963*c*]: 'Three Views Concerning Human Knowledge', in H. D. Lewis (ed.): *Contemporary British Philosophy*, 1957, pp. 355–88; reprinted in Popper [1963*a*], pp. 97–119.

Popper, K. R. [1968*a*]: 'Epistemology Without a Knowing Subject', in B. Rootselaar and J. Staal (eds.): *Proceedings of the Third International Congress for Logic, Methodology and Philosophy of Science*, pp. 333–73.

Popper, K. R. [1968*b*]: 'On the Theory of the Objective Mind', in *Proceedings of the XIV International Congress of Philosophy*, vol. 1, pp. 25–33.

Price, D. J. [1959]: 'Contra Copernicus: A Critical Re-estimation of the Mathematical Planetary Theory of Ptolemy, Copernicus and Kepler', in M. Clagett (ed.): *Critical Problems in the History of Science*, pp. 197–218.

Priestley, J. B. [1968]: *The Image Men.*

Scheffler, I. [1967]: *Science and Subjectivity.*

Shapere, D. [1964]: 'The Structure of Scientific Revolutions', *Philosophical Review*, **63**, pp. 383–84.

Shapere, S. [1967]: 'Meaning and Scientific Change', in R. G. Colodny (ed.): *Mind and Cosmos*, pp. 41–85.

van der Waerden, B. [1967]: *Sources of Quantum Mechanics.*

Watkins, J. W. N. [1952]: 'Political Tradition and Political Theory: an Examination of Professor Oakeshott's Political Philosophy', *Philosophical Quarterly*, **2**, pp. 323–37.

Watkins, J. W. N. [1958]: 'Influential and Confirmable Metaphysics', *Mind*, **67**, pp. 344–65.

Watkins, J. W. N. [1963]: 'Negative Utilitarianism', *Aristotelian Society Supplementary Volume 37*, pp. 95–114.

Watkins, J. W. N. [1967]: 'Decision and Belief', in R. Hughes (ed.): *Decision Making*, pp. 9–26.

Watkins, J. W. N. [1970]: 'Against Normal Science', in I. Lakatos and A. Musgrave (eds.): *Criticism and the Growth of Knowledge*, pp. 25–38.

Williams, L. P. [1970]: 'Normal Science and its Dangers', in I. Lakatos and A. Musgrave (eds.): *Criticism and the Growth of Knowledge*, pp. 49–50.

Atomism versus thermodynamics*

PETER CLARK

LONDON SCHOOL OF ECONOMICS AND POLITICAL SCIENCE

1 *Introduction*

Throughout the nineteenth century there were two quite separate approaches to the problems posed by thermal and thermochemical phenomena. The first, the mechanical theory of heat, which developed into a fully-fledged phenomenological thermodynamics, was based upon two very general empirical laws, independent of any hypothesis as to the ultimate nature of matter.[1] The second, the kinetic theory, on the

* My greatest personal and intellectual debt is to the late Imre Lakatos. His methodology of scientific research programmes was the crucial philosophical influence on this paper. I gratefully acknowledge the helpful criticisms of earlier drafts of this paper made by Jon Dorling, Adolf Grünbaum, Erwin Hiebert, Colin Howson, David Lavis, Alan Musgrave, Heinz Post, John Stachel, Peter Urbach, John Watkins, John Worrall and especially Elie Zahar, none of whom should, however, be taken as agreeing with all I say. This is a very much revised version of a paper read before the British Society for the Philosophy of Science, on 29 April 1974 and at the Nafplion Colloquium, September 1974.

[1] Clausius for instance regarded the independence of the mechanical theory from any hypothesis as to the nature of matter as a special merit of the theory. Indeed he informs us that he had taken 'especial care to base the development of the equations which enter into the mechanical theory of heat upon certain general axioms, and not upon particular views regarding the molecular constitution of bodies, and accordingly should be inclined

contrary began with specific assumptions as to the constitution of matter, viz. that it was discrete, molecular, ultimately atomic, and that heat was a 'concealed' form of motion associated with the molecules of a substance.

The kinetic theory is now regarded (rightly) as one of the greatest achievements of nineteenth century physics. However, in the last decade of that century it was subject to severe attacks from some of the leading scientists of the day. Planck, for example, regarded the theory as faced with 'insurmountable obstacles' such that 'every attempt at elaborating the theory has not only not led to new physical results but has run into overwhelming difficulties'.[2] Similarly Ostwald saw in the theory 'a superficial habit to cover up rather than promote actual scientific tasks by arbitrary assumptions about atomic positions, motions and vibrations', which in his opinion did 'great harm to science'.[3] Duhem poured scorn on the various atomic models proposed as a basis for the kinetic theory and remarked of the latter theory's relationship to thermodynamics that 'Thermodynamics had reached maturity and constitutional vigour when the kinetic hypothesis came along to bring it the assistance it did not ask for, with which it has nothing to do and to which it owed nothing.'[4]

This opposition to the kinetic theory is often attributed by modern commentators to the existence of a fashionable philosophy of science which dismissed as unacceptable *any* theory based upon speculative, unobservable entities, such as atoms. For instance, Einstein remarked of (in particular) Ostwald and Mach that:

> the prejudices of these scientists against the atomic theory can be undoubtedly attributed to their positivistic philosophical views. This is an interesting example of how philosophical prejudices hinder a correct interpretation of facts even by scientists with bold thinking and subtle intuition.[5]

The same sentiment is echoed by Brush when he writes,

> Those scientists who did suggest that the kinetic theory be abandoned in the later 19th Century did so not because of empirical difficulties but because of a more deep seated *purely* philosophical objection. For those who believed in a positivist methodology, any theory based on invisible and undetectable atoms was unacceptable.[6]

to regard [this] treatment of the subject as the more appropriate one...' (Clausius [1863], pp. 273–4; cf. Clausius [1850]). Here Clausius makes a sharp distinction between general thermodynamic principles and particular hypotheses as to the nature of heat.

[2] Planck [1891], p. 650.

[3] Ostwald [1927], pp. 178–9.

[4] Duhem [1906], p. 95.

[5] Quoted by Suvorov (Suvorov [1966], p. 578). Cf. Einstein [1949], pp. 21, 49.

[6] Brush [1974], p. 1169, my italics. Nye in her [1972] repeats the thesis that those who supported and used the methods of thermodynamics 'began from an experimentalist positivist viewpoint'. The opposition to the kinetic theory here again is attributed not to that theory's inadequacies but to a blanket application of a particular philosophy of science.

Similarly Jaynes argues that:

> the rise of the school of 'Energetics' championed by Mach and Ostwald, represents an early attempt of the positivist philosophy *to limit the scope of science*. This school held that to use modern terminology the atom was not an 'observable', and that physical theories should not, *therefore*, make use of the concept.[7]

This thesis of philosophical prejudice is introduced to explain away objections to the kinetic theory, in terms of an extra-scientific, external influence, viz. the dominance of an anti-atomistic, sensationalist philosophy. Other *external explanations*[8] have been employed to account for the opposition to kinetic theories at the turn of the century. For example D'Abro writes of those who preferred thermodynamics to kinetic theories,

> their criticisms appear to have been dictated by preference rather than by reason. At all events their views found favour with many physicists who, though of first rate ability in the experimental field, never evidenced any particular liking for mathematics...one cannot help suspecting that the more difficult mathematical techniques of the microscopic theory may have influenced their philosophy.[9]

I shall argue on the contrary, that the history of what is now called the kinetic theory must be appraised as the development of a powerful research programme, which after some early notable successes, in the last decade of the nineteenth century, was *degenerating*. In contrast the research programme of pure thermodynamics was *progressive* from its inception. Thermodynamics and the atomic–kinetic research programme were two quite *distinct* research programmes. Many scientists (e.g. Clausius, Maxwell, Helmholtz and Boltzmann) contributed to the development of both programmes in separate publications, but both programmes retained their separate identity. Each possessed a distinct hard core and employed quite different heuristic techniques. Part of the problem situation of the atomic–kinetic programme was, of course, to provide an explanation of the two laws of thermodynamics in terms of the aggregate behaviour of systems involving an enormous number of molecules.[10] However, Boltzmann's entropy theorem actually contradicted the phenomenological second law. My claim is that it is precisely these two factors, the degeneration of the atomic–kinetic programme and the fact that one of its central achievements (the entropy theorem) was incompatible with the phenomenological second law, which formed the basis of the objections to the programme.[11]

[7] Jaynes, [1967], p. 80, my italics.

[8] For the crucial distinction between 'Internal' and 'External' history, cf. Lakatos [1971], pp. 1–9.

[9] D'Abro [1939], p. 92.

[10] Clausius, Szily and Helmholtz attempted to do so from mechanics alone, i.e. from the principle of least action alone (cf. especially Bryan [1891], pp. 87–121). Maxwell and Boltzmann employed both mechanics and statistics, cf. below, pp. 50–7.

[11] I do not argue for the supersession of one research programme by another here. My thesis does *not* entail that those scientists (the vast majority both in Britain and on the Continent) who stuck to the atomic-kinetic programme acted irrationally. It would be

My second thesis concerns the research programme of thermodynamics. I shall show that the heuristic of thermoydnamics was limited, in a quite well-founded sense: that it was *fact dependent*. That is, the heuristic was limited to methods for drawing consequences from low-level empirical generalisations about substances in conjunction with the second law and to resolving anomalies by the adjustment of parameters to fit the experimental facts. This quite marked limitation of the heuristic of thermodynamics meant that there was no way of systematically improving the theory.[12] This I think captures objectively the rationality of those scientists like Planck and Ostwald who while rejecting the kinetic programme as an inadequate, now *stagnant* attempt to provide a foundation for thermodynamics (which would provide a means for systematically extending it), tried to develop alternative research programmes to do just that.

However, as a matter of fact, the attempts to develop a research programme to supersede thermodynamics in the late nineteenth century were unsuccessful. Like the kinetic programme they degenerated. It was the degeneration of these attempts to provide a foundation for thermodynamics in terms of some deeper theory, compared with the empirical progress of the phenomenological programme, which led to the elevation of thermodynamics into a 'paradigm' of great science. For Duhem and Mach, for instance, it was precisely the economical classification of the facts which characterised thermodynamics. This they elevated into a general demarcation criterion. In my view it was the degeneration of the kinetic programme compared with the empirical progress of thermodynamics which accounts for the rise of scientific positivism.

In the remainder of my paper I shall argue for the conclusions sketched above. I shall first show that the early kinetic programme was progressive (§2). Then that thermodynamics, though a progressive research programme, had a limited heuristic (§3), and that the kinetic programme degenerated after 1880 (§4). Finally I shall show that the kinetic programme became empirically progressive after 1905, with the prediction, for example, of the existence and magnitude of the Brownian motion.

Throughout this paper I shall employ the explicit criteria of appraisal provided by the methodology of research programmes. In locating the hard core and heuristic of a programme what is important is not what working scientists *say* about what they were doing, nor their *beliefs* about the truth or falsity of the specific theories they espoused, but what principles they *adopted and followed in practice* in developing their theories.

wrong to argue that sticking to a degenerating research programme is irrational, for one may well decide to work on such programmes in order to try to *improve* them. Cf. below, pp. 88–93. Indeed the atomic-kinetic programme provides a paradigm case of a research programme which became empirically progressive (the prediction for example of density and thermal fluctuations) after a degenerating phase.

[12] Cf. *below*, pp. 74–7.

2 *The atomic–kinetic research programme*

2.1 *The hard core and positive heuristic*

The *hard core* of the atomic–kinetic research programme consisted simply of the proposition that:

> the behaviour and nature of substances is the *aggregate* of an enormously large number of very small and constantly moving elementary individuals subject to the laws of mechanics.

This is a very general, metaphysical statement about the constitution of matter. Testable versions of it were generated as a sequence of particular kinetic theories by the positive heuristic of the programme. This heuristic consisted of four methodological directives:

(i) Make specific assumptions as to the nature of the elementary individuals and as to their available degrees of freedom, such that *all* interactions among them are subject to the laws of mechanics.

(ii) Since the motions to be treated are 'aggregate', although that motion is chaotic assume that for every property of that motion a mean value determined by the *distribution* of that property among the molecules exists.[13]

(iii) Try to weaken or if possible eliminate the simplifying assumptions introduced to facilitate calculation once the specific assumptions of (i) and (ii) have been introduced so as to simulate, as far as possible, conditions obtaining in a 'real'[14] gas.

(iv) Use the specific assumptions introduced to investigate the *internal* properties of gases (e.g. the viscosity) while the macroscopic (hydrodynamic) and equilibrium properties should be derivable as *limiting cases*.[15]

[13] The heuristic of the programme thus directs that all interactions shall be treated according to the laws of mechanics, while distribution of the properties of the molecular motion among the molecules shall be treated according to the laws of statistics.

[14] 'Real' here is meant to indicate no more than that the model of the gas should at least take into account properties of molecules provided by (often unarticulated) touchstone or observational theories. (Cf. Lakatos [1970], p. 135.) For example, gases have weight, so their molecules must. Maxwell showed that in the case of monatomic gases in the absence of external forces, molecular collisions do not affect the steady state velocity distribution once it is attained. (Maxwell [1866], pp. 45–50.) However, the effects of some external forces, e.g. gravity, cannot be eliminated, and from elementary chemical atomism, gases at ordinary temperatures are polyatomic. Now in this case two of Maxwell's assumptions obviously fail, viz. that all directions of velocity are equally likely (for now the downward is preferred), and that the number of molecules in a unit volume can be assumed to be the same in all parts of the gas. Equally the polyatomic molecule need not be spherical. The heuristic, however, directs that these must be taken into account and gives clear hints as to how this could be done. The Maxwell theorem for polyatomic molecules was deduced by Boltzmann in his [1868], and his [1871*b*] (cf. Maxwell [1873], p. 351). The difficulties here in carrying out the programme are purely mathematical. (Cf. Lakatos [1970], p. 136.)

[15] This requirement is a stringent one. In modern terms if $f(\mathbf{r}, \mathbf{v}, t)$ is the distribution function such that $f(\mathbf{r}, \mathbf{v}, t)\, d^3r\, d^3v$ is the number of molecules which at time t have positions lying within the volume element d^3r and within the velocity element d^3v, then the form of $f(\mathbf{r}, \mathbf{v}, t)$ as $t \to \infty$ must contain all the equilibrium properties of the system.

Thus each successive theory as the programme developed consisted of a particular *model* of a gas, constructed in accordance with (i) and (ii), each designed to be a closer approximation to the conditions known to obtain in a gas than the one before. Wherever anomalies arose, the refuting instance was attributed to the set of auxiliary assumptions constituting the model in question. The heuristic contained suggestions as to how to fill in, elaborate and draw consequences from each theory within the programme. It thus set out a *research policy* and gave hints and suggestions as to how it might be carried out.[16]

In identifying the hard core and positive heuristic of the programme I do not mean to assert that the scientists working on the programme did so for the same reasons or with the same aims in view. For instance, Clausius's main interest in developing the kinetic theory lay in the problem of the physical interpretation of his entropy function in terms of the molecular disgregation, itself a function of the arrangement of the constituent particles of the body and of its thermal content.[17] Maxwell, however, was principally concerned with a theory of matter, while Boltzmann's problem situation was to provide a rational foundation for thermodynamics. What the identification of hard core and heuristic presupposes is that the working scientists *utilised a common approach to, and set of techniques for, solving their problems*. In general, in locating the hard core and heuristic of a programme what is important is not what working scientists said about what they were doing, nor their beliefs about the truth or falsity of the specific theories they espoused, but what principles they *adopted and followed in practice* in developing their theories.[18]

[16] Cf. Zahar [1973], this volume, p. 215. Lakatos in his [1970] gave an example of just such a sequence of 'models' constructed in accordance with the heuristic in his sketch of the development of the Newtonian programme. In this case each model (successive theory) took into account more and more of the 'real' planetary situation: first the planets were treated as point masses, with fixed sun and only heliocentric forces, then as spherical masses revolving about a common centre of gravity, and then as spinning balls with interplanetary forces and perturbations. Only at this advanced stage after the solution of innumerable dynamical problems did the testing of the latest version of the programme become important. Cf. Lakatos [1970], p. 135.

[17] Clausius's form of the second law of thermodynamics for reversible cycles is

$$\int \frac{\mathrm{d}Q+\mathrm{d}H}{T}+\int \mathrm{d}Z = 0.$$

For Z, the molecular disgregation, and H, the thermal content of the working substance. (Clausius [1862*b*], pp. 220–8, Clausius [1863], Clausius [1865], esp. pp. 354–8. For the history of the disgregation function see, particularly, Gibbs [1889], Daub [1967] and Klein [1969*b*].)

[18] Cf. Urbach [1974], pp. 108–11. It might be objected that the heuristic of the kinetic programme is a historical fabrication arrived at purely by hindsight. I think this would be quite wrong for there were, as is clear from the original papers, unpublished manuscripts and letters (cf. particularly Brush and Everitt [1969] on Maxwell's heuristic route to the solution of the radiometer problem) of those scientists working on the programme, generally accepted clear guide-lines as to how problems in the programme should be tackled. Maxwell, for instance, sketched the general form of approaching such problems

In the next section I shall show how the positive heuristic played a vital role in the development of the programme, while itself being further articulated and strengthened. In so doing I shall appraise the development of the early kinetic programme, according to the explicit criteria provided by the methodology of research programmes.

2.2 *The empirical progress of the early kinetic programme*

(*a*) *The elementary theory*[19]

In the original Krönig–Joule theory a very simple statistical hypothesis was adopted, viz. 'the path of each molecule must be so irregular that it will defy all calculation. However, according to the laws of probability theory, one can assume a completely regular motion in place of this completely irregular one.'[20] More specifically all the molecules can be regarded as traversing rectilinear paths with the same speed, in arrays parallel to and normal to the walls of the containing vessel.[21] (This amounts to a particular specification of (ii) in the heuristic.) Furthermore, the molecules are *smooth* elastic spheres (specification of (i)) from which it follows that only the motion of translation is present. Impacts are perfectly elastic. From this very simple theory the equation of state (Boyle–Charles law) for an ideal gas follows, when the vis viva of the molecules is assumed to be proportional to the absolute temperature of the gas.

Clausius[22] extended the simple Krönig model by attempting more realistic assumptions about molecular mechanics and a more realistic account of the state of motion of the molecules. On the basis of translatory

in his famous letter to Stokes: first assume the molecular model, the model for interaction, deduce the laws of motion and investigate the resulting 'molecular' quantities. (Cf. Brush [1965], p. 27.) Boltzmann gave a retrospective sketch of the heuristic he had followed in his development of the programme in his [1897*a*]: 'the chief end and purpose of science [is] to shape the constructs representing one group of facts that from them may be predicted the behaviour of other similar facts...the prediction is afterwards to be checked by experiment. Probably it will be only in part confirmed. There is then hope of so modifying and supplementing the constructs that they will conform also to the new facts; which is tantamount to making new discoveries with respect to the constitution of the atoms.' (Boltzmann [1897*a*], p. 75.) He regarded 'the present atomism as nothing more than a body of directions for the construction of a theory and view of the world'. (*Op. cit.*, p. 79.)

[19] Isolated, single kinetic theories like that of Bernoulli, Herapath and Waterson had, of course, appeared long before Clausius's first paper of 1857. However, it is with this latter paper that the atomic–kinetic *research programme* began, embodying systematic techniques for going beyond the ordinary gas laws to investigate the crucial internal properties of gases, e.g. viscosity, diffusion and heat conduction.

[20] Krönig [1856], p. 316.

[21] To calculate the pressure arising from the transfer of momentum at the walls of the container Krönig assumed that since all directions are equally likely exactly $\frac{1}{3}$ of the number of molecules within a unit volume $u\Delta t$ would be moving normal to a unit area of the wall. The net transfer of momentum per unit area per unit time would then be

$$-2mu\cdot\tfrac{1}{2}\cdot\left(\frac{1}{3}\frac{nu\Delta t}{\Delta t}\right)=\frac{-mnu^2}{3}.$$

[22] Clausius [1857].

motion alone, Clausius could not account for the observed specific heat of gases. This anomaly was attributed by him to the auxiliary hypothesis employed in Krönig's theory that the molecules were smooth elastic spheres. However, independent considerations suggested that they were not, and since not all molecular collisions could be rectilinear and central, a rotatory motion would ensue. Furthermore, the collisions could not be regarded as perfectly elastic, for if a molecule was a combination of atoms one would expect vibrations to occur among the atoms during and after impact. Thus any motion of translation alone would gradually become distributed among the other available degrees of freedom of motion.[23] Once a steady state had been reached it would be possible to neglect the irregularities occurring as the result of these inelastic collisions and to 'assume that in reference to the translatory motion, the molecules follow the common laws of elasticity'.[24] Also, since a steady state would be attained, a distribution of molecular speeds would be assumed which had a very small dispersion about a most frequent value; thus in this state all the molecules could be assumed to have a constant (the mean) speed.[25] Most important, however, the assumption of motion in regular *arrays* was abandoned (in accordance with the directives of the heuristic (iii) above). Clausius adopted the method of calculating the probability of finding a molecule with speed u with a *direction* within an element of solid angle dw and then integrating over all directions. This innovation, together with the concept of a distribution of speeds possessing a definite spread, constituted important heuristic progress as we shall see when we consider the later theories of Clausius and Maxwell. First, however, let us examine the *novel* consequences of the elementary theory (i.e. the genuine empirical support, if any, for the theory).

Here many of the *novel* consequences were *not temporally novel*; they were novel in the sense that the theory was not specifically designed to accommodate them. Clausius's theory in effect predicted Dalton's law of partial pressures for a mixture of gases[26] *and* Gay-Lussac's law of equivalent volumes.[27]

[23] Clausius [1857], pp. 131–4. This result later became a major anomaly for the research programme. Cf. below, pp. 82–3. [24] *Ibid.*, p. 114. [25] *Ibid.*, p. 127.

[26] The total pressure of a gas mixture, P, is related in Clausius's theory to the total kinetic energy of translation by $P = \frac{2}{3}E/V$ for V the total volume of the mixture. Each gas (i) if it alone filled the space available would exert a pressure $P_i = \frac{2}{3}E_i/V$. But $E = \sum_i E_i$, from which it follows that $P = \sum_i P_i$, which is Dalton's law of partial pressures (Dalton [1808], pp. 154–5).

[27] This prediction, of course, requires the *assumption* of Avogadro's law that when pressure and temperature are the same, equal volumes of gas contain the same number of particles. Maxwell later eliminated the assumption by deriving it from the equipartition of energy theorem (cf. below, pp. 55–6). However, as Clausius put it, Avogadro's law was 'a hypothesis which for *other* reasons is very probable'. For this reason I regard the deduction of Gay-Lussac's law as an example of the empirical progress achieved by the kinetic programme.

Furthermore, it gave a novel qualitative explanation of the phenomenon of evaporation, and the equilibrium between liquid and gaseous phases.[28] Thus, Clausius's first theory constituted a progressive shift over the Krönig–Joule theory since it predicted a series of novel facts which did not follow from the Krönig–Joule theory. However, a major anomaly was soon pointed out in the new theory by Buijs-Ballot[29] to the effect that on this theory gases should instantaneously inter-diffuse, since the molecules travel rectilinearly between successive collisions of infinitesimal duration compared with the entire duration of the rectilinear motion.[30]

In his second paper[31] Clausius utilised the heuristic of the programme in an attempt to resolve the anomaly. The heuristic directs that more and more of the 'real' properties of molecules be taken into account, one of these is clearly the finite size of the molecules and the effect of that finite size upon the dimensions of the mean free path (i.e. the *average* distance a molecule moves before colliding with another molecule). By taking into account the finite size of the molecule Clausius constructed its 'action-sphere'.[32] Then the number of collisions which take place in a time interval Δt between two groups of molecules coming toward each other can be computed as (volume swept through by the action-spheres of the first group in time Δt) $\times$ (number of molecules of the second group per unit volume).[33] The mean free path can then be computed as the mean speed divided by the collision frequency given as above. Clausius had, of course, no way of estimating the size of the action-sphere of the molecules; he actually assumed that the ratio of the mean free path to the action-sphere was 1000 to 1[34] which gave a mean free path length 'very small compared to our usual units of length'. The number of collisions per cm^3 would then be enormous and the rate of diffusion equally slow, which at least qualitatively resolved the Buijs-Ballot anomaly.

[28] Evaporation results on Clausius's theory from the escape of molecules from the liquid whose velocity is in excess of that required to overcome the cohesive forces of the liquid phase. In particular it contradicts the background theory of equilibrium of the liquid–vapour phase, for it had been assumed that equilibrium was a state of *no* evaporation and *no* condensation, whereas on Clausius's theory evaporation and condensation continally take place but exactly compensate each other. (Cf. Clausius [1857], p. 119.)

[29] Buijs-Ballot [1858], p. 240.

[30] On Buijs-Ballot's account 'the molecules of a gas in a room must traverse the room many hundred times in one second'. *Op. cit.*, p. 240.

[31] Clausius [1858].

[32] If the molecular *diameter* is d, the 'action-sphere' is a sphere of *radius* d about the centre of the molecule. No other molecule can have its centre within this sphere. The problem then becomes how far on average the molecule moves before its centre of gravity comes 'within' the sphere of action of another molecule. (Clausius [1858], p. 139.)

[33] This is in fact the *Stosszahlansatz* which played so crucial a role in Boltzmann's original proofs of the H-theorem. (See below, pp. 78–82. Cf. Ehrenfest [1911], pp. 4–16 and Klein [1969*a*], pp. 101–2.) Clearly this holds only if the number of molecules per unit volume is the same in the space traversed by the action-spheres as in any other part of the space available to the gas. This is an assumption of equal frequencies, the specific assumption invoked by Clausius satisfying (ii) in the heuristic. Furthermore, all the molecules are assumed to move with the same (mean) speed.

[34] Clausius [1858], p. 145.

What I wish to emphasise is the vital role played by the heuristic of the programme in providing a general policy for the resolution (digestion) of such anomalies and in the extension of each successive theory to a more and more detailed account of the internal motion of gases. By direct application of the heuristic Clausius was led to the concept of the mean free path, which in turn, at the hands of Maxwell, led to enormous empirical progress.[35]

(*b*) *The mean free path and the law of velocities*

The introduction of the mean free path technique was crucial in the development of the kinetic programme because it enabled an investigation into the internal properties of gases to be undertaken (i.e. an investigation of irreversible processes, of transport phenomena).[36] In order to apply the method of the mean free path to the internal properties of gases, some determination of the law of the distribution of molecular velocities was required. Maxwell considered the problem from the point of view of molecular dynamics rather than simply postulating the existence of a mean speed and that dispersion about this mean speed should be small. He realised that pure dynamics would not alone yield the result; certain statistical assumptions and a particular molecular model would need to be introduced, the point being that these assumptions should be as innocuous[37] as possible.

In fact he made three assumptions, first that the molecules be regarded as elastic spheres, second that after a collision all directions of rebound are equally likely, and third that the probability distribution for each component of velocity was independent of the values of the other components.[38] The distribution law then obtained must be of a form such that

$$f(v_x)\,f(v_y)\,f(v_z) = \phi(v_x^2+v_y^2+v_z^2),$$

i.e. $f(v_x) = C \exp(A\,v_x^2)$ (A must be negative otherwise the number of particles would *increase* with the velocity). Thus the number of particles whose velocity lies between $\boldsymbol{v}$, $\boldsymbol{v}+\mathrm{d}\boldsymbol{v}$ is obtained as a function of $\boldsymbol{v}$,[39] and the mean velocity as $\int_0^\infty \boldsymbol{v} f(\boldsymbol{v})\,d\boldsymbol{v}$. All that remained to complete the analysis was to determine the value of a transport quantity (A) across an arbitrary plane in the gas. The mean free path is proportional to the average distance from the plane at which a molecule makes its last collision before

[35] Maxwell's [1860] was the first kinetic theory to make *predictions* going beyond the known gas laws.

[36] For example, the transport of momentum, i.e. viscosity; the transport of vis viva, i.e. heat conduction; the transport of mass, i.e. diffusion.

[37] The three assumptions (above) were not as innocuous as had been suspected. In particular the third would intuitively be expected to be false, for the collisions would presumably introduce a functional dependency between the velocity components. That this is in fact not so, was shown by Maxwell in his [1866], i.e. the assumption was rendered redundant.

[38] Maxwell [1860], p. 379–80.

[39] Cf. Maxwell [1860], p. 381.

crossing. It is this *last collision*[40] which determines the value of the property (A) transported across the plane in Maxwell's theory. The net transport per unit area per unit time across the plane, is then

$$\Gamma_A \propto \text{(average number of molecules crossing the plane per unit time per unit area)} \times \text{(mean free path)} \times \frac{\partial A}{\partial z}$$ [41]

i.e.

$$\Gamma_A = -b_A n\bar{v}\lambda \frac{\partial A}{\partial z} \qquad (1)$$

Once $\bar{v}$ has been determined, (1) can be compared with the hydrodynamic equations. This is a quite general method for treating the internal properties of gases, applicable to any transported property of the molecular motion. This is a clear example of the heuristic progress of the early kinetic programme, for by employing this general technique in particular cases, a sequence of theories was developed, each describing the transport of a particular property. In the case of viscosity, the property transported is momentum. If η is the coefficient of viscosity then the hydrodynamic equation assumes the form $\Gamma_A = -\eta \frac{\partial A}{\partial z}$ which gives on comparison with (1), $\eta \propto \lambda\bar{v}n$.[42] Since, however, λ (the mean free path) is inversely proportional to the density n, the viscosity of a gas should be, according to Maxwell's theory, independent of the density. *This was a quite startling novel prediction for it contradicted background theories* (the previously available *experiments* contradicted it). Maxwell wrote of the prediction,

> A remarkable result here presented to us in [the] equation is that if this explanation of gaseous friction be true, the coefficient of friction is independent of the density. Such a consequence of a mathematical theory is very startling, and the only experiment I have met with on the subject does not seem to confirm it.[43]

Further, the application of the general equation to heat conduction (transport of vis viva) and diffusion led to further *novel* predictions: first

[40] This is the assumption used in Maxwell's theory; however, previous collisions also play a role, and correction must be made for this 'persistence of velocities'. (Cf. Jeans [1925], pp. 260–7.)

[41] Maxwell [1860], pp. 384–90. A detailed analysis of Maxwell's deduction can be found in Hopley [1956], Part 2, section 1 and Brush [1962], pp. 246–7. In general the mean free path method for transport phenomena amounts to the following. Consider the molecular transport of a property A which varies uniformly in a given direction (the z-direction). The transport coefficients can then be computed by considering the transport across an imaginary plane located at $z = z_0$ in the gas. Cf. Loeb [1961], pp. 31–58.

[42] Maxwell [1860], p. 390.

[43] Maxwell [1860], p. 391. Boltzmann wrote of the prediction that, 'all previous observations counted against it, resistance had always been found to be much greater in dense air than in thin. Besides, the result seemed to be *a priori* unlikely, for if resistance were independent of density it would have to remain the same at zero density, when there would be no gas at all!' (Boltzmann [1886], p. 18.) There are some quite unexpected physical consequences of this law, for instance that the terminal velocity of a sphere falling through a gas would be independent of the density of the gas. (Cf. Jeans [1925], p. 227.)

that gases have a thermal conductivity coefficient (and an estimation of its value $\frac{3}{4}p\lambda\bar{v}/T$),[44] and second a deduction of the rate of diffusion through a porous plug and a narrow tube.[45] *From the law of the velocity distribution alone Maxwell deduced Avogadro's law.* This latter follows trivially from the theorem that if two systems of particles move in the same vessel, the mean kinetic energy of each particle will be the same in the two systems. The prediction that the viscosity is independent of the density was confirmed in a series of experiments carried out by Maxwell, Meyer and Loschmidt.[46] However, since all the exact predictions involve λ, the mean free path, at least one of the three relations must be used to determine its value, and the remaining two can then be used as test cases. Maxwell actually used Graham's experimental data on diffusion to determine λ.[47]

Thus Maxwell's application of the mean free path and his law for velocities resulted in empirical progress: the prediction of temporally novel phenomena (e.g. the viscosity independent of the density, the existence of a thermal conductivity coefficient for gases).[48] However, two further predictions of exactly the same theory were in clear *conflict* with experimental results and as such constituted major anomalies for the kinetic programme. They were first, the prediction that the viscosity should vary with the root of the temperature,[49] and second, that the ratio of the whole kinetic energy of the molecules to the kinetic energy of translation should be 2.[50] Maxwell's own experiments indicated, however, that the viscosity varied with the *temperature* and that the ratio of vis viva was 1.634.

These two refutations which were explicitly stated by Maxwell had, however, little effect upon the *appraisal* by the scientists of the time of the value of the kinetic approach. Meyer, Loschmidt and Stefan *all regarded as crucial the confirmation of the novel prediction that the viscosity be independent of the density* in their decision to work on the kinetic theory in an attempt to improve it.[51] The reason for this is clear. Firstly, Maxwell's prediction gives

[44] Maxwell [1860], pp. 403–4. The prediction of a thermal conductivity coefficient for gases was confirmed experimentally by Magnus.

[45] Maxwell [1860], pp. 398–400.

[46] Maxwell [1865], pp. 11–13; Loschmidt [1870].

[47] Loschmidt in 1865 used the results of Maxwell and Clausius to make an estimate of Avogadro's number and the sizes of atoms and molecules. He obtained for the carbon dioxide molecule a diameter of 3.0×10^{-8} cm and for Avogadro's number 10×10^{23}. (Loschmidt [1865].)

[48] Avogadro's law is, of course, not temporally novel, but it does constitute a novel fact relative to Maxwell's theory, for the theory was not specifically designed by adjustment of parameters to yield it.

[49] Since $\eta = \frac{1}{3}n\bar{v}\lambda$, and from the elementary theory $\bar{v} \simeq (\overline{v^2})^{\frac{1}{2}} \propto T^{\frac{1}{2}}$.

[50] Maxwell [1860], p. 409. Cf. below pp. 82–8.

[51] For example see Meyer [1877], author's preface: Loschmidt [1865] and van der Waals [1873]. Indeed, van der Waals begins his doctoral thesis with the sentence: 'The view according to which the molecules of a body considered to be in molecular equilibrium remain at rest, and the invariability of intermolecular distances is ascribed to a repulsive force has generally been given up.' (Quoted by Klein in his [1974], p. 34.)

rise to a 'crucial' experiment between theories involving a static molecular model of the gas and kinetic theories; in the former case, the viscosity should be proportional to the density while for the latter it is independent of it.[52] Secondly, the refutations like ($\eta \propto T^{\frac{1}{2}}$) instead of ($\eta \propto T$)[53] could, at least in principle, be thought resolvable by adopting and refining the molecular model. This I think provides a clear illustration of the *importance of confirmations as opposed to refutations in the development of the programme*, for what was learned from experience by the early kinetic theorists was provided by the confirmation of the viscosity–density relation. In this case it was, to quote Lakatos, 'the verifications which [kept] the programme going, recalcitrant instances not withstanding'.[54]

(*c*) *Maxwell's second theory – the transfer equations*

Maxwell first tried to resolve the anomaly of the viscosity–temperature dependency by refining the mean free path method, taking into account the effect of collisions prior to the last one before the molecule crossed the arbitrary plane in the gas. However, to do this required a deduction of the law of velocities for a gas in a non steady state; this after two attempts Maxwell could not do and the manuscript which contained the corrected transport equations was left unpublished.[55] He then changed his tack and returned to the problem of the molecular model and the resulting dynamics of collision. If the molecule could be represented as a centre of force, the distance of approach of two colliding molecules (and hence the apparent molecular diameter) would depend upon the velocity of approach and thus upon the temperature. If the temperature increased the apparent molecular diameter would decrease, the mean free path increase, and thus the viscosity increase with the temperature. *Thus, in the light of the experimental result that the viscosity varies with the temperature, Maxwell assumed a molecular model comprising a centre of repulsive force* ($f(r) = -G/r^n$); and he regarded $n = 5$ as 'the only case which is consistent with the laws of viscosity of gases'.[56]

[52] Cf. Brush [1958], pp. 247–8 and Brush [1965], p. 28. This 'crucial' experiment cannot be reconstructed as one which decisively refutes the static molecular theory, while corroborating Maxwell's first theory, for it actually refuted *both* theories. 'Crucial' here only refers to the fact that static molecular theories systematically failed to account, in a non-*ad hoc* way, for the observed viscosity–density relationship.

[53] These refutations of Maxwell's first theory render its degree of corroboration -1, the least possible value, for in Popper's theory all refutations have equal (devastating) weight. Cf. Popper [1963], pp. 231–5 and pp. 388–91.

[54] Lakatos [1970], p. 137. Cf. especially Lakatos [1974] and Grünbaum [1963].

[55] A detailed account of Maxwell's unpublished paper is given by Garber [1970], especially pp. 310–15. In a memoir published in 1862 Clausius attacked the assumptions involved in Maxwell's deductions in his [1860] as being inconsistent with the molecular model assumed. He pointed out that on Maxwell's theory heat conduction implies the mass motion of the gas, even when the flow of heat is steady, but the transport equations are calculated on the assumption that the gas is in an equilibrium state.

[56] Maxwell [1866], p. 47. Cf. also Maxwell [1865], p. 11 and [1866], pp. 32–3. Since the molecular model was constructed precisely to account for the experimental result

Thus having in accordance with the heuristic of the kinetic programme succeeded in obtaining a better (more realistic) molecular model, Maxwell was directed by the same heuristic to attempt (*a*) to obtain a more rigorous deduction of the law of velocities in a steady state by weakening the assumptions needed to obtain the law, and (*b*) to attempt a better approximation of the effect of molecular collisions upon the values of the transport coefficients (i.e. to reduce the number of simplifying assumptions required in order to carry out the calculation). To do this he introduced a quite novel method of considering molecular transport: the so-called transfer equations. In each element of the velocity space $d\tau$ there is a quantity of the transport property A in question. The method, which is quite generally applicable, is to compute the changes arising from (i) the influx and efflux of A as a result of the transport of A through the bounding surfaces of the element $d\tau$, and (ii) the changes in A produced by collisions inside the element. When this is done components of the net transfer assume the form[57]

$$\iiint AV^{(n-5)/(n-1)} f(V_x, V_y, V_z)\, dV_x dV_y dV_z \qquad (2)$$

where V is the relative velocity of approach of the colliding molecules. Since Maxwell had already assumed $f(r) = -G/r^5$, i.e. $n = 5$, the term in (2) in V vanishes and (2) reduces to $N\bar{A}$ (for N the number of colliding molecules). Once $f(V_x, V_y, V_z)\, dV_x\, dV_y\, dV_z$ is known, $\bar{A}$ can be calculated. In accordance with (*a*) above, Maxwell did give a more rigorous deduction of the distribution function (f) *for a gas in a steady state* (this time by 'proving' what he had formerly assumed: that distribution for each component of velocity was independent of that of the other components[58]). To find the distribution of velocities in a non-steady state, in the case of streaming, he simply added the motion of translation to the steady state distribution.[59] Yet one more simplifying assumption was required, however, for in the cases Maxwell considered, i.e. those involving diffusion and conduction, the gas is not in a steady state, the distribution of velocities is not spherically symmetric. However, the distribution function was unknown in these cases and Maxwell assumed that the steady state distribution would apply, the 'lack of symmetry being very small in most actual cases'.[60] In this way $\bar{A}$ could be calculated to a good approximation.

($\eta \propto T$), the fact that from Maxwell's second theory ($\eta \propto T$) follows cannot be accounted empirical support for this second kinetic theory. In the improved version of the methodology of scientific research programmes due to Zahar, 'A fact will be considered novel with respect to a given hypothesis if it did not belong to the problem situation which governed the construction of the hypothesis.' (Zahar [1973], this volume p. 218.)

[57] Maxwell [1866], p. 42. Cf. Brush [1958], p. 248.

[58] Maxwell [1866], pp. 43–7. Actually Maxwell deduces this for any law of force between the molecules. (Cf. Jeans [1925], pp. 231–8.)

[59] Maxwell [1866], p. 46. 'When the gas moves in mass the velocities now determined are *compounded* with the motion of translation of the gas.'

[60] *Ibid.*

In the second section of his paper 'On the Theory of a Medium Composed of Moving Molecules', Maxwell first re-deduced the crucial transport coefficients comparing the predicted values with the ordinary hydrodynamic equations. The predicted values for interdiffusion and viscosity were in much better agreement with the experimental values determined by Graham, Stokes (and later Loschmidt). Maxwell further deduced the Gay-Lussac law of combining volumes[61] (a novel prediction since the previous deductions had all involved the assumption of Avogadro's law, e.g. that of Clausius, or the assumption of a particular molecular model, e.g. Maxwell's original deduction – Maxwell now showed that Avogadro's law followed independently of the molecular model assumed); Poisson's adiabatic formula[62] (the relation between pressure and density for an adiabatic process); and the *in*dependence of temperature from height for a gas column in equilibrium.[63]

However, in Maxwell's second theory a major anomaly remained unsolved: the ratio of specific heats presented exactly the same difficulty as it had in the first theory. The importance of Maxwell's second theory was twofold: first, the prediction, *independent of a molecular model*, of the major properties of gases in equilibrium (e.g. the Gay-Lussac law, the adiabatic law, the independence of the temperature from the height), and second, the corrected treatment of the transport properties (model dependent). The transition from Maxwell's first theory to his second theory can, within the methodology of research programmes, be seen as a quite natural one. Maxwell had available, firstly, the observed dependency of the viscosity upon the temperature, which given the form of central force laws from mechanics determines the power of repulsion; secondly, the distribution law for molecular velocities (f) which, together with the mechanical principles of conservation of kinetic energy and conservation of motion of centre of mass, gives rise to a partial differential equation for the change in f in a very short time; and thirdly, standard statistical assumptions, e.g. the assumption of molecular chaos. In particular, the two central components of the heuristic of the programme – the demand that all interactions obey the laws of mechanics and that the molecular motion

[61] Maxwell [1866], pp. 63–4.

[62] *Ibid.*, pp. 64–5; Poisson [1823].

[63] *Ibid.*, pp. 75–6. Maxwell regarded the fact that a vertical column of gas in equilibrium has the same temperature throughout as an elementary consequence of the second law of thermodynamics. He wrote of the prediction of this result from the kinetic theory that: 'this [was] derived from the law of the distribution of velocities to which we were led by *independent considerations*. We may therefore regard this law of temperature, if true, as in some measure a *confirmation* of the law of the distribution of velocities.' (*Ibid.*, p. 76.) However, what Maxwell's derivation from the second law shows is that if there is a dependence of the temperature upon the height, then that dependence is the same for all substances and *not*, as Maxwell believed, that there is functional independence. (Cf. Maxwell [1872], p. 300 and Maxwell [1866], p. 76.) Guthrie pointed out the discrepancy in Maxwell's reasoning and tried a series of experiments to settle the matter. His first results seemed to indicate that Maxwell's law was incorrect; in a second more delicate series of experiments Maxwell's law was confirmed.

be assumed to be 'chaotic' – provided quite clear guidelines as to how the anomalies of the first theory might be resolved.

The crucial role played by the heuristic in the generation of new theories in a research programme, illustrates the superiority of the methodology of research programmes over its major rival, falsificationism. On falsificationist grounds the characteristic feature of scientific method is 'trial and error'.[64] On this view, the transition from one theory to the next, given that the first has been refuted, is unanalysable, the result of an 'irrational' creative intuition. As Popper puts it, 'there is no such thing as a logical method of having new ideas, or a logical reconstruction of this process'.[65] In particular he writes, for instance, of the transition from Kepler's to Newton's theory:

> It is important to note that from Galileo's or Kepler's theories we do not obtain *even the slightest hints* of how these theories would have to be adjusted...should we try to proceed from these theories to another and more generally valid one such as Newton's.[66]

Thus, within falsificationism the continuity of theoretical science is inexplicable. Given the methodology of research programmes, one would expect that within a powerful research programme successive theories are constructed on the basis of certain very general principles, the heuristic principles, which can be articulated and appraised. It is the existence of such principles which accounts for the continuity of the succession of theories within such a research programme. Further, whether or not the sequence of theories is constructed in accordance with the heuristic plays a major role in the *appraisal* of the programme.[67]

(*d*) *Boltzmann and the transport equation*

The resolution and digestion of anomalies is only one aspect of the role played by the heuristic of a programme, it also provides hints and suggestions as to how the programme can be *elaborated independently of the existence of refutations of specific theories within a programme.* This process is clearly illustrated in Boltzmann's early work on the programme. It was

[64] Cf. Popper [1934], pp. 31–2; [1963], pp. 312–35; [1972], p. 253. On Popperian grounds, how a new theory is arrived at is relegated *entirely* to the real of empirical psychology. He writes: '...the act of conceiving or inventing a theory seems to me neither to call for logical analysis nor to be susceptible of it. The question how it happens that a new idea occurs to a man...may be of great interest to empirical psychology, but it is irrelevant to the logical analysis of scientific knowledge.' (Popper [1934], p. 31.)

[65] Popper [1934], p. 32.

[66] Popper [1972], p. 201 (my italics). This particular claim of Popper's would seem to be false. According to Zahar (below, p. 254), 'Kepler created the programme which culminated in the Newtonian system; Kepler's method consisted in trying to discover the law of force responsible for the periodic motion of the planets round the sun.' Within the Keplerian programme there are hints as to how to proceed to a more general theory, for example, the requirement that whatever the force is, it must be central. Cf. also Lakatos and Zahar [1976].

[67] Cf. Lakatos [1970], pp. 173–7; and Zahar [1973], this volume, pp. 237–62.

the statistical techniques introduced by Maxwell which stimulated Boltzmann's interest in the theory after he read Maxwell's second paper. He undertook to *extend* and 'correct' many of Maxwell's deductions, but again precisely in accordance with the heuristic of the programme.[68] *That is, by attempting to eliminate the need for those simplifying physical assumptions adopted to expedite the analysis of the very complicated interactions.* He derived Maxwell's law for the distribution of velocities for *polyatomic* molecules in an *external field of force* and even obtained an exact expression for the velocity distribution in a layer of gas which is not in a steady state,[69] thereby eliminating the need for Maxwell's assumption that the streaming velocity can simply be added to the steady state distribution. The result was a general non-linear integro-differential transport equation,[70] but one which proved impossible to satisfy except for the perfectly elastic spherical molecules of a dilute gas.

Boltzmann had derived a general equation whose solution for any given form of molecular interaction would generate the exact values of the transport coefficients. However, since the transport equation proved in practice impossible to satisfy, a whole range of alternative kinetic theories were developed, each differing in the simplifying assumptions used to calculate the transport coefficients. Most notable among these were the theories of Stefan, Tait and Meyer.[71] These theories were in general unsatisfactory since in arriving at the transport equations they neglected certain terms expressing the variation of the collision frequency, although those terms were of the same order of magnitude as the terms retained. Strictly speaking, the values of the transport coefficients were therefore *not* logical consequences of the theories. This fact greatly impressed Planck.[72] Although they arrived at different transport coefficients which were better approximations to experimental data in some cases and worse in others, essentially they did not go beyond Maxwell's results. They did not lead to genuine empirical progress.

(e) *The equation of state and laws of stress in rarefied gases*

Before concluding this appraisal of the early development of the kinetic programme, mention should be made of two more examples of how following the positive heuristic of the programme led to empirical progress. The two examples are the van der Waals equation of state and the laws of stress in rarefied gases. We have seen how the heuristic of the kinetic

[68] Boltzmann [1868] and Boltzmann [1871*a*]. Cf. Maxwell [1873], p. 351.

[69] Boltzmann [1872].

[70] See below, pp. 78–82.

[71] They each differ in the means employed to calculate the mean free path. A full account of each approach with their inadequacies is given in Jeans [1925], pp. 250–67 and pp. 290–301. Depending upon the method employed five different ratios for the coefficient of heat conduction to the coefficient of diffusion could be obtained.

[72] The proliferation of such theories played a major role in the later degeneration of the kinetic theory. Cf. Planck's retrospective remarks in his [1931], p. 353. See below, p. 87.

programme suggests ways of elaborating particular kinetic theories to take account of more of the 'real' properties of gases. Clausius, for instance, in his deduction of the 'ideal' gas law, showed that in order for this idealised case to be deducible (gases, of course, were known empirically to deviate from this law significantly) three conditions had to be satisfied.[73] They were:

(i) that the space actually filled by the molecules of the gas must be infinitesimal in comparison to the whole space occupied by the gas itself;

(ii) that the duration of impact (i.e. change of direction) of the molecules must be infinitesimal compared with the time interval between collisions;

(iii) that the influence of the molecular forces between the molecules must be infinitesimal.

But of course the molecules, whether elastic spheres or centres of force, are presupposed to have a finite 'size' and thus the ratio of molecular volume to the space available to the gas must be finite. The duration of interaction (especially if the molecules are centres of force) is not in infinitesimal ratio to the mean free path duration.

Clausius further suggested that the inter-molecular forces which account for the cohesion of the liquid phase *must act throughout* the range of temperature and pressure; in particular their effect would still be appreciable in the gaseous phase, when the molecules closely approached each other at collision. *That is, any property of the molecular model should be present under all temperatures and pressures*, since the molecular model, once determined, is a fixed characteristic of the *substance*. In his Nobel prize lecture van der Waals specifically refers to these considerations as being responsible for his theory of the continuity of the liquid and gaseous phase. He remarks:

> Thus I conceived the idea that there is no essential difference between the gaseous and the liquid state of matter – *that the factors which*, apart from the motion of the molecules, act to *determine the pressure* must be regarded as quantitatively different when the density changes and perhaps also when the temperature changes, but that they must be the very factors which exercise their influence throughout. And so the idea of continuity occurred to me.[74]

Thus, what was crucial in the genesis of van der Waals's theory were *the clues provided by the assumption of a 'realistic' molecular model into the properties of matter.* Given the idea of the continuity of the liquid and gaseous phase, all that remained was to find the general form of the equation of state which expressed this continuity. To achieve this van der Waals employed a general mechanical theorem deduced by Clausius from

[73] Clausius [1857], p. 116.

[74] van der Waals [1910], pp. 254–5 (my italics). It is interesting to note that van der Waals always adopted a firmly realistic stance, as to the existence of molecules. He wrote of his attitude to the elaboration of molecular theories, '...in all my studies I was quite convinced of the real existence of molecules, ...I never regarded them as a figment of my imagination, nor even as mere centres of force.' Quoted by Klein [1974], p. 28.

simple kinetic assumptions: the virial theorem.[75] The theorem asserts that for a system of material points, in which the coordinates and velocities of all the particles are bounded, the mean kinetic energy of the system is equal to its virial.[76] For the particular case of material points confined in a volume V by an internal pressure (from the containing walls) P, Clausius showed that the theorem reduces to

$$\Sigma \tfrac{1}{2}m\overline{u^2} = \tfrac{3}{2}PV + \tfrac{1}{2}\overline{\Sigma r\phi(r)} \tag{3}$$

where *central* forces $\phi(\mathbf{r}).\mathbf{r}$ act between *pairs* of molecules. Since the form of $\phi(\mathbf{r})$ was unknown, van der Waals assumed that the second term in the right hand side of (3) could be replaced by an effective inter-molecular pressure P', i.e.

$$\Sigma \tfrac{1}{2}m\overline{u^2} = \tfrac{3}{2}(P+P')\,V$$

However, since the molecules are not point masses but particles of finite size, the effective volume available to the molecules must have the form $(V-b)$, the volume of the container minus the sum of the molecular volumes. That is

$$\Sigma \tfrac{1}{2}m\overline{u^2} = \tfrac{3}{2}(P+P')\,(V-b). \tag{4}$$

The problem remaining for van der Waals, was to determine the inter-molecular pressure, P', arising from the attractive force between pairs of molecules. That the forces were very short range followed from continuity considerations. As such, the inter-molecular attraction which produced the decrease in the pressure that the molecular bombardment exerted on the walls of the containing vessel could arise only from a thin layer adjacent to the containing walls. The number of interacting pairs of molecules in this layer would, in equilibrium, be proportional to the square of the density, that is, inversely proportional to the volume of the container (i.e. $P' = a/V^2$). Finally, using the standard kinetic assumption that $\Sigma \frac{1}{2}m\overline{u^2} = \frac{3}{2}RT$ for one mole of the gas, van der Waals arrived at the general equation of state as (from (4) above)

$$(P+a/V^2)\,(V-b) = RT.$$

The two parameters a and b must be determined from experiment (i.e. two sets of values for P, V, T). Once this is done, however, the isotherms (constant T) can be predicted for all (P, V). The van der Waals equation, for example, gave the first theoretical determinations of the critical points (the change of state). The van der Waals isotherms and the predicted values for the critical points were confirmed in tests carried out using the empirically determined isotherms for carbon dioxide.[77] The agreement

[75] Clausius [1870].

[76] The virial of the system is the *mean* value of the sum $-\frac{1}{2}\sum_i (X_i x_i + Y_i y_i + Z_i z_i)$, where the ith particle at position $\langle x_i, y_i, z_i\rangle$ is acted upon by the force $\mathbf{F}_i = (X_i, Y_i, Z_i)$. The steps in van der Waals's application of the virial theorem to the problem of the equation of state are given in detail in Klein [1974], pp. 38–40.

[77] Andrews [1869].

between the van der Waals isotherms and the experimental ones was, however, by no means complete; anomalous results abounded, expecially at high compression. *What is important is that the novel predictions were found to be in good agreement with experiment. Application of the heuristic of the kinetic programme had resulted in empirical growth*, in this case the discovery of a *new general law*, which for the very first time allowed for the theoretical determination of some of the intractable properties of non-ideal gases. This discovery of a quite general law was arrived at, using the heuristic of the programme, quite independently of the detailed experimental investigations of Andrews into the isotherms of carbon dioxide and the numerous experimental deviations from the perfect gas law (i.e. refutations of earlier kinetic theories).[78]

As a final example of the empirical progress of the early kinetic research programme there is Maxwell's explanation of the Crookes radiometer effect[79] and thermal transpiration which gave the first account of gas surface interactions in which he

> simply extend[ed] to the surface phenomena the method which I think most suitable for treating the interior of a gas.[80]

The whirling of the radiometer vanes was first attributed to the pressure of light. Schuster, however, showed that if the vanes were held the radiometer case would rotate! Maxwell's explanation of the effect, was that the vanes were subject to a tangential force (if a constant temperature gradient was attained, as in the vane case, no net pressure normal to the surface of the vane would result) in a rarefied gas at their edges, produced by the creep of molecules of the rarefied gas from colder to hotter parts of the vanes. When a hypothesis as to the nature of the surface of the vanes was assumed, the slipping force could be calculated using the transfer equations. The kinetic theory provided the only viable explanation of the phenomena. Its prediction as to the force on the vanes was qualitatively confirmed. Exact predictions were impossible because of the need to assume some 'arbitrary' hypothesis as to the gas–vane surface interaction.

I have attempted in this section to appraise the early development of the kinetic research programme. I should now like to emphasise three important points:

First: what we have appraised *was* a research programme. That is, it was a sequence of theories, each of which consisted of specified forms of a set

[78] The discovery of the law of corresponding states (a consequence of the van der Waals equation) constitutes a refutation of the Popperian historiographical thesis that *all* major scientific discoveries are made in attempts to refute existing theories. (Cf. Agassi [1963], pp. 64–74.) Van der Waals discovered the law by way of a systematic attempt to find the consequences for the properties of matter of elementary kinetic assumptions. Cf. Klein [1974], pp. 29–30.

[79] Maxwell [1879]. Cf. Brush and Everitt [1969], esp. pp. 118–20.

[80] Quoted by Brush and Everitt [1969], p. 114.

of hypotheses held constant (in effect regarded methodologically as unfalsifiable) and a set of auxiliary hypotheses differing in each theory in the sequence. Further, each successive theory can be seen to have been constructed in accordance with a general, overall plan.

Second: the research programme was *progressive.* That is, until the early 1880s each version of the research programme predicted a *novel fact* (moreover, as we have noticed, each version predicted a *temporally*[81] novel fact).

Third: the overwhelming importance of the heuristic in the development of the kinetic research programme. For instance, many of the difficulties faced by the programme were purely mathematical.[82] If the research policy were to be put into effect, clearly defined mathematical problems would require solution; for example, the deduction of a law for the distribution of velocities for a gas in thermal equilibrium (Maxwell originally developed his law, as he put it, 'as an exercise in mechanics... I have taken to the subject for mathematical work'[83]) and in Boltzmann's extension of the law to cases involving the presence of an external field of force and polyatomic molecules.[84]

One can also see in the development of the programme the importance of confirmations of the novel predictions as opposed to the numerous 'refutations' each version had to contend with. Maxwell (see his letter to Stokes) developed his theory anticipating that it would be completely knocked out as it led to 'absurd' (the term is his own) consequences. And indeed it did, for the ratio of specific heats problem seemed to him decisive against the view that gases are systems of hard elastic spheres.[85] But he did not abandon work on the programme for an equally 'absurd' prediction (that viscosity was independent of density) was confirmed.[86] The heuristic

[81] The temporal novelty in these cases was not spurious or 'accidental' in the sense envisaged in Zahar (see below, p. 218, note 17). The magnitude of, and the relations existing among, the transport coefficients were quite unsuspected before the advent of the Clausius–Maxwell theory of the mean free path, as was the relationship between the viscosity and the density; indeed in this latter case, the exact opposite was regarded as being true.

[82] Cf. Lakatos [1970], pp. 135–7. He wrote, 'Indeed if the positive heuristic is clearly spelt out, the difficulties of the programme are mathematical rather than empirical'. This is clearly true to a large extent of the development of the kinetic programme. In this case the fact that the solution of some abstruse analytic problems in the relation between mechanics and statistics and the gas model was *not* forthcoming, retarded the development of the programme; see below, p. 82.

[83] Maxwell's letter to Stokes, quoted by Brush [1965], p. 27.

[84] One could continue the list, e.g. the dynamics of collision for molecules as centres of force, the solutions of non-linear transport equations for differing molecular models.

[85] Maxwell [1860], p. 409.

[86] Cf. Lakatos [1970], p. 137. This bears out a major thesis of Lakatos, viz. 'it is the "verifications" which keep the programme going, recalcitrant instances notwithstanding'. It is clear that what was 'learned from experience' by the early kinetic theorists was provided by the confirmation of the viscosity–density result. This provided the impetus for developing the programme, for the model dependent refutations like ($\eta \propto T^{\frac{1}{2}}$) could at least in principle be thought resolvable by adopting some suitable molecular model.

of the programme suggested a whole series of ways (principally involving refinements of the molecular model) in which the difficulty over the specific heat anomaly might be avoided. Indeed the choice of problems which Maxwell and Boltzmann adopted, in their development of the programme, was in my view determined by the principle that the kinetic theory should make predictions about the *internal properties of gases*, it should go beyond hydrodynamic laws, and investigate the transport *properties*.[87] The anomaly of the ratio of specific heats was listed but largely ignored for empirical progress was forthcoming in other directions. It was not until it became quite clear that the heuristic could not provide any hope of resolving the difficulty and all that could be gained in other directions by applying the heuristic had been achieved that the specific heat anomaly became important. Brush makes this point concerning the anomalous behaviour of the specific heat ratio. He remarks

> The only way to make progress in the kinetic theory was to ignore these difficulties and hope someone would eventually work out a better theory of atoms and of the propagation of electromagnetic waves...[88]

It seems to me a very important point in favour of the methodology of research programmes that it can provide a *methodological rationale* (internal explanation) for such episodes in the history of physics. It is, I think, impossible to reconcile with a falsificationist methodology, for all refutations in such a methodology have equal (devastating) weight. Popper seems now to suggest that some refuting instances are more important than others.[89] However, the decision as to which ones are, seems to be left to a Polanyiite *Fingerspitzengefühl* on the part of the working scientists.[90] It was, however, by employing quite general systematic methods, that Maxwell, faced with the anomaly of the dependence of viscosity upon too low a power of the temperature, introduced a new molecular model from which it followed that $\eta \propto T$. He tried to immunise his programme against the refutation. The immunisation was successful, and the *problem shift* (elastic spheres to centres of force) was progressive because from the new hypothesis a series of novel facts were deducible. Thus the early kinetic research programme consisted of a series of 'immunising stratagems',[91] the point being that the methodology of research programmes

[87] For Boltzmann's preoccupation with transport phenomena, cf. Boltzmann [1897*a*], pp. 74–5. According to the methodology of research programmes the heuristic of the programme decides which are important problems and which are not. Cf. Lakatos [1970], p. 137: 'which problems scientists working in powerful research programmes rationally choose, is determined by the positive heuristic of the programme rather than by psychologically worrying (or technologically urgent) anomalies'.

[88] Brush [1965], p. 29.

[89] Popper [1972], p. 38, footnote 5.

[90] Cf. Lakatos and Zahar [1976].

[91] I have adopted the term 'immunising stratagems' in the light of Popper's reply to Lakatos in the former's [1974]. Popper begins his reply to Lakatos by 'disregarding the possibility of immunising stratagems' (Popper [1974], p. 1004), but it is precisely the

gives us a method of appraising these strategems as to whether they lead to empirical growth or not, as to whether the research programme was progressive or degenerating.

Having shown that the early kinetic programme was indeed a progressive one, let us turn now to the development of thermodynamics.

3 *The research programme of thermodynamics*

3.1 *The hard core*

Lakatos regarded the history of physics as the history of rival, competing research programmes.[92] Indeed, he specifically mentions Einstein's prediction of the Brownian motion as a major defeat for the phenomenological research programme which, to use his phrase, turned 'the war in favour of the atomists'. Elkana too, in his recent case study, described phenomenological thermodynamics as an 'unusually ambitious' programme, one of 'great depth and daring'.[93] In fact the research programme consisted of a sequence of four theories. They were Carnot's theory, the mechanical theory of heat, fully-fledged phenomenological thermodynamics and, finally, the system of the four laws of thermodynamics. Throughout this sequence a fundamental assumption, the hard core of the programme, was held constant, viz. *that there was a definite relation between a quantity of heat and the work which in any way could be produced by it.*[94] The aim of the programme was to determine this relation and so to deduce the laws of heat. As Carnot put it, the goal was the discovery of a 'complete theory' of heat, in which 'all cases were foreseen' and 'all imaginable movements referred to general principles firmly established and applicable under all circumstances', which would 'make known beforehand all the effects of heat acting in a determined manner on any body'.[95]

3.2 *The heuristic[96] of thermodynamics*

The heuristic techniques of thermodynamics were articulated and extended as the programme developed. However, three standard requirements were

importance of the methodology of research programmes that it provides criteria upon which the sequences of 'immunising stratagems' which constitute a research programme may be appraised. Popper's ignoring of this point from the beginning means that his whole reply simply begs the question.

[92] Lakatos [1970], p. 155.

[93] Elkana [1973], p. 60.

[94] The statement that there is a definite relationship between heat and work is a metaphysical one. For the regulative role performed by such metaphysical propositions see especially Watkins [1958], pp. 345–8, Watkins [1975], and Lakatos [1970], pp. 144–5.

[95] Carnot [1824*a*], p. 6.

[96] In an earlier version of this paper I claimed that classical thermodynamics was not a research programme because it had *no* heuristic. That this conclusion was incorrect was pointed out to me by Erwin Hiebert and John Stachel. This, of course, in no way commits them to my present characterisation of the programme.

introduced in Carnot's original paper.[97] (i) First, always investigate cyclic processes in the particular systems under consideration. In such cases the system attains its original state and thus the complicated (unknown) internal changes occurring in the working substances can be neglected. (ii) Further, all transformations considered shall be *reversible* ones, for in these cases the path of the transformation will consist of a succession of equilibrium states, for which an equation of state always applies. (iii) The third requirement introduced by Carnot was simply the directive to employ particular empirical laws (constitutive relations) to discover using (i) and (ii) the functional relationships persisting among empirically determinable parameters.

Carnot's theory consisted of two propositions: that work is produced by the 'fall' of heat from a higher to a lower temperature, and that no heat is *consumed* in the production of work. Carnot arrived at his theory from a principle of the conservation of *substance*. Since heat, the caloric fluid, was a substance, it must be subject to a conservation law. Thus when work or motive power was produced in an engine its production was *not* due to the consumption of caloric but 'to its transportation from a warm body to a cold one, that is, to the re-establishment of equilibrium in the distribution of caloric fluid'.[98] In particular, the 'motive power produced would be found exactly proportional to the fall of heat'.[99] By employing an infinitesimal *cycle* (the first heuristic technique) consisting of two isothermal transformations and two 'adiabatic' transformations, Carnot derived an expression for the amount of work performed as a function of the heat added in the first isothermal transformation of the cycle.[100] Carnot's theory

[97] Since Carnot held that heat was a substance, while within general thermodynamics the nature of heat is untreated, it might be thought that Carnot's theory cannot be included in the thermodynamic programme. However, a research programme is identified by its heuristic, that is the principles working scientists adopt in practice in developing their theories. The principles introduced by Carnot remained (in more elaborate formulations) a central component in the subsequent elaboration of thermodynamic theories. To this extent Carnot's theory can be regarded as the foundation of the thermodynamic programme as it developed.

[98] Carnot [1824*a*], p. 7.

[99] From his unpublished manuscripts, it is clear that Carnot abandoned the caloric theory in favour of the hypothesis that heat is 'the movement of particles of bodies'. (Carnot [1824*b*], p. 67.) He remarks that heat is equivalent to work, and even calculates a conversion factor. He continues: 'whenever there is destruction of motive power, there is at the same time production of heat in quantity exactly proportional to the quantity of motive power destroyed. Reciprocally, wherever there is destruction of heat, there is production of motive power.' He lists five experiments involving the production of heat by friction and radiant heat. Both sets of phenomena he regards as being characterised by motion, and then asks 'Could a motion (e.g. that of radiant heat) produce matter (caloric)? Undoubtedly no; it can only produce motion. Heat is then the result of motion.' (Carnot [1824*b*], p. 63.)

[100] Since caloric is a substance, the amount of caloric, or heat, in a body is, in Carnot's theory, a function of the volume and temperature, i.e. $Q = Q(v, t)$; $\mathrm{d}Q$ the heat added is thus a perfect differential. In adiabatic processes, $\mathrm{d}Q = 0$, and thus on Carnot's assumption no work can be performed. The assumption of an *infinitesimal* cycle is therefore crucial; had Carnot considered a finite cycle, the argument would have failed. (Cf. especially Truesdell [1971], pp. 16–20.)

was empirically successful. From the function connecting heat and work and the equation of state for an ideal gas, he derived a general expression for the heat function of a perfect gas, from which it followed that the difference between the specific heat of a (perfect) gas at constant volume and constant pressure, was a constant.[101]

The importance of the theory, however, lay in the three central techniques it introduced for the analysis of the relation between heat and work.[102] However, by 1850 it was clear that Carnot's theory was faced by a major anomaly, for although the production of work required the fall of heat it was not generally true that whenever heat was transferred work was done. As Clausius put the problem, 'if the mere transfer of heat were the true equivalent of work, there would be a loss of working power in nature, which is hardly conceivable'.[103] That is, the principle of the conservation of energy and Carnot's theory were incompatible.[104] To Kelvin this difficulty seemed insurmountable. He wrote of the problem posed by Joule's experiments indicating the existence of a *mechanical equivalent* of heat that, 'if we abandon Carnot's fundamental axiom, a view which is strongly urged by Mr. Joule, we meet with innumerable other difficulties, insuperable without further experimental investigation, and an entire reconstruction of the theory of heat, from its foundation'.[105] That this drastic action was not necessary, was shown by Clausius who resolved the inconsistency simply by dropping Carnot's assumption that the quantity of heat remained undiminished when work was performed, while retaining the assumption that work results from the diminution of heat.[106] Clausius's theory (the mechanical theory) thus consists of two propositions: first the equivalence of heat and work, and second a reformulated version of Carnot's principle that: 'heat can never pass from a colder to a warmer body without some other change, connected therewith, occurring at the same time'.[107] Thus, *the transition from Carnot to Clausius's theory involved simply dropping from Carnot's theory that part of it seen to be inconsistent with the conservation of energy.*

[101] Carnot [1824*a*], p. 24.

[102] This highlights an open problem in the history of thermodynamics. If Carnot's theory predicted a whole series of novel facts, and provided techniques of quite general significance, why was it ignored by working scientists (with the exception of Clapeyron) from 1824 to the late 1840s? Why the irrational slowness in accepting such a powerful theory?

[103] Clausius [1850], p. 111.

[104] For the discovery of the conservation of energy cf. especially Hiebert [1962*a*], Elkana [1974], Kuhn [1957] and Mach [1872].

[105] Kelvin [1849], p. 119, footnote *; and Joule [1845*a*], [1845*b*].

[106] Clausius [1850], p. 112. Clausius states that, 'a careful examination shows that the new method [the conservation of energy] does not stand in contradiction to the essential principle of Carnot, but only to the subsidiary statement that *no heat is lost*, since in the production of work it may very well be the case that at the same time a certain quantity of heat is consumed and another quantity transferred from a hotter to a colder body, and both quantities of heat stand in a definite relation to the work that is done.'

[107] Clausius [1854], p. 117. Cf. Kelvin [1849], p. 118; Kelvin [1852]; Kelvin [1856].

It is at this point that a fourth general heuristic principle was added to methods of thermodynamics, namely, the principle that *whenever it appears that heat is converted entirely into mechanical effect in some process, 'some other change connected therewith must be occurring at the same time'*, which renders such a process impossible. Furthermore, since the path of the process is given independently, the number of candidates for the compensating effect is *determined* and *finite*. Anomalies could thus be systematically tackled by searching the finite number of steps in the cycle, for the compensating process. There is a clear example of this technique in operation in the discovery by James Thomson, of the effect of pressure in the lowering of the freezing point of water.[108] Thomson actually based his deduction upon the exclusion of a *perpetual source* of mechanical work which operated at single temperature a principle which follows from *Carnot's* theory. Thomson noticed that it appeared possible using a particular cycle of operations to convert water at 0° C into ice without the net expenditure of any mechanical work.[109] However, given the fact that water expands while freezing, and is thus capable of doing work, it would appear that a perpetual source of mechanical work was made available, simply by the continuous conversion, using the cycle, of water at 0° C into ice. Thomson resolved this anomaly by inspecting the cycle[110] for the compensating effect. The only way in which the net work done in the cycle can be non-zero was for the work of expansion to be less than the work of compression; this could only be the case if the freezing point of the water was lowered by the pressure induced during the compression. Thomson thus discovered a quite novel relationship between the freezing point of water and the pressure to which it was subject.[111]

The mechanical theory of heat superseded Carnot's theory in the sense that all the *corroborated* content of Carnot's theory[112] followed from the mechanical theory as well as a series of quite novel consequences.[113] The consideration of cyclic transformations in Clausius's theory takes on particular significance, since in that theory the heat added in a trans-

[108] Thomson [1849].

[109] *Ibid.*, pp. 157–8. The cycle consists of the isothermal compression of air at 0° C in a cylinder placed in contact with an indefinite mass of water at 0° C; heat is then transferred from the air to the water. The cylinder is then placed in contact with the water at 0° C to be converted into ice, followed by isothermal expansion (heat being transferred from freezing water to the expanding air) until the original volume is attained when the cycle is repeated. The work of compression is thus apparently compensated by the work done by the air in the expansion.

[110] *Ibid.*, pp. 158–60.

[111] Thomson remarks having used a simple infinitesimal cycle to calculate the depression of the freezing point that, 'the variation to be appreciated is extremely small, so small in fact as to afford sufficient reason for its existence never having been observed by any experimenter'. *Ibid.*, p. 160.

[112] For example, Clapeyron's equation, Clapeyron [1834], pp. 88–91, and Clausius [1850], p. 136.

[113] Cf. below, p. 68.

formation is not the differential of an equation of state, but depends upon the path of the transformation. The work done in the performance of a *cycle* can be identified with the heat added, for at the end of the cycle no other changes have been effected.[114] The analysis of cyclic transformations was crucial to the proof of his fundamental theorem that for all reversible cycles (those for which an equation of state always applies, the process proceeding infinitely slowly through successive equilibrium states) $\oint dQ/T = 0$.[115] It is important to note that Clausius arrived at both theorems by employing the *heuristic techniques* introduced by Carnot, *in drawing out consequences* from his second fundamental axiom. The fundamental axiom states that, 'heat can never pass from a colder to a warmer body without some other change, connected therewith, occurring at the same time.' The problem was to specify unambiguously what the 'other change, connected therewith' actually was. Clausius noticed that if the transformation of heat into work proceeded infinitesimally slowly, then it would be 'reversible without any permanent change'. In such a case the problem reduces to finding the law[116] 'according to which the transformations must be expressed as mathematical magnitudes in order that the equivalence of two transformations may be evident from the equality of their values'.

However, whatever the value was, it must for any transformation be proportional first to the heat produced and secondly to the temperature potential, that is to $Qf(t)$[117] or Q/T, for T some unknown function of the empirical temperature.[118] Hence, $Q(1/T_2 - 1/T_1)$ represents the required value for the transformation of heat Q from temperature T_2 to T_1. If there are N transformations operating in a cycle, the total transformation value is $\sum_i^N Q_i/T_i$. If the cycle is *reversible* then there can be no 'permanent' change and the algebraic sum of the transformation values must be zero. Hence, $\sum_i^N Q_i/T_i = 0$ in the limit $\oint dQ/T = 0$. If the cycle is irreversible then $\oint dQ/T < 0$. Since $\oint dQ/T = 0$ around a reversible cycle, the integral $\int dQ/T$ for any reversible transformation is independent of the path, depending only on the initial and final states of the transformation.[119] This fundamental theorem enabled Clausius to define a function of state S

[114] In Carnot's theory $Q = Q(v, t)$, hence in a complete *cycle*,

$$\Delta Q = Q(v_0, t_0) - Q(v_0, t_0) = 0.$$

Q in Carnot's theory was the *total* heat in the body. Cf. Clausius [1850], pp. 112–15.

[115] Clausius [1854], p. 129.

[116] *Ibid.*, p. 122.

[117] *Ibid.*, pp. 123–7.

[118] T is the absolute thermodynamic temperature. By comparison with Clapeyron's equation for a perfect gas undergoing isothermal transformation, $T = \alpha° + t$. That is, 'T is nothing more than the temperature counted from $\alpha°$, or about 273 °C below the freezing point; and considering the point thus determined as the absolute zero of temperature, T is simply the absolute temperature.' (Clausius [1854], p. 135.)

[119] Clausius [1862*b*], pp. 218–19.

(the entropy) given by $S(A) \equiv \int_0^A \frac{dQ}{T}$, (for an arbitrary fixed reference state O) where the integration is along any reversible path between O and A.[120] Thus the entropy of a thermally isolated system can never decrease ($dQ \equiv 0$). Since the universe is thermally isolated, the entropy of the universe must tend to a maximum.[121] The transition from Carnot's to Clausius's theory was a progressive problem shift in that from the mechanical theory a sequence of quite novel facts followed, in particular, the relationship between the moduli of compression and thermal expansion, the change of the freezing point with pressure, the change of specific volume at solidification, and between the latent heat of vaporisation and the vapour pressure at any temperature and in the case of electrochemical cells the relationship between the heat of reaction, the electromotive force and its functional dependency on the temperature. The progress achieved by the mechanical theory *consisted in establishing relations among properties*, or empirically determinable parameters, which had been thitherto quite unsuspected.[122]

The third theory in the thermodynamic research programme can be regarded as a fully-fledged phenomenological theory. It was developed independently by Gibbs and Planck in the late 1870s. The theory consisted of two laws: the laws of the conservation of energy and of increasing entropy. Planck gave them their most general formulations as:

(i) Every physical or chemical process in nature takes place in such a way as to leave the sum of the total energy of *all* bodies taking part in the process constant.

(ii) Every physical or chemical process in nature takes place in such a way as to increase the sum of the entropies of all bodies taking part in the process. The limiting case of neutral (reversible) processes is only an ideal case, one can consequently only say: 'The Entropy of a material system can in Nature only increase, never diminish or remain constant.'[123]

[120] Clausius [1863], and Clausius [1865], p. 355.

[121] Clausius [1865], p. 365. For an earlier application of the second law to the universe, cf. especially Kelvin [1852]. Here the *universal* tendency is a dissipation of available energy.

[122] The phenomenological character of thermodynamics had already appeared in the mechanical theory, for the theory makes no assertions about the physical causes of the increase in entropy, or the properties of matter from which the increase in entropy results. Clausius developed his physical interpretation of the entropy function, the molecular disgregation, as a quite separate enterprise, independent of his elaboration of the mechanical theory.

[123] Planck [1887], pp. 199–200, Planck [1893], p. 438 and Planck [1897*a*], p. 103. The empirical content of the law consists in the assertion that in any natural process, the entropy of all bodies taking part in the process increases. Given a particular process, it is assumed that a set of bodies can be found, which take part in the process and whose interaction with the rest of the universe may be rendered arbitrarily small. Once this set is fixed, the total entropy of the bodies in the set must increase, for any process in which they take part. A refutation of this conclusion *may* be attributed to the exclusion of some body actually taking part in the process, from the specified set.

Planck regarded the second law as a universal empirical law which determined the direction of all processes in nature. All transformations were *in fact* irreversible and proceeded in the direction of increasing total entropy of the bodies involved, the measure of the irreversibility being provided by the change in entropy.[124] Planck's major concerns in developing his theory were with the foundation of the second law and with the extension of the mechanical theory into the problems of thermochemistry. He regarded Clausius's attempt to formulate an unrestricted[125] form of the second law, by referring to the entropy of the universe as tending to a maximum, as devoid of content. The 'universe' was left undefined and there was no conceivable way in which the entropy of such a system could be determined.[126] Similarly, Planck found inadequate the attempt by Kelvin to provide a universal form of the second law in the principle of the dissipation of energy since there were, he noted, irreversible processes which involved *no transference of heat*, or change in energy (e.g. diffusion at constant temperature).[127] Planck also sought to expunge from thermodynamics the anthropomorphic interpretation of the second law based upon the impossibility of *constructing by any means* a perpetual motion machine of the second kind; that is, an engine which working in a cycle would produce no effect other than the performance of work and the cooling of a heat reservoir. Since perfectly reversible cycles are central to thermodynamic methods, it was possible to interpret the second law as merely asserting that because of the *limitations of experimental methods*, practically available in the laboratory, perpetual motion machines were unavailable to us. The second law would then assert an experimental limitation, but would put no limit on nature. This Planck regarded as an absurdity, considering that

> It would be absurd to assume that the validity of the second law depends in any way upon the skill of the physicist or chemist in observing or experimenting...The limitations to the law, if any, must lie in the same province as its essential idea, in the observed nature, and not in the observer.[128]

[124] For Planck the advantage of his formulation of the second law lay in its applicability to *any* process whatsoever, providing a universal measure of the irreversibility of all transformations. He wrote: 'the form given here is the only one of unrestricted applicability to any finite process, no other universal measure of the irreversibility of processes exists than the amount of the increase of entropy to which they lead' (Planck [1897*a*], p.103).

[125] The second law, as formulated by Clausius in the mechanical theory, refers *only* to transformations in which there is a *transfer* of heat. Cf. above, p. 65.

[126] Planck [1897*a*], p. 104. The first law Clausius gave as 'The energy of the world is constant'. Here too the same difficulty emerges, for the energy of the universe is similarly undefined. Planck reformulated Clausius's first law as, if E denotes the energy contained in a very large space S, then

$$\lim_{S\to\infty} \frac{1}{E}\frac{dE}{dT} = 0 \quad (\textit{ibid.}, \text{ p. 105, footnote 2}).$$

[127] Cf. Planck [1897*a*], p. 106.

[128] Planck [1897*a*], p. 106. Cf. especially Hiebert [1970] for the controversy between Planck and Mach as to the correct interpretation of the second law.

The fact of nature being precisely that all processes *were* irreversible. Planck was quite clear as to what would refute the law: the observation that any single process occurring in nature be reversible. As Planck put it, in such a case 'the entire edifice of the second law would crumble', for

> if in a single instance, one of the processes should be found to be reversible, then all of these processes must be reversible in all cases. Consequently, either all or none of these processes are irreversible. There is no third possibility.[129]

For Planck and Gibbs the content of the second law lay in the entropy theorem, which established a relation between the final and initial states of any process. In the mechanical theory the theorem is proved by consideration of processes involving the *transfer* of heat and mass *motion*. Planck and Gibbs simply generalised the theorem so that it applied to *all* processes whatsoever, the second law of the mechanical theory being then regarded as a special case of the more general entropy law.[130] The advantage of such a problem shift is that the generalised second law (the entropy law) could be applied to *all* processes, in particular to the problems of thermochemistry and thermochemical equilibria where in general no transfer of heat or mass motion takes place.[131]

Having sketched the sequence of theories which ended with the most general form[132] of the thermodynamic laws I shall first indicate some of the novel consequences which followed from the phenomenological theory, and illustrate in those cases how the kinetic programme was only able to provide *ad hoc* explanations of results already achieved by application of the thermodynamic laws.

(*a*) *The theory of dissociation*

A clear example of a *novel* prediction of thermodynamics is provided by Gibbs's analysis of dissociation and the anomalous vapour densities and their temperature-dependent deviations from the ordinary gas laws. The

[129] Planck [1897*a*], p. 86. The paradigm case of an irreversible process is, of course, the production of heat by friction. Cf. especially below pp. 81–2.

[130] From Clausius's postulate (the second law of the mechanical theory) that a transformation whose only final result is to *transfer* heat from a body at a given temperature to a body at a higher temperature is impossible, the result that for a cycle $\oint dQ/T \leqslant 0$ follows as a logical consequence, for processes involving the transfer of heat. However, from the principle of increasing entropy (i.e. that for *all* cyclic transformations whatsoever $\oint dQ/T \leqslant 0$) then the postulate of Clausius follows as a special case.

[131] Cf. Planck [1897*a*], p. 103 and Gibbs [1875], p. 59.

[132] The principle of increasing entropy was used directly by Planck in his four papers on thermochemistry. However the form of the law actually employed depended upon the specific aim in view and the nature of the problem under consideration. For instance, van't Hoff repeatedly makes use of the reversible isothermal cycle and the extremely simple form of the generalised second law applied to such processes, viz. that the sum of the work done throughout the cycle is zero. Gibbs, on the other hand, based his entire development of the theory on what he called the 'prime' equation: $dU = TdS - p dv$, the combined first and second law for an infinitesimal reversible variation (Gibbs [1875], pp. 55–62).

standard explanations for the great deviation of the vapours in question was the supposition that the vapours consisted of two or more different gases formed by chemical combination or dissociation, the degree of which depended upon the temperature and pressure. Each component would of course obey the perfect gas law at a single temperature and pressure. The problem then was to derive a general law for determining the proportions of the component gases (their partial pressures) in an equilibrium condition under varying conditions of temperature and pressure. Gibbs attacked the problem first in 1875,[133] using the second law in the form that in an equilibrium state when the energy and volume do not vary, the entropy is at a maximum. On the assumption that the entropy of the gas mixture is the sum of the entropies of the component gases,[134] Gibbs considered the effect of an infinitesimal variation in temperature. From the condition that the energy does not vary, if m_1, m_2 are the proportions by weight of the two gases, c_1, c_2 their specific heat at constant volume, and t the absolute temperature,

$$(m_1c_1+m_2c_2)\,\mathrm{d}t+(c_1t+E_1)\,\mathrm{d}m_1+(c_2t+E_2)\,\mathrm{d}m_2 = 0.$$

By the second law the variation in entropy is zero for all equilibrium states and so:

$$\left(\frac{m_1c_1+m_2c_2}{t}\right)\mathrm{d}t+\left(\frac{\partial S_1}{\partial m_1}\right)\mathrm{d}m_1+\left(\frac{\partial S_2}{\partial m_2}\right)\mathrm{d}m_2 = 0.$$

Two further conditions hold: the first arises from the fact that the mixture is isolated (thus $\mathrm{d}m_1 = -\mathrm{d}m_2$) while the second derives from the chemical formula for the dissociation of the two gases (e.g. $N_2O_4 \rightleftharpoons 2NO_2$). These two equations are easily transformable into terms of the partial pressures of the dissociated gases which gives D, the mean density, (and thus the degree of dissociation) as a function of the absolute temperature and pressure. The form of the deduction is not important since it is simply algebraic, what is important is that the mean density so determined fitted beautifully with the observed densities for all temperatures and pressures found from the experimental investigations of Deville and Troust.[135] The mean density is actually given by, $f(D) = A+c/t+\ln P$, once the two constants are determined for a particular gas, D follows at *all* other temperatures and pressures and, from D, the degree of dissociation.

Not only were the gaseous dissociation laws *not* derived from the kinetic heory but they were extremely difficult to reconcile with it. Boltzmann gave an elaborate deduction of the dissociation law in his lectures on gas

[133] Gibbs [1875], pp. 176–84, Gibbs [1879], pp. 374–95.

[134] The energy of perfect gas is given by c_1t+E_1; the entropy by

$$S = m(H+c\ln t+a\ln v/m),$$

where H is an arbitrary constant.

[135] Cf. Gibbs [1879], p. 378.

theory in 1896.[136] This deduction depends essentially upon three assumptions:

(i) That chemical attraction is exerted by one atom upon another only over a very small surface region of each atom; they attract each other only when their sensitive regions are in contact – as would be expected from valency considerations.

(ii) If two single atoms do collide so that the sensitive regions touch, the time during which the sensitive regions overlap will be very short compared to the collision time (this is natural since the atoms will be rotating when they collide) and only when *more* than two atoms are overlapping will two atoms be united long enough for combination until the next collision.[137]

With these two assumptions Boltzmann calculated the probability of chemical binding of an atom with a similar one in a given volume element for a given relative velocity of approach and then integrated over all directions and velocities, from which the total number of bound and unbound atoms follows, provided (and this is the third assumption): the dependence of *the force of chemical attraction on the sensitive region can be given.* This force must increase the probability of binding, that is: probability of binding = (probability of overlapping of sensitive regions) × (force of attraction once the regions overlap). The third assumption that is required, is that at the *instant* when the sensitive regions overlap the force is a strong attraction *but* as soon as they inter-penetrate the force of attraction drops to zero. With these three assumptions Gibb's equation is deduced and in the case of the last two assumptions *no independent evidence can be added in their favour; they serve merely to attain the previously known result, the formula of Gibbs. They were ad hoc.*[138]

(*b*) *The theory of affinity*

The second example will serve to illustrate how thermodynamics solved fundamental problems in thermochemistry in a way in which the kinetic theory was not able to match. The problem in question was one of chemical 'affinity', viz. how to predict the direction in which a particular chemical reaction would proceed and to determine reliably the strengths of the affinities involved.[139] Such a measure was apparently provided by the heat

[136] Boltzmann [1896*a*], pp. 376–411.

[137] Boltzmann's deduction of the statistical form of the second law relies upon considering only binary collisions. Tri-molecular collisions (which are used to explain dissociation) occur with such low frequency as to have negligible effect, on the way H changes with collisions. Yet it is precisely these which are needed to explain dissociation.

[138] No novel facts were deduced by Boltzmann in his account of the phenomenon of dissociation. Boltzmann's theory was thus *ad hoc*$_2$ (cf. Zahar [1973], p. 216, note 12). Exactly the same number of parameters is determined from experiment in both formulae.

[139] Chemical affinity is simply a measure of the relative ease by which certain substances combine or dissociate with each other. Thus, some molecules in combining with others must perform work in overcoming the potentials or forces binding them together; when the reaction occurs at constant volume this work appears as heat. Cf. Hiebert [1962*b*] for the history of the concept of affinity in thermodynamics.

of reaction, for this heat (which was always constant for a given reaction) was taken as a measure of the sum of the chemical and physical changes accomplished.[140] This was the Thomsen–Berthelot law of molecular work: 'The quantity of heat disengaged in any reaction measures the sum of the chemical and physical changes accomplished in that reaction.'[141] Further if the reaction was spontaneous (i.e. taking place without the intervention of any external agency) it would tend 'to the production of that body, or system of bodies *which disengages most heat*'. However, by 1880 a considerable number of reactions had been discovered which were spontaneous, yet took place with the absorption of heat (e.g. the freezing mixtures of hydrochloric acid and sodium sulphate to common salt and sulphuric acid). Further, some *exothermic* reactions did not proceed spontaneously in the required sense.[142]

Helmholtz considered the general question of the direction in which a chemical reaction would take place from the point of view of the second law which provided a necessary condition for all natural processes, viz.: that the total entropy of all bodies taking part in the process shall increase. From the first law, considering an infinitesimal change of state, (1) $dU = Q+W$ where U is the total energy of the system, Q the heat absorbed by the system, and W the work done on the system. From the second law, (2) $dS+dS_0 \geqslant 0$ where S is the entropy of the system and S_0 the entropy of all surrounding bodies. If the reaction proceeds isothermally $dS_0 = -Q/T$, whence (2) becomes $dU-TdS \geqslant W$ or $d(U-TS) \leqslant W$. The function $F = U-TS$ is a function of state, whence for reversible isothermal changes $dF = W$. Thus, for a finite reversible isothermal change the total work performed by the system is equal to the decrease of F. Three important novel consequences follow from Helmholtz's thermodynamic analysis:

(i) For a finite reversible change the change in free energy is the maximum work obtainable or the work equivalent of the heat released.

(ii) If the external work is zero during an isothermal process then the free energy decrease measures the work done by the *forces of chemical affinity*.

(iii) *It explained the deviations* from the Thomsen–Berthelot law, for if T is very high or S is very great, as in a gas or dilute solution, then the term $(T\Delta S)$ is not negligible and the reaction may absorb heat: at constant temperatures the reaction will only proceed spontaneously if the free energy of final state is *less* than the free energy of the initial state. Finally, it gave van't Hoff a reliable measure of the forces of chemical affinity, for by (2) the chemical affinity is the maximum work obtainable by an isothermal *reversible* transformation from the initial to the final state.

[140] The empirical law of Hess (Hess [1846], p. 423).

[141] Berthelot [1879], pp. 28–9.

[142] See van't Hoff [1903], p. 32 for a summary of the known anomalies to the Berthelot–Thomsen law.

van't Hoff wrote in 1904 of the progress in the theory of 'affinity' that the 'progress due to thermodynamic methods most curious of all enable us to treat the problems of affinity in an absolutely trustworthy way...without having to admit anything concerning the nature of affinity or the matter wherein the affinity is supposed to reside'.[143]

It was only in the final section of his 1882 paper that Helmholtz developed a *kinetic* interpretation of his results which had been achieved by thermodynamic considerations alone: he wrote

We require finally an expression in order to be able to distinguish clearly... between actual energy from the work equivalent of heat, which are indeed likewise to be regarded as the *vis viva* of invisible molecular motions. I would suggest that the former should be called the *vis viva* of orderly motion and the latter kind (disorderly motion) one might in this sense regard as the entropy or the measure of disorder, for only orderly motion can be converted into other forms of mechanical work.[144]

The new results[145] were achieved by direct application of the two laws of thermodynamics, the molecular account providing only a *post hoc* interpretation, which did not go beyond the thermodynamic results.

Thermodynamics in the nineteenth century was a *progressive research programme*. From each particular theory (Carnot's theory, the mechanical theory and phenomenological thermodynamics) a sequence of novel predictions followed which were subsequently confirmed. It was precisely the enormous empirical progress achieved by the programme which was the crucial factor in the appraisal of the programme by Planck, Ostwald and Duhem at the turn of the century. For instance, Planck regarded as the real merit of thermodynamics[146] that 'from the two fundamental principles...by pure logical reasoning, a large number of *new* physical and chemical laws [were] deduced, which are capable of extensive application, and have hitherto stood the test without exception'. Planck's point here is an important one: the empirical progress achieved by the programme consisted of the *drawing out of consequences* of two very general empirical laws. One can express this condition sharply by saying that the heuristic of thermodynamics is *fact dependent*, in the sense that the applica-

[143] van't Hoff [1904], p. 84.

[144] Helmholtz [1882], p. 972.

[145] Another clear example of the empirical progress achieved by the thermodynamic programme is Gibbs's deduction of the phase rule (Gibbs [1875], pp. 96–100) from the condition of equilibrium given by the second law $\delta S = 0$. Cf. Planck [1897*a*], pp. 181–200.

[146] Planck [1897*a*], p. ix. The statement that the thermodynamic laws were confirmed without exception is false. There were persistent anomalies to the second law, for example, the combination of hydrogen and oxygen. On the basis of the second law alone, because the free energy of water is less than the sum of the free energies of hydrogen and oxygen, the reaction should proceed spontaneously. It does not, however, for it requires heat (the spark of ignition) to begin the reaction. This anomaly was systematically dismissed on the grounds that the energy of the spark was negligible compared with the reaction energy (Cf. Planck [1897*a*], p. 116). Only Duhem took the problem seriously; cf. his long series of papers on 'false' equilibrium, Duhem [1900].

tion of thermodynamics to the solution of specific problems relies upon sets of auxiliary hypotheses which are not constructed in accordance with general overriding principles, but are taken either piecemeal, as low-level empirical generalisations, or are simply borrowed from other theories. For example, from the point of view of classical thermodynamics it is experience which shows that the three coordinates (P, V, T) which uniquely characterise a system in equilibrium are not all independent. Given a particular system we derive from *experience* which equation of state holds, e.g. $(P+a/V^2)(V-b) = RT$.[147] That is, thermodynamics was dependent upon the existence of particular *constitutive relations*, which had to be arrived at as empirical generalisations or borrowed from other theories. As an example of the latter in the theory of black body radiation, the deduction of Stefan's law[148] from thermodynamics requires that the law connecting the energy density and the radiation pressure ($p = \frac{1}{3}\omega$, for ω the energy density and p the radiation pressure) be borrowed from Maxwell's theory. Thus, the 'large number of *new* physical and chemical laws', to which Planck refers, were arrived at by using the heuristic of thermodynamics to draw out the *content* of the two laws when applied to differing situations characterised by different constitutive relations.

Since the heuristic techniques of thermodynamics are limited to the drawing of consequences from the two laws, given particular constitutive relations, and to suggestions as to how to deal with refutations (i.e. the fourth heuristic technique), it is clear that there was no possibility of employing the heuristic to go beyond the two laws, *to supersede them*. This situation contrasts sharply with the kinetic programme, for there the heuristic laid down quite general guidelines[149] for developing, independently of refutations, new kinetic theories of ever greater empirical content. For example, the heuristic of the kinetic programme led both Clausius and van der Waals to try to reduce the number of simplifying assumptions involved in considering molecular interactions, by taking into account, using the results of mechanics, more of the properties of the molecules (e.g. their finite size and the existence of short range cohesive forces). These attempts led to quite new universal laws (not mere logical consequences of auxiliary hypotheses and previous kinetic theories), e.g. the non-ideal equations of state. The kinetic programme consisted of a

[147] The existence of such an equation of state is a *consequence* of the kinetic theory.

[148] Cf. Boltzmann [1884].

[149] The fact that a heuristic does not provide an *algorithm* for generating new theories does not mean that such heuristic principles are *empty*. The heuristic of the kinetic programme did *not* provide an algorithm for generating new kinetic theories, but it did provide quite specific hints as to how this could be done. The distinction between strong and weak heuristic is independent of the distinction between effective and non-effective procedures. Indeed, in the case of the Ptolemaic research programme, the single heuristic principle – *save the phenomena by a combination of uniform periodic motions* – provides an algorithm for incorporating recalcitrant orbits given the observed data; but the heuristic is *weak* since elaboration of the programme depends crucially upon the existence of anomalies and data derived therefrom. The heuristic is *fact dependent*.

sequence of theories of ever greater empirical content, each taking account more and more of the 'fine structure' of the world by the *systematic* improvement of initial conjectures. A second example of a research programme which incorporates systematic guidelines for the construction of *new* theories is provided by the relativity programme. According to Zahar[150] the heuristic of Einstein's programme was based upon two distinct requirements. First, a new law should be Lorentz-covariant and second, it should yield some classical law as a limiting case. These two stringent requirements did indeed guide the construction of the revolutionary relativistic laws like Planck's law of motion and Einstein's equation relating mass and energy. The heuristic of thermodynamics was, however, limited[151] to the drawing of consequences from the two laws and providing suggestions as to how anomalies might be resolved given particular constitutive relations. In this sense the heuristic of thermodynamics is *weak*.[152]

It might be objected to this thesis that Nernst's discovery of the third law of thermodynamics (i.e. that the curves of free energy and total energy are tangential at absolute zero) indicates that the heuristic of thermodynamics did at least provide some guidelines for the discovery of new laws of nature.[153]

However, the two central clues in Nernst's route to the third law were both results based on *experience*. First, that the heat effect of a chemical

[150] See Zahar's paper below, especially pp. 237–62.

[151] The fact that the research programme consisted of a sequence of three theories, each successive theory having greater empirical content than the former, might appear to refute my conclusion as to the limitations of the heuristic. However, if we examine the transition from Carnot's theory to the mechanical theory, as given above, we see that the latter was arrived at simply by dropping from Carnot's theory the assumption that *only the fall* of heat generates work and adding the law of the conservation of energy. The transition from the mechanical theory to the phenomenological theory was simply achieved by the *generalisation* of a theorem of the mechanical theory.

[152] This is a particular application of a general criterion for distinguishing between programmes with strong and weak heuristics, given that the heuristic can be fully articulated. A strong heuristic is one which guides the construction of new laws independently of the existence of anomalies and of the determination of the values of crucial parameters from experiment. It is otherwise weak. Of course, progress in science is measured by *empirical* support and not by heuristic power: the possession of a powerful heuristic, in no way guarantees *empirical* success. Indeed, in the examples of the kinetic programme and the ether programme, both possessed powerful heuristics but later degenerated, while for example the thermodynamic programme, possessing a weak heuristic, achieved enormous empirical success in the prediction of a host of novel phenomena.

[153] Nernst [1906], pp. 167–73. Cf. Planck [1910], chapter 6. The third law states, that the difference between the free energy and the total energy of chemical reactions between pure solid or liquid bodies tends to zero as the temperature tends to absolute zero (cf. Nernst [1906], p. 168). This implies that the entropy of all systems at absolute zero is a universal constant which may be taken to be zero. From the second law the Gibbs–Helmholtz equation $F = U + T\partial F/\partial T$ is a consequence, for isothermal processes. From F, U may be calculated, but from U, F is not determined since in the integration of the equation there occurs an arbitrary function of T of the form $(\alpha T + b)$. Given the third law, however, the value of the constant α can be determined and consequently the change in F can be calculated from purely thermal data; i.e. from change in the total energy U.

process was not altered very greatly when the temperature was altered was an empirical generalisation. Second, that in galvanic combinations involving highly concentrated solutions or solid bodies, the *difference* between the free energy and the total energy was very small,[154] was simply observed to hold in a great number of cases.

In this case, then, the two vital clues in the discovery of the third law were not provided by general thermodynamic principles but were based upon two generally *observed* characteristics of reactions.

The fact that thermodynamics possessed only a weak heuristic, in the sense specified above, led to an attempt to strengthen the heuristic principles of thermodynamics by conferring on the programme an ontological basis.[155]

3.3 *Energetics*

The ontological basis was energy, the new strengthened research programme, *energetics*. According to Ostwald, pure thermodynamics is characterised by the 'analytic procedure', that is, 'the essential way in which results flow from the initial assumptions remains in the dark from the point of view of physical intuition'.[156] That is, many steps in thermodynamic deductions have no physical interpretation. Energetics arose as an answer to the need for giving a physical interpretation to those purely mathematical operations and so 'to connect more tightly from the point of view of physical intuition the premises with the conclusion and so make the deduction more transparent'.[157] The *heuristic* of the energetics programme *was* a single overriding requirement: the construction of *all* concepts as well as *all* calculations should take 'as their starting point the magnitude of the energies present'.[158] The first testable version of energetics simply reinterpreted the two laws of phenomenological thermodynamics as relations between forms of energy. Thus the law of the conservation of energy became the first law of energetics. The second law of energetics was that in any thermodynamic process the heat absorbed (or evolved) could be reduced to the change in two quantities of energy, the capacity (or volume energy, $p\,\mathrm{d}V$) and the intensity factor, T. However, as Planck and Boltzmann pointed out, if the volume energy were actually to be representative of a substance then it must be a function of state. That is, that the change in volume energy is determined by the change of state, but the $\int p\,\mathrm{d}V$ is dependent on the path and as such cannot be a function of simply the state of the substance.[159] Despite the claims of Ostwald and Helm for the heuristic value of energetics, they were forced in each separate case

[154] Nernst describes the *observed* difference as 'astonishingly' small. Cf. Simon [1956].

[155] The connection between 'ontology' and 'heuristic' is spelled out by Zahar in his [1973]: see below, especially pp. 238–9.

[156] Ostwald [1896], p. 157. Cf. also Helm [1896], [1898], and Ostwald [1891], [1892].

[157] Ostwald [1896], p. 158.

[158] *Ibid.*, p. 159.

[159] Cf. Planck [1896], p. 74, and Boltzmann [1896c], pp. 63–71.

to introduce new kinds of energy, *ad hoc* as the facts required. They could point to no result which followed from the principles of energetics which did not follow the ordinary thermodynamic principles; indeed the energeticists became bogged down in attempts to *reconcile* the energetic laws with those of thermodynamics.[160] Ostwald finally conceded that energetics won an 'epistemological' advance rather than a scientific one.[161] Energetics became a philosophical fashion but not a scientific one.[162]

We have appraised classical thermodynamics as an eminently successful programme, in which each successive theory predicted a whole range of novel facts. Furthermore, these novel facts followed from the application of the second law, in one form or another, of the principle of the *absolute increase* in entropy. A major problem situation for the kinetic theorists, if (as Clausius put it) the kinetic theory was to provide 'an adequate foundation for the theory of heat', was to provide a derivation, from kinetic assumptions, of the second law. That is, to demonstrate that an entropy function S exists satisfying $dS = dQ/T$, and that this function could *only* increase for irreversible processes.

3.4 *The kinetic programme and the second law*

By 1871 there were two important results available to Boltzmann, first the *general* form of the distribution law for molecular velocities and second the equipartition of energy theorem. Boltzmann considered the mean energy of a system of molecules specified as a function of the position and momentum of the molecules. Then the average total energy $\bar{E}$ is given by $\bar{E} = \bar{E}(\text{kinetic}) + \bar{E}(\text{potential})$ or $\bar{E} = \bar{T} + \bar{V}$, and thus

$$\delta\bar{E} = \delta\bar{T} + \delta(\bar{V}),$$

for a reversible variation. By the first law the heat added in a reversible process may be identified with the difference between the energy change and the work done on the system that is $\delta Q = \delta\bar{E} - \delta W$. But the work done on the the system is given by $\overline{\delta V}$, whence

$$\delta Q = \delta\bar{T} + \delta(\bar{V}) - (\overline{\delta V}). \tag{5}$$

Given the general distribution function (from which each of the terms on the right hand side of (5) may be evaluated), Boltzmann was able to show that dQ/T is an exact differential of a function of state S, the entropy of the

[160] Planck regarded it as demonstrable that 'energetics' would be unable to accomplish its avowed task of adding new heuristic principles to thermodynamics, since it merely *re-defined* crucial concepts, e.g. the internal energy, without adding any new physical principles. He remarks perceptively of energetic programme that, 'it is evident that no definition, however ingenious, although it contain no contradiction in itself, will ever permit the deduction of a new fact.' (Planck [1897*a*], p. 82.)

[161] Cf. particularly Ostwald [1907]. Ostwald's *anti-atomistic* arguments are summarised in his famous [1895]. Cf. below, p. 89.

[162] Cf. particularly Post [1968] and Hiebert [1972]. For the philosophical reception of energetics in England, cf. especially Hibben [1903].

system. In effect Boltzmann had derived from the kinetic theory, without appeal to a particular molecular model, the restricted form of the second law for reversible cycles, viz. $\oint dQ/T = 0$.[163] This is a remarkable novel fact for the kinetic theory, for essentially the result follows from the distribution function above, and the identification of $(\overline{\delta V})$ with the work done on the system.

In order to effect a complete reduction it remained to show that for irreversible processes $\int dQ/T < 0$. To do this Boltzmann introduced the famous single valued function H, such that $H = \int f(\boldsymbol{v}, t) \log f(\boldsymbol{v}, t)\, d\boldsymbol{v}$ for any distribution function f such that $f(\boldsymbol{v}, t)\, d\boldsymbol{v}$ is the number of molecules per unit volume which have velocity within the range $\boldsymbol{v}$ to $\boldsymbol{v}+d\boldsymbol{v}$. The rate of change of f with time, which is effectively the transport equation, can be given as

$$\left(\frac{\partial f}{\partial t}\right) = (B-A)\, dt,$$

where B = the number of molecules which as a result of collisions acquire in the time interval dt a velocity in the range $\boldsymbol{v}$ to $\boldsymbol{v}+d\boldsymbol{v}$ and A is the number of molecules that as a result of collisions in the time interval acquired new velocities outside this range. In order to calculate the number of collisions taking place in the time interval dt Boltzmann introduced a fundamental assumption which was essentially a generalisation of a method used by Clausius[164] in determining the mean free paths of molecules. That is that the number of collisions taking place in time dt between two groups of molecules moving toward each other can be computed as the product:

[Volume swept through by the 'action' spheres (cylinder of radius equal to twice the molecular radius) in time dt by the first group]
$\times$ [Number of molecules of the second group lying in that volume]

which in Boltzmann's deduction effectively amounts to the assumption that the number of molecules per unit volume is the same in the space to be traversed (by the first group of molecules) as in any other part of the space. On this assumption,[165] (without going into the analytic form of the argument[166]) Boltzmann was able to show that $\partial H/\partial t \leqslant 0$, that is that H could never increase, and further that when $\partial H/\partial t = 0$, the distribution function f has the Maxwell form. *That is, the acquisition by the gas of the Maxwell–Boltzmann distribution is a necessary as well as a sufficient condition for equilibrium.* Boltzmann then identified the entropy with $-kH+C$ which

[163] Boltzmann [1871*a*]. Cf. especially Klein [1969*a*], pp. 97–100; Klein [1972], pp. 62–5; Daub [1969].

[164] See above, p. 49.

[165] Boltzmann's proof was *not* model independent in this case. The molecules are assumed to be spherical and perfectly elastic; only binary collisions are considered, the gas being regarded as sufficiently dilute; further the gas is spatially homogeneous and has no external forces acting upon it.

[166] Cf. Boltzmann [1872], Boltzmann [1896*a*], pp. 49–62.

completes the deduction of the second law in its *absolute* form. Since H can never increase, the entropy proportional to $-H$ must always increase until it reaches a maximum value, when H is a minimum.

This apparent victory for the kinetic programme, that H can be shown to monotonically decreasing with time (the final reduction of thermodynamics), was reversed by a devastating argument of Loschmidt. Loschmidt's thesis amounts to this.[167] Consider a gas in an initial non-equilibrium state Γ_1; by the monotonic decrease of H, it will evolve through a series of states $\Gamma_1, \Gamma_2, \Gamma_3, \ldots, \Gamma_i, \ldots, \Gamma_n$ with the corresponding H values $H_1 \geqslant H_2 \geqslant H_3 \geqslant \ldots \geqslant H_n$. At state Γ_n, say, let all the molecules have the same positions and let the velocities of each molecule be reversed. Then since the gas molecules form a mechanical system they will now pass through the reverse states in order $\Gamma'_n, \Gamma'_{n-1}, \ldots, \Gamma'_{n-2}, \ldots, \Gamma'_1$, such that in Γ'_i the molecules will then have the *same positions* but velocities the reverse of those they had in Γ_i. From the form of the H-function $H_i = H'_i$ then, the corresponding H values will be $H'_n \leqslant H'_{n-1} \leqslant \ldots \leqslant H'_2 \leqslant H'_1$. Thus, for every motion of the model in which H decreases from H_1 to H_n, there exists a motion in which H increases from H_n to H_1, and hence the fact that entropy increases in our part of the universe must always be due to the initial conditions prevailing. Thus any attempt to prove from the nature of bodies and their interactions the law of entropy increase without taking into account initial conditions, which Boltzmann's did not, must be 'necessarily futile'.[168]

Loschmidt's argument is very important, for it effectively showed the *essentially* statistical nature of the H-theorem. Boltzmann re-interpreted the collision assumption referred to above so that that assumption gives for each time interval dt only 'the most probable' values of the number of collisions occurring, and that correspondingly the H-theorem gives for each dt only the most probable value of the change in H. The actual number of collisions therefore fluctuates about this most probable value assuming other values with a *small* but non-zero probability. Boltzmann wrote in his reply to Loschmidt:

> Loschmidt's theorem seems to me to be of the greatest importance, since it shows how intimately connected are the second law and probability theory, whereas the first law is independent of it. In all cases where $\int dQ/T$ can be negative, there is also an individual but very improbable condition for which it may be positive and the proof that it is almost always negative can only be carried out by means of probability theory.[169]

To illustrate this process exactly he developed the theory of thermodynamic probability[170] in which the increase in the entropy is simply the passage

[167] Cf. Ter Haar [1954].

[168] Loschmidt [1876], p. 173.

[169] Boltzmann [1877*a*], p. 193. Cf. Post [1968], especially, pp. 3–4.

[170] Boltzmann [1877*b*]. The possibility of providing the required probability metric is guaranteed by Liouville's theorem.

from initial states of low intrinsic probability to states of higher thermodynamic probability.

Thus Boltzmann's theorem flatly contradicts the phenomenological second law. The absolute increase of entropy is on kinetic principles reduced to the statistical law that the entropy *almost always* (and with overwhelming likelihood) increases with time. Boltzmann's theorem thus predicts the *existence* of *fluctuations*, for example in a state of thermal equilibrium away from the minimal value of H, such that the probability of a fluctuation occurring is inversely proportional to the magnitude of the fluctuation. *This was an entirely novel prediction.* However, it was not confirmed.[171] No independent evidence for the existence of such fluctuations existed and it seemed likely to both Maxwell and Boltzmann that the fluctuations were of so small an order that they would never be of an observable magnitude.[172]

Nevertheless Boltzmann's 'reduction' proved unacceptable to the phenomenologists.[173]

My thesis is that their attitude in rejecting the 'reduction' was perfectly rational, for first the deduction was *model dependent* (in fact the theorem was proved for an ideal case – dilute monatomic gas, with spherical molecules homogeneously distributed) given for a very restricted number of cases, *second* it contradicted the absolute entropy law[74] upon which so much empirical progress had been achieved (see above, pp. 68–77), and *third* (and principally) subsequent attempts to elaborate and develop the kinetic programme degenerated. The crucial difficulty was the irreconcilability

[171] Boltzmann's 1877 theory which predicts the fluctuation phenomenon was *ad hoc*$_2$, i.e. its novel prediction had not been confirmed. (Cf. Zahar's paper, below, p. 216. Zahar remarks that *ad-hocness* in this sense becomes a demerit of a research programme only if it is a lasting feature. But of course, the non-confirmation of the fluctuation prediction did last for years, from 1877–1905.)

[172] Cf. Boltzmann [1896*a*], p. 449. Deviations from the average value are such that on Boltzmann's view the relaxation times are of the order of millionths of a second. Maxwell also considered the fluctuations predicted by the kinetic theory, to be in principle unobservable. His demon would never 'crash the gate of the laboratory'. He wrote 'the second law is being continually violated, and that to a considerable extent, in any sufficiently small group of molecules belonging to a real body. As the number of molecules in the group is increased the deviations from the mean of the whole become smaller and less frequent; when the number is increased till the group includes a sensible portion of the body, the probability of a measurable variation from the mean occurring in a finite number of years becomes so small that it may be regarded as practically an impossibility.' (Maxwell [1878], p. 670.) When confronted with the Brownian motion he suggested that it was merely apparently discontinuous and 'random': if looked at under a more powerful microscope it would exhibit a 'more perfect repose'. (See below, footnote 238.)

[173] For instance, Plank wrote: 'Boltzmann's theorems are an unsuccessful attempt to show how a system governed by conservative interactions can proceed irreversibly to a final state of thermodynamic equilibrium' (Planck [1897*b*], p. 493).

[174] Cf. Feyerabend [1966]. Given the reversibility and recurrence objection, Zermelo saw the choice before physicists as a starkly simple one: either accept the Carnot–Clausius principle and abandon the kinetic approach, or accept the mechanical interpretation of nature and abandon the second law. Cf. Zermelo [1896*b*], p. 230, and Poincaré [1893], p. 206–7.

of the equipartition of energy theorem and the specific heats of gases. The plethora of *ad hoc* atomic models designed to resolve the persistent specific heat anomaly not only systematically failed to do so, but also possessed such properties as to render the *approach* to equilibrium given by the H-theorem impossible.[175]

4 *The degeneration of the kinetic research programme after 1880*

We have seen how the early kinetic programme was empirically progressive (see above, pp. 47–63), each successive kinetic theory predicted *novel* facts. However, none of these theories was free of anomalies, the problem of the ratio of specific heats proving particularly intractable. The anomaly was listed but set aside,[176] for Maxwell's heuristic directed his research into other problems (the transport properties),[177] where it was clear that the heuristic provided hints as to how they might be resolved. However, the heuristic of the kinetic programme could suggest no ways out of the specific heat ratio anomaly. In order to substantiate this claim, we must examine the problem situation in detail. It arises, essentially, as a consequence of a fundamental theorem common to all kinetic theories, the theorem of the equipartition of energy. This theorem allots to each degree of freedom of motion of a molecule the same amount of energy as is found for a degree of freedom of translation. Thus, if the total mean energy E of a sample of gas is

$$\overline{E} = \overline{E}(\text{translation}) + \overline{E}(\text{rotation}) + \overline{E}(\text{vibration})$$

then

$$\overline{E} = 3(RT/2) + r(RT/2) + v(RT/2)$$

whence the ratio of specific heats at constant volume and constant pressure becomes

$$\gamma = \frac{C_p}{C_v} = \frac{5+r+2v}{3+r+2v}.$$[178]

If the molecules are 'perfect spheres' then $r = v = 0$, so $\gamma = 1.66$. But this ratio was too great for any real gas then known.[179] For polyatomic molecules, $r = 3$ and thus $\gamma = 1.33$. But here the result was equally

[175] See below, pp. 85–6.

[176] It is the heuristic of the programme which directs which problems scientists work on, rather than the existence of 'psychologically' worrying anomalies.

[177] Cf. Maxwell [1860], pp. 406–9, Maxwell [1866], pp. 33–5 and Maxwell [1875], p. 433. Boltzmann's major theoretical interest lay in the relation between the kinetic theory and the second law of thermodynamics.

[178] Where r is the number of degrees of freedom of rotation, and v the number of degrees of freedom of vibration.

[179] Maxwell [1866], p. 78. In 1876, Kundt and Warburg ([1876]) found for mercury vapour that $\gamma = 1.66$, which confirmed the prediction for a monatomic gas. However $\gamma = 1.66$ follows only if it is assumed that $v = 0$. For a monatomic gas Boltzmann proposed that the molecules were elastic spheres, composed of 'inner' ether atoms. The 'atoms' would vibrate, but play no role in the thermal capacity of the gas. See below, p. 84. Further monatomic gases, argon and neon, were discovered by Ramsay and Rayleigh. Cf. Hiebert [1963] for a history of the discovery of argon.

anomalous, for the observed ratio for 'atmospheric air and other simple gases' was found to be 1.41.[180] This highlights a crucial difficulty for the research programme, *for in both these cases the degrees of freedom of vibration* (v) *have been ignored, and according to the heuristic of the programme such simplifying assumptions should be eliminated as there is independent evidence for the existence of spectral lines indicating that molecules do perform vigorous vibrations as they move between collisions.* However as v increases γ becomes smaller and nearer to 1, but already for $v = 0$ and $r = 3$, γ is too small.[181] The qualitative kinetic theory of spectral lines and bands was summarised by Maxwell in relation to the specific heat problem, in his *Theory of Heat*.[182] When a molecule moved along its free path it vibrated, since the parts of the molecule (the atoms) were capable of relative motion about some equilibrium position. The vibrations could be resolved into simple vibrations which were roughly like the oscillations of a pendulum. The amplitude was proportional to the violence of the last collision undergone by the molecule, while the period was dependent only on the constitution of the molecule. The vibrations were communicated to the luminiferous ether, giving rise to the luminous sharp spectral lines. If the mean free path were very long (as in rarified gases) the molecules would transmit all their oscillatory energy to the ether and cease to vibrate until another collision. If the density were very high the mean free path would be very short, each collision would superimpose new vibrations upon the regular ones and a continuous spectrum with bright bands (the regular oscillations) would result.[183] On these grounds Maxwell regarded the molecule as a system 'which must be capable of changing its form'.[184] Since, however, for $v = 0$, γ was already too small, in general the degrees of freedom of vibration of the molecules (v) would have to be excluded from contributing to the heat capacity of the gas, although their energy occurred as a result of intra-molecular collisions. The heuristic of the programme gave no suggestion as to how to account for the absence of the effect of vibration (particularly since from the laws of mechanics one would suspect quite the opposite effect). Even if the energy of vibration played no role in the thermal capacity of the gas, this would not completely solve the difficulty, for the value of r (at 3) would still be too great.

Boltzmann in 1876[185] proposed a molecular model for diatomic gases

[180] Boltzmann [1871*b*], p. 258.

[181] Maxwell [1866], p. 33.

[182] Maxwell [1872], p. 306.

[183] Cf. Clausius [1857], p. 113.

[184] Maxwell [1875], p. 433.

[185] In Boltzmann's earlier attempts to resolve the specific heat problem he assumed that in every gas, elementary or compound, at least two 'atoms' were associated with every 'molecule'. These 'atoms' would play no role in the distribution of energy during collisions but served merely to produce the spectral line by their oscillations about an equilibrium point (Boltzmann [1866]). In his [1871*b*] he changed his tack, attributing the entire difficulty to the effect of the ether 'drag'. As the gas molecules move through the ether, he argues, they lose kinetic energy to it via heat radiation, but he gives no clear account of the mechanism involved apart from a calculation as to the amount precisely by means of the deviation of the observed from the predicted values.

designed precisely to account for the fact that the observed ratio was 1.41 instead of 1.33. If a diatomic molecule was made up of two elastic spheres rigidly connected (such that collisions would not affect rotation about the axis of symmetry joining the two atoms), r would be equal to 2, and $\gamma = 1.4$ in accordance with the observations. However, this model does predict that if the molecule is *not* a solid of revolution (i.e. a polyatomic molecule) r must equal 3 and $\gamma = 1.33$.[186] Since Boltzmann's hypothesis was designed to account for $\gamma = 1.4$, its prediction of the correct value of γ when the molecules are polyatomic (the 'heavy' gases) counts as a novel fact.[187] It is, however, *ad hoc in the sense that it was obtained from the previous models in a way which did not accord with the heuristic of the programme* (*i.e. it is ad hoc*$_3$), for it did not comply with the principle that interactions should obey the laws of mechanics. In this case, the superiority of the system of appraisal provided by the methodology of research programmes over that of falsificationism emerges. Boltzmann's solution satisfies Popper's third requirement, for it explains the successful Kundt–Warburg confirmation for a monatomic gas ($r = 0$, $v = 0$), it explains the falsifying instance $\gamma = 1.4$ and predicts a novel fact, i.e. for *poly*atomic gases $\gamma = 1.33$. Nevertheless it was regarded by leading scientists of the time, e.g. Maxwell, Kelvin, Loschmidt, and Poincaré as *ad hoc*, as arbitrary and unsatisfactory because it ran counter to ordinary dynamical considerations.[188] This appraisal is captured objectively by the methodology of research programmes in the criterion of *ad hoc*$_3$, i.e. violation of the spirit of the heuristic of the programme.[189] Maxwell objected to Boltzmann's solution precisely on these grounds. He realised that if Boltzmann's solution were taken seriously from the point of view of dynamics, the specific heat of a gas would be infinite. The atoms of diatomic molecules are elastic, but they cannot be perfectly elastic, since the 'internal' atoms must vibrate upon collision to give rise to spectral lines. As such they must absorb some thermal energy (i.e. $v > 0$), *but on Boltzmann's account they do not*. According to the conservation laws more and more translational energy would be converted into vibrational energy, and this *without limit*. In Maxwell's opinion, there was no escaping the fact that the internal vibrators must take up their due proportion of the thermal energy according to their number of degrees of freedom. Even supposing only that the atoms of the molecule be elastic, although of nearly perfect rigidity, the specific heat is still infinite, for as Maxwell pointed out:

It will not do to take a body formed of continuous matter endowed with elastic properties, and to increase the coefficients of elasticity without limit until the body becomes practically rigid. For such a body, though apparently rigid, is in reality

[186] Boltzmann [1876].
[187] Cf. Zahar [1973]; see below, pp. 216–9.
[188] Cf. Maxwell [1877], p. 245; Kelvin [1886], p. 616; Loschmidt [1886], pp. 434–6.
[189] Cf. Lakatos [1970], pp. 182–3; Zahar [1973] (below, p. 217).

capable of internal vibrations, and these of an infinite variety of types so that the body has an infinite number of degrees of freedom.[190]

Exactly the same conclusion was reached by Kelvin, who regarded the Boltzmann molecules as a wholly unsatisfactory solution, since they would affect the permanent distribution provided by the Maxwell law of velocities. He thought it 'rigidly demonstrable that repeated mutual impacts must gradually convert *all* the translational energy into the energy of shriller and shriller vibrations of the molecule'.[191] True to dynamical principles, Kelvin regarded the problem as insoluble. He wrote:

There is in fact no possibility of reconciling the Boltzmann–Maxwell doctrine with the truth regarding the specific heats of gases.[192]

Rayleigh, too, saw that Boltzmann's solution was incompatible with general dynamical principles; he said of its relation to the equipartition theorem:

We are brought face to face with a fundamental difficulty relating not to the theory of gases merely, but rather to general dynamics...However great may be the energy required to alter the distance of the two atoms in a diatomic molecule, practical rigidity is never secured, and the kinetic energy of the relative motion in the line of junction is the same as if the tie were of the feeblest. The two atoms, however related, remain two atoms, and the degrees of freedom remain six [not five as in the Boltzmann solution] in number.[193]

The *ad hoc* (in the sense specified above) nature of the Boltzmann solution urged Maxwell to look elsewhere for a basic molecular model compatible with both the thermal and spectral properties of gases. He thought he had found such a model in the vortex atom, for in this case the behaviour of the atoms followed from purely mechanical principles. He wrote of the attractiveness of the proposed model:

The success in explaining phomena does not depend on the ingenuity with which its contrivers 'save appearances' by introducing first one hypothetical force and then another. When the vortex atom is first set in motion all its properties are absolutely fixed and determined by the laws of motion of the primitive fluid, which are fully expressed in the fundamental equations.[194]

The analytic difficulties, however, in solving the equations of vortex motion were enormous. Kelvin calculated the complete equation of motion only for a single infinitely long straight cylindrical vortex.[195] The deduction of even the perfect gas law proved possible only after great difficulty.[196] In general it proved impossible to show that a complex vortex ring did *not*

[190] Maxwell [1877], p. 245.

[191] Kelvin [1886], p. 616.

[192] Kelvin [1901], p. 504.

[193] Rayleigh [1900], pp. 117–18. Cf. Kelvin [1901], pp. 493–527.

[194] Maxwell [1876], pp. 471–2.

[195] Kelvin [1880], p. 165. Tait wrote of the complexities of the theory, 'to investigate what takes place when one circular vortex atom impinges upon another, and the whole motion is not symmetrical about an axis, is a task which may employ perhaps the lifetimes, for the next two or three generations, of the best mathematicians in Europe.' (Tait [1876], pp. 302–3.)

[196] Kelvin [1888].

have an infinite number of vibrational modes and therefore an infinite specific heat.[197] The vortex theory, despite its early promise and non-arbitrariness, did not succeed in providing an adequate foundation for the theory of molecules.

Watson, the author of a highly influential monograph on the kinetic theory, attempted to reconcile the equipartition theorem and the available degrees of freedom of the molecules by allowing that it take a 'sensible', even *long*, interval of time for the addition of an amount of energy added to the gas in the form of heat to be distributed equally among the available degrees of freedom. At first, all the energy would go into translational energy and only very slowly would equipartition be established. This would seem to increase the observed specific heats in the required manner. Watson, however, could give no means of independently calculating the postulated relaxation time apart from backwards in the light of the observed deviation from the predicted value of the specific heat ratio. The method simply achieved a reconciliation with the experimental values.[198]

Thus a succession of molecular models[199] were proposed, each designed specifically to account for the anomalous ratio of the specific heats, none succeeding in predicting any novel facts. The only model which accounted for the anomalous values, while possessing some independent support, the Boltzmann–Bosanquet solution, was in *violation of the heuristic of the programme*, that is the interactions based upon the model were, in an essential way, incompatible with the laws of mechanics. There is an analogy here with the situation, at the turn of the nineteenth century, of the ether programme. According to Zahar, 'The ether programme developed rapidly in certain respects, yet towards the end of the nineteenth century its positive heuristic was running out of steam.'[200] The problem for the ether programme was to provide a *mechanical* model for the ether, yet in order for it to be a carrier of *electromagnetic* phenomena as well the ether had to be at rest and in constant motion 'simultaneously'. Similarly, the molecule had to be elastic, not perfectly rigid, so as to allow for the vibrations of its internal 'atoms', yet the energy of vibration played no role in the *thermal capacity* of the gas. The molecular model possessed contradictory properties. Zahar continues:

Lorentz had reached a point where the behaviour of the electromagnetic field dictated what properties the ether ought to have, no matter how implausible these

[197] Cf. above, p. 84.

[198] Cf. Watson [1876], pp. 84–7; Watson [1894]; Fitzgerald [1895], pp. 221–2; and especially Bryan [1894].

[199] Boltzmann even grew irritated with the plethora of *ad hoc* hypotheses designed to account for the specific heat anomaly. He wrote: 'It must further be admitted that the form of research frequently led to extravagancies against which some sort of reaction was imperative. Every Tom, Dick and Harry felt himself called upon to devise his own special combination of atoms and vortices and fancied having done so that he had pried out the ultimate secrets of the creator'. (Boltzmann [1900], p. 247.)

[200] Zahar [1973]; see below, p. 255.

properties might be... *This involved a reversal of the heuristic of Lorentz's programme: instead of learning something about the field from a general theory of the ether, he could only get at the ether post-hoc by way of the field.*[201]

Exactly the same situation existed for the kinetic programme in the late nineteenth century. Originally the heuristic laid down that by making assumptions as to the nature of molecules and their interactions, the internal properties of gases (e.g. viscosity, diffusion) could be discovered. However, for Boltzmann it was the internal properties of the gas which dictated the nature of the molecules and their interactions.[202] *Thus, although the kinetic programme had developed very rapidly during the 1860s and 1870s, towards the end of the nineteenth century its heuristic had 'run out of steam'.* This can be seen in connection with the transport properties too. Since *no general solution*, even for a simple gas model, could be given for Boltzmann's transport equation, various approximate methods had to be used to arrive at values for the transport coefficients. Each method of making the approximation (e.g. those of Meyer, Stefan, Tait)[203] depended upon finding an expression for the mean free path, but involved neglecting certain terms which were of the same order of magnitude as the terms retained. Each different method of computation produced a different value for the mean free path and thus for the viscosity and diffusion coefficients. Planck remarks: 'Each of some six or more investigators in this field obtained different values for the ratio of the coefficient of diffusion to the coefficient of the conduction of heat, according to the particular method of calculation.'[204]

Each attempt at improving the treatment of the free path led to better approximations to experimental values for some gases at a given range of temperatures and pressures, while for the same gas within another range the predicted values showed larger deviations from the experimental values.

To summarise, the heuristic of the kinetic programme in the last decade of the nineteenth century had run out of steam: *First*, because Boltzmann's solution of the specific heat anomaly, the molecular model, was arrived at by postulating that degrees of vibratory freedom played no role in the thermal capacity of the gas; *this violated the heuristic principle of the programme that the molecular interactions* (the collisions giving rise to the molecular vibrations) *should be treated according to the laws of the mechanics.*[205] And *second*, the absence of a general solution for the Boltzmann transport equation necessitated the use of arbitrary approximations in the calculation of the transport coefficients.

[201] Below, p. 256.

[202] Cf. Kelvin [1901]. His first 'cloud' over nineteenth-century physics was the notion of the ether, his second, the equipartition theorem.

[203] Cf. above, p. 87.

[204] Planck [1931], p. 353.

[205] Boltzmann in his letter to *Nature* expressed his own conditions on the adequacy of a theory in mathematical physics, which his solution to the specific heat problem clearly violates. He states, 'Every hypothesis must derive indubitable results from mechanically well-defined assumptions by mathematically correct methods.' (Boltzmann [1895], p. 413.)

The kinetic programme was a degenerating programme from 1880 to 1905 for the following reasons.

(i) The series of molecular hypotheses put forward to account for the specific heat ratio were *ad hoc*. For example the Fitzgerald–Watson hypothesis of the equipartition time lag was designed purely to account for the observed ratio, by re-interpreting the experimental situation in such a way that the heat added to the gas would take so long to be equi-distributed that the predicted ratio based upon equipartition could never be observed. No novel consequences were drawn for this hypothesis.[206] The only hypothesis with independent support, the Boltzmann–Bosanquet model, was *ad hoc*$_3$, i.e. it was arrived at by violating an important heuristic principle.

(ii) The novel prediction of the existence of thermal and density fluctuations within a gas in thermodynamic equilibrium was not confirmed.

(iii) In a number of important instances the kinetic programme lagged behind thermodynamics: the kinetic programme could reconcile the novel predictions by introducing a series of *ad hoc* hypotheses which did not lead to any new facts.

5 *Thermodynamics and the kinetic research programme* (*1880–1905*)

In the last section, the kinetic programme in the late nineteenth century was shown to have been a degenerating research programme. From this conclusion it would be incorrect to infer that the kinetic programme should have been abandoned, or that those scientists like Boltzmann, Meyer, Rayleigh and Kelvin who continued to work on the programme acted irrationally. Indeed, the kinetic programme provides a paradigm case of a degenerating programme, which later became empirically progressive as a result of efforts to develop it. What the appraisal does indicate is that the kinetic programme was in need of a 'creative shift'.[207]

The methodological appraisal that the kinetic programme was degenerating was reflected in the judgements of the leading scientists of the time. Planck for instance commenting upon the state of the kinetic programme at the Halle conference in 1891 referred both to the progressive and degenerating phases of the programme. He concluded his appraisal of the development of the programme with the judgement that, 'Despite

[206] This explanation does not appear to be *ad hoc*$_1$, i.e. such that *no* novel consequences *can* be deduced from it. It is clear for instance that a long equipartition time (i.e. the time before the heat added becomes equally distributed among the available degrees of freedom) would make the transport coefficients time dependent. Cf. Lakatos [1970], pp. 123–4.

[207] This shift was provided by Planck and Einstein in the quantum theory (a shift in 'mechanics') and its application to the problems of specific heats (cf. Planck [1900], Einstein [1905*b*] and Einstein [1907*b*]); and by Einstein and Smoluchowski in their statistical investigations of thermal and density fluctuations (cf. Einstein [1905*b*] and Smoluchowski [1906]).

a short meteoric rise in the early sixties, every attempt at elaborating the theory has not only not led to new physical results but has run into overwhelming difficulties.'[208] Similarly Ostwald writing in retrospect of the state of the programme at the end of the last century states that he saw in the atomic–kinetic hypothesis, 'a superficial habit to cover up rather than promote actual scientific tasks by arbitrary assumptions about atomic positions, motions and vibrations.'[209]

Even scientists more sympathetic to the programme of providing a rational foundation for thermodynamics, like Gibbs, concluded that the *atomic* approach had degenerated. In a now famous passage of his *Statistical Mechanics* he summarised the problems facing a *dynamic* theory of molecular action, and concluded:

> In the present state of science, it seems hardly possible to frame a dynamic theory of molecular action which shall embrace the phenomena of thermodynamics, of radiation, and of electrical manifestations which accompany the union of atoms. Yet any theory is obviously inadequate which does not take account of all these phenomena. Even if we confine our attention to the phenomena distinctly thermodynamic, we do not escape difficulties in as simple a matter as the number of degrees of freedom of a diatomic gas. It is well known that while theory would assign to the gas six degrees of freedom per molecule, in our experiments on specific heat we cannot account for more than five. Certainly one is building on an insecure foundation who rests his work on hypotheses concerning the constitution of matter.[210]

Boltzmann, too, was aware of the degenerating state of the kinetic programme. He responded by abandoning his former strongly realistic stance as to the existence of atoms and molecules, and their interactions and developed a methodology based upon regarding theories as mental pictures[211] or classificatory devices to be compared as to their fruitfulness

[208] Planck [1891], p. 650. In the preface to his textbook on thermodynamics (Planck [1897*a*], p. viii) he gave a similar appraisal of the state of the research programme referring to 'essential difficulties...in the mechanical interpretation of the fundamental principles of thermodynamics'. The crucial difficulty was the recurrence theorem of Poincaré ([1890]). Poincaré showed that for any system of particles confined in a finite volume acting under forces that depend only on position in space and possessing a finite energy, any state of motion will recur to within arbitrarily specified limits infinitely often. Hence, the entropy of such a system would have to possess a periodic character which contradicted the irreversible approach to equilibrium. (Cf. Zermelo [1896*a*], Zermelo [1896*b*].) Boltzmann's replies (Boltzmann [1896*b*], Boltzmann [1897*c*]) re-emphasised the statistical nature of the *H*-theorem, to the effect that if an initial state deviates from the Maxwell distribution, it is overwhelmingly likely that it will tend toward the Maxwell distribution and subsequently deviate from it by vanishingly small amounts, the recurrence times themselves for systems involving large (10^{18}) numbers of molecules being inconceivably long.

[209] Ostwald [1927], pp. 178–9. This is a remarkable intuitive definition of heuristic and theoretical degeneration.

[210] Gibbs [1902], p. x. For the relationship between the statistical mechanics of Gibbs and Boltzmann, see Ehrenfest [1911], pp. 43–69 and Jaynes [1967], pp. 84–92.

[211] Cf. Popper [1974], p. 129. According to Popper, Boltzmann was a realist in respect of atoms throughout his work.

and to their ability to provide economical classifications. The *ideal* of a physical theory was then 'the most perfect which represents all appearances in the simplest and most useful way'.[212]

That Boltzmann gave up his realism in regard to atoms and their interactions is a corroboration of the prediction made on the basis of the methodology of research programmes that scientists *believe* the hard cores of their programmes just to the extent that those programmes are objectively *progressive*. In this case Boltzmann was forced into abandoning his realistic position on atoms and molecules, because his own hypothesis as to the constitution of molecules and their interactions was incompatible with the laws of mechanics, as was clearly pointed out by Maxwell.[213]

The programme of classical thermodynamics was, on the other hand, a progressive research programme from its inception. Using thermodynamic methods a whole range of quite novel facts were deduced as consequences of the second law. However, as was noted above, the heuristic of the thermodynamic research programme was *weak*, for it contained no hints or guidelines as to how the thermodynamic laws could be systematically extended and improved.

This account of the state of the thermodynamic research programme captures objectively the rationality of scientists like Planck, Ostwald and Gibbs who attempted to develop a programme which would supersede thermodynamics (yielding the second law as a limiting case), while rejecting the kinetic programme as an unsuccessful, now stagnant, attempt to do just that. Planck, for instance, attempted to derive the phenomenological second law as a consequence of radiation damping in which irreversibility was to be explained by the conversion of incident plane electromagnetic waves to out-going spherical waves. He realised that such an attempt would, *if* it succeeded, supersede thermodynamics in a progressive way. It would, for instance, predict the

[212] Boltzmann, [1899], p. 253. Also Boltzmann [1897*b*]; and Boltzmann [1897*a*].

[213] Cf. above, p. 84. For a discussion of Boltzmann's methodology, cf. Feyerabend [1967]; Broda [1955]; Dugas [1959], pp. 101–40; Elkana [1973]; Brush [1968]. Elkana states in his [1973] that 'the very question whether a probabilistic argument is "scientific" or not depends on the image of science and not on the existing body of knowledge'. But it seems to me that the use of a given argument (probabilistic or not) is *scientific* if the argument results in the prediction of a novel fact, not derivable without the (probabilistic) premise. Boltzmann's programme used statistics in the form of the distribution law right from his second paper in 1868. The introduction of his theory of mental pictures stems, in my view, from the impossibility of producing a model of the atoms not endowed with contradictory properties, and not as Elkana seems to suggest as resulting from a need to introduce probabilistic laws (cf. Elkana [1973], p. 48). I completely agree, however, with Elkana in his account of how rival theories are appraised in Boltzmann's methodology, i.e. as to their fruitfulness in the prediction of novel facts (cf. Elkana [1973], pp. 148–9). Boltzmann's reply to the accusation that the heuristic of the kinetic programme had run out of steam is contained in his [1897*a*], in which he tried to show atomism was presupposed by continuum mechanics and that limiting processes were a purely mathematical fiction!

spectral distribution law for black body radiation.[214] He wrote of the programme:

The study of conservative damping seems to me to be of the greatest importance, since it opens up the prospect of a possible general explanation of irreversible processes by means of conservative forces – a problem that confronts theoretical physics more urgently every day.[215]

Similarly, Ostwald attempted in the research programme of energetics to derive the two laws of thermodynamics as consequences of the properties of the fundamental substance *energy*, the *aim* of the programme being to *divest* thermodynamics of its phenomenological character and thus to arrive at new laws, inaccessible by pure thermodynamic techniques.

In summary, the situation at the end of the nineteenth century with respect to the two major research programmes, the kinetic programme and thermodynamics, was one of genuine scientific uncertainty. The diversity of appraisals of the relative merits of the two programmes can be accounted for on the grounds that the two components of appraisal, *empirical success* and *heuristic power*, here diverge. The programme possessing (in principle) the *strong* heuristic, the kinetic programme, was (after a progressive period) now *degenerating* – in particular, its heuristic had run out of steam; while the programme possessing the *weak* heuristic, thermodynamics, remained *empirically progressive*. Given this appraisal of the state of the research programmes, there were two paths open to working scientists of the time, each equally legitimate. First, to stick to the kinetic programme in the hope that by utilising hitherto unexplored consequences of the theory (e.g. the existence of fluctuations[216]) new facts would be discovered, or by attempting to add to the programme's heuristic new principles (i.e. a creative shift in the heuristic) which would lead to the discovery of new laws. Second, to try to systematically strengthen the heuristic of thermodynamics to develop a programme which would supersede it in a progressive way, while avoiding the empirical and heuristic problems of atomism. This latter path was adopted by Ostwald, Planck and Gibbs, who by employing admittedly quite different methods attempted to super-

[214] Planck abandoned this approach when Boltzmann pointed out that the equations of electromagnetism are invariant under time reversal. Cf. Klein [1962], and [1969*a*], pp. 218–34.

[215] Planck [1897*b*]. The heuristic of Planck's programme, the requirement that irreversibility be explained by the action of conservative forces, was a prescription which corresponded to a metaphysical principle held by Planck: determinism. The kinetic programme essentially involved statistical laws which introduced into classical physics an element of *indeterminism*, in the sense that spontaneous reversals of apparently irreversible processes were predictable on kinetic grounds. Even long after Planck had accepted a modified form of Boltzmann's analysis of thermodynamic probability, he regarded it as necessary to introduce a separate hypothesis excluding large scale fluctuations, such as the spontaneous separation of two inter-diffused gases. (Cf. Planck [1909], pp. 13–17, and Planck [1914].)

[216] Cf. below, pp. 93–8.

sede thermodynamics. Only Gibbs's programme, employing statistical but non-atomistic methods, achieved any lasting success.[217]

There was a third possibility open which seems to have been adopted only by Duhem. That is to regard pure thermodynamics as a closed system, the business of the physicist being to investigate the logical consequences of the three laws as applied to differing initial conditions. Physics would then consist of the system of classification of experimental laws provided by the principle of least action for mechanics, Maxwell's equations for electromagnetism, and the laws of thermodynamics for heat and physical chemistry. In this case thermodynamics is elevated into a 'paradigm' of great science, providing as it does a superb classification of experimental data (and laws). Duhem seized upon this characteristic, 'deriving' from it a universal demarcation criterion, viz: that one set of theoretical principles, T_1, is better than another, T_2 say, when T_1 gives a more economical, simple codification of the relationships persisting among observable facts (or of the set of empirical laws) than does T_2.[218] In elevating thermodynamics into a paradigm case of 'great' science Duhem eliminated an essential component in the growth of knowledge, the ability to learn more about the 'fine structure' of the world by the *systematic* improvement of initial conjectures. The 'initial conjectures' are simply the first naive 'models' of the physical situation investigated. *New laws*, as was illustrated in the early development of the kinetic programme, were arrived at precisely by improving in a systematic way (according to the principles laid down in the heuristic of the programme) the original naive models. These ever-improving models were thus not, as Duhem would have them, 'parasitic growths fastening themselves on a tree already robust and full of life',[219] but *essential components* in the discovery of new laws. In this respect the methodology of research programmes identifies a crucial weakness of scientific positivism at its source – the overestimation of the superiority of thermodynamics over the kinetic

[217] Cf. Schrödinger [1944].

[218] Cf. Duhem [1906], particularly pp. 165–79 and pp. 219–70. Duhem's methodology imposes a hierarchy upon physical theories. At the bottom of the hierarchy is the potentially infinite list of experimental facts. These are classified by (and are consequences of) low level experimental laws, which have content. Purported laws are either true or false. The experimental laws are classified by, and are deductive consequences of, the physical theories. These physical principles, like the principle of least action which accounts for, in a simple economical fashion, all purely mechanical phenomena, constitute the highest level of physical theory. However, the general physical principles are purely systems of classification, decisions between rivals being based upon the convention of simplicity and economy. They are contentless, neither true nor false, yet they imply (have as logical consequences) laws which have content. But deduction is content decreasing, which requires that if the general principles have zero content, so must the physical laws which follow from them. Cf. also Mach [1882].

[219] Duhem [1906], p. 95. Cf. also his [1902] and [1895]. Duhem here argues for the redundancy of atomism. What he in fact shows is that the physics of the time could be reconstructed without appeal to the atomic hypothesis; what he often claims but does not substantiate is the *heuristic* valuelessness of atomism in general.

programme.[220] In opting for the programme which possessed only a weak heuristic, the vital role played by a strong well articulated heuristic (based possibly upon 'vague' metaphysical beliefs[221]) in the discovery of new laws, was overlooked.[222]

6 *Brownian motion and the empirical progress of the kinetic programme after 1905*

As noted above, the kinetic programme predicts that for a system composed of molecules (a gas, say) in thermodynamic equilibrium significant fluctuations from the mean thermodynamic quantities can be expected in a sufficiently small volume element of the gas.[223] These fluctuations were, however, regarded by both Maxwell and Boltzmann to be in principle unobservable.[224] If, however, such fluctuations would produce effects which were experimentally determinable, *then their existence would count as a novel fact for the kinetic programme.* In his autobiography Einstein reveals that his motivation for investigating the effects of molecular motion was 'to find facts which would guarantee as much as possible the evidence of the existence of atoms of definite size'.[225] He noticed an *asymmetry* in the thermodynamic analysis of the suspension of small particles in a fluid. According to classical thermodynamics, small particles suspended in a fluid will *not* produce any osmotic pressure due to their concentration gradient, whereas substances which are dissolved will. Thus the phenomenological theory makes an artificial distinction which Einstein regarded

[220] Lakatos first pointed out the possibility that a research programme could be so progressive as to set itself up as a standard of *scientificness*. But this is exactly what Duhem tried to do for thermodynamics. In this case the arbiter was, as Lakatos conjectured, an empirically successful research programme. He remarked, 'one must never allow a research programme to become a *Weltanschauung*, or a sort of *scientific rigour*, setting itself up as an arbiter between explanation and non-explanation, as mathematical rigour sets itself up as an arbiter between proof and non-proof' (Lakatos [1970], p. 155).

[221] Cf. especially Zahar [1973]: see below, p. 239. Vague metaphysical beliefs can be transformed into powerful tools for the invention of new theories when applied to specific problem-situations: for example, in the discovery of the continuity of liquid and gaseous phase, cf. above, pp. 57–60.

[222] The basis of Mach's criticism of the kinetic programme was that programme's heuristic and empirical degeneration (cf. above, pp. 82–8). His early defence of the programme, for example in his [1863], relies upon the range and number of new facts deduced by employing the theory. In the preface he wrote of the theory, 'One may accept the atomic theory, ...as a formula which already has led to many results and will continue to do so in the future.' (Quoted by Hiebert in his [1966] p. 87.) Mach's opposition to atomism stems from its subsequent failure as a heuristic tool, e.g. in its persistent failure to deal with the specific heat problem. He came to regard it as a model for representing the facts, which would in accordance with the principle of the economy of thought be subsequently replaced by a complete theory. (Cf. particularly Hiebert [1966] and Brush [1968] for a detailed account of Mach's changing attitude to atomism.)

[223] Cf. Maxwell [1872], p. 308.

[224] Maxwell [1878], p. 670 and Boltzmann [1896*a*], p. 449; cf. above, p. 81, footnote 172.

[225] Einstein [1949], p. 47.

as dependent upon the arbitrary fact that we can *observe* suspended particles while we *cannot observe* the molecules of a substance in solution.[226]

Thermodynamics predicts that the osmotic force produced by the suspension should be zero. On the kinetic programme, however, 'a dissolved molecule is differentiated from a suspended body solely by its dimensions'.[227] Thus the suspended particles should, according to the kinetic theory, experience an osmotic pressure due to the concentration gradient as they diffuse through the medium. This pressure would be resisted by the hydrodynamic viscous force of the fluid. Since dynamic equilibrium exists in the fluid the number of particles per unit area per unit time passing as a result of the concentration gradient should exactly balance the number of particles passing as a result of the remaining forces acting on the particles. Providing Stokes's law for the velocity of a particle acted upon by a force in a viscous medium holds for a suspended particle, the coefficient of diffusion could be found as a function of the temperature, the viscosity and the radius of the particles.[228] In the second part of his paper Einstein developed the theory of the *motion* of the suspended particles. On the assumption that each single particle executes a movement which is independent of the movement of all other particles, and that the movements of one and the same particle at different intervals of time are independent so long as the intervals are not too small, Einstein showed that if $f(x, t)$ was the distribution for the particles then f satisfied the equation $\partial f/\partial t = D\,\partial^2 f/\partial x^2$ which has the solution,[229]

$$f(x, t) = \frac{n}{(4\pi D)^{\frac{1}{2}}} \frac{e^{-x^2/4Dt}}{t^{\frac{1}{2}}}$$

From this equation and the value determined for D, Einstein deduced a series of novel predictions concerning the magnitude of the motion: for instance, that the displacement of the particle in time τ is proportional to $\tau^{\frac{1}{2}} (\bar{d}_\tau = (2D\tau)^{\frac{1}{2}})$, that the particles *rotate* with mean angular displacement in time τ proportional to $\tau^{\frac{1}{2}}$ and that the number of particles in a suspension performing the motion decays *exponentially* with height. One can sum up the content of Einstein's paper as a prediction of the *existence* and *magnitude* of the *Brownian motion* from the kinetic *theory* of thermal equilibrium.

226 Einstein [1905*a*], p. 3.

227 *Ibid.*

228 Cf. Einstein [1905*a*], section 3. Einstein's use of Stokes's law in this situation is very problematic, first because the law holds only when the object acted upon is subject to a *continuous* steady force but the molecular bombardment is *discontinuous*, and second, the radius of the object subjected to the viscous drag must be large compared with the mean free path of the molecules of the surrounding fluid, which would not be the case for Brownian particles suspended in a gas. Perrin had to re-interpret Einstein's application of the law so that it applied not to single particles but actually represented that the departures from the law expected because of the irregular nature of the Brownian movement would on average cancel out. Cf. Perrin [1908].

229 Einstein [1905*a*], p. 16.

The Brownian motion, or the random agitation of tiny particles suspended in a fluid in thermal equilibrium, was actually first noticed by Brown in 1827. On observing pollen grains in water he noticed 'many of them very evidently in motion; their motion consisting not only of a change of place in the fluid, manifested by alterations in their relative positions, but also not infrequently of a change of form in the particle itself.... In a few instances the particle was seen to turn on its longer axis.'[230] Throughout the nineteenth century various hypotheses were put forward to explain the motion, ranging from agitation of the fluid by external sources of vibration to the influence of *light* upon the conditions of equilibrium around a given particle.[231]

Thus, for almost a hundred years scientists *apparently* ignored what Gouy described as 'a direct and visible proof of the modern hypothesis of the nature of heat'.[232] In fact, however, a number of investigations of the Brownian motion were undertaken from the point of view of the kinetic theory, but all with negative results. For example, in 1877 Delsaulx published a paper in which he gave a qualitative explanation of the motion: if the diameter of the Brownian particles or 'bubbles' in the case of a gas were of the order of the dimensions of the molecules of the surrounding fluid, the pressure on the surface of the given particle would be non-uniform. He claimed 'that from the laws of chance alone it would be possible to derive that the small particles would be agitated to a degree which was observable'.[233]

However this suggestion was refuted as soon as a quantitative calculation was attempted. Nägeli took up Delsaulx's suggestion first by considering a single impact between a Brownian bubble and a molecule of the surrounding gas. In such a case the increase in velocity acquired would be of the order of vm/M (where m = mass of a molecule, M = mass of the particle). From the elementary kinetic theory $v \simeq (3kT)^{\frac{1}{2}}/M$, whence the increase in velocity of the 'sun-mote' would be $\simeq (3kT)^{\frac{1}{2}}/M$ at ordinary temperatures, 0.002 mm s^{-1}. He further assumed that in the case of a particle suspended in water, v would be smaller (than in the case of a gas) and the density of the medium much greater; thus for a starch particle to acquire the *observed velocity* of motion, a billion such collisions must occur per second in the direction of motion. That that number of collisions did occur per second he conceded, but that the irregular pressure developed in Delsaulx's hypothesis should be such as to produce such an excess number of collisions in *one given direction* he refused to accept. William

[230] Brown [1827]. Perrin says of the motion: 'It would be difficult to examine for long preparations in a liquid medium without observing that all the particles situated in the liquid instead of assuming a regular movement of fall or ascent according to their density, are on the contrary animated with a perfectly irregular movement.' (Perrin [1910] pp. 1–2.)

[231] For a detailed account of the various theories of the motion in the nineteenth century, cf. Nye [1972], pp. 21–9.

[232] Gouy [1895], pp. 1–2.

[233] Delsaulx [1877], p. 6.

Ramsay in 1892, on the basis of his own and Nägeli's analysis of the motion, introduced in order to overcome just that difficulty a hypothesis completely alien to kinetic theory, the hypothesis of coagulate motion 'that the bombardment be produced from water molecules existing in complex groups of *considerable mass*, and *some stability*'.[234] This *ad hoc* adjustment was never really successful for from it no quantitative evaluation of the motion would follow. In a series of papers from 1888 to 1895 Gouy returned to the problem of the kinetic theory of fluctuation and the Brownian motion, but again he could supply no quantitative analysis of the motion apart from citing Bernoulli's 'law of large numbers', such that the ensemble of molecules in the Brownian collisions was so great that 'there is not any contradiction [in assuming random fluctuations] no more than with other phenomena governed by chance alone'. He reproduced a whole series of experiments which he had conducted on the motion (including watching the motion in an apparatus in a ploughed field in the country late at night in the 'total' absence of vibration) to show that it never settled down. If the kinetic theory in its explanations were correct, this of course should follow, since indeed the motion would be *eternal*. Given that it persisted in a state of thermodynamic equilibrium he even suggested a design for *a perpetual motion machine of the second kind* based upon it. In the absence of any quantitative analysis of the motion Gouy felt that the Brownian motion nevertheless constituted a *crucial experiment* which went conclusively against the second law, showing that 'Carnot's principle is essentially false in small volumes'.[235] However, the crucial experiment when performed refuted the kinetic theory's explanation. Exner in 1900[236] returned to the fundamental *dynamic analogy* presented by van't Hoff's law for solutions. He argued that if the kinetic theory as an explanation of the Brownian motion was correct, a connection must exist (namely one of equality) between the kinetic energy of the particles and the kinetic energy of the surrounding liquid molecules, regarding the entire suspension as one fluid. The calculation of the kinetic energy of the liquid molecules was easily performed and gave $v = 0.2\ \text{cm s}^{-1}$ which was very much greater than the observed velocity of the Brownian particles.[237] Exner's 'refutation' and the unavailability of any quantitative analysis of the motion seemed to throw explanation back to the older convection current,

[234] Ramsay [1892].

[235] Gouy [1895], p. 23.

[236] Exner [1900].

[237] This method of obtaining the kinetic energy of a Brownian particle, which relies upon following under the microscope a *grain path*, gives a velocity very much smaller than the true velocity. Einstein's theory from its inception was 'fact correcting' in that it showed precisely why the observational theory employed by Exner – the path-following method – was inadequate. Einstein concluded that the 'observed' velocity corresponds 'to no objective property of the motion whatsoever', (Einstein [1907*a*]): for an observer 'can never perceive the actual path transversed in an arbitrarily small time, a certain mean velocity will always appear to him as an instantaneous velocity'. Using Einstein's theory it first became possible to verify Exner's relation.

non-equilibrium explanations. (An explanation which, incidentally, Maxwell preferred for he felt that if the motion were looked at under a *more* powerful microscope it would be seen as *continuous*; indeed after a while the particles would exhibit 'a more perfect repose'.[238])

What I wish to emphasise is that until Einstein's 1905 paper all attempts to give an adequate account of the Brownian motion using the kinetic theory had failed. The Brownian motion did not amount as Gouy had suggested to a direct and visible proof of the kinetic theory. Quite the opposite, as Einstein pointed out, for the 'observed' velocity of the Brownian particles corresponded 'to no objective property of the motion whatsoever'.[239]

Initial attempts to test the predictions went against Einstein's theory. In 1906 and 1907 Svedberg undertook a series of experiments which he hoped would verify Einstein's predictions, but when his results were analysed they gave the displacements of the particles as 4 to 6 times as large as the predicted value. Indeed, Svedberg claimed to have found that, quite contrary to Einstein's theory, the trajectories of the Brownian particles were 'lines regularly undulated, of well-defined amplitude and wavelength'.[240] He even calculated the value of the amplitude of the oscillation. Similarly, Henri found displacements three times greater than Einstein's formula predicts.[241]

Einstein's predictions were finally confirmed in a series of experiments carried out by Perrin and Chaudesaigues. The confirmation of Einstein's theory constituted *empirical progress* for the kinetic programme. It did so because although the Brownian motion was well known (it did not constitute the prediction of a temporally novel fact), Einstein's theory was in no way 'accommodated' to arrive at it.[242] In essence, Einstein showed that the motion can be predicted directly from the kinetic model of thermal equilibrium.[243] The importance of the theory lies in the fact that

[238] Quoted by Gouy in his [1889], p. 102.

[239] Cf. above p. 96, footnote 237.

[240] Svedberg [1907], p. 138.

[241] Henri [1908]. Perrin's later confirmations of Einstein's theory have led historians into underplaying the role of the refutations of the theory. For example, Fürth states that Henri's experiments confirm Einstein's theory. The displacements measured by Henri were three times greater than the formula allows, which would imply (if we accept the formula) that the granules were in fact 8 times smaller than Henri stated them to be. This is impossible, since the granules would then be invisible, contrary to Henri's statement that he worked with visible granules. Perrin claims that Henri's refutations were well known at the time (Perrin [1910], p. 58). When Perrin undertook to re-do Henri's experiments he failed to get the same results; the new ones confirmed Einstein's theory. Indeed, Henri's results seem later to have been dismissed as *occult* effects. Perrin wrote of them: 'The method was quite correct and had the merit of being then used for the first time. I do not know what source of error falsified the results.' (Perrin [1916] p. 121, footnote 4.)

[242] In the appraisal of Einstein's prediction of the Brownian motion and the Smoluchowski explanation of critical opalescence, Zahar's amended criterion of 'novelty' is essential, for both *facts* were well known. Cf. Zahar [1973], p. 218.

[243] The series of experiments and the results are fully documented in Perrin [1910] and [1916], and Chaudesaigues [1908]. Cf. also Nye [1972], especially pp. 97–142.

it confirms the kinetic programme's account of the second law of thermodynamics as being an *essentially statistical* law.[244]

Einstein's theory did not, of course, solve all the difficulties facing the kinetic programme. This latter programme did not *supersede* phenomenological thermodynamics (i.e. predict *progressively* all that thermodynamics predicted and more besides) with Einstein's theory. What it did show was that the utilisation of the statistical approach of the kinetic programme was the only way in which *empirical* progress could be achieved.

References

Agassi, J. [1963]: *Towards an Historiography of Science.*

Agassi, J. [1969]: 'Popper on learning from Experience', in N. Rescher (ed.): *Studies in the Philosophy of Science*, pp. 162–72.

Andrews, T. [1869]: 'On the Continuity of the Gaseous and Liquid States of Matter', *Philosophical Transactions of the Royal Society*, **159**, pp. 575–90.

Berthelot, M. [1879]: *Essai de Mécanique Chimique.*

Boltzmann, L. [1866]: 'Über die mechanische Bedeutung des zweiten Hauptsatzes der Wärmetheorie', *Wissenschaftliche Abhandlungen*, **1**, pp. 9–33.

Boltzmann, L. [1868]: 'Studien über das Gleichgewicht der lebendigen Kraft zwischen bewegten materiellen Punkten', *Wissenschaftliche Abhandlungen*, **1**, pp. 49–96.

Boltzmann, L. [1871*a*]: 'Analytischer Beweis des zweiten Hauptsatzes der mechanischen Wärmetheorie aus den Sätzen über das Gleichgewicht der lebendigen Kraft', *Wissenschaftliche Abhandlungen*, **1**, pp. 288–308.

Boltzman, L. [1871*b*]: 'Über das Wärmegleichgewicht zwischen mehratomigen Gasmolekülen', *Wissenschaftliche Abhandlungen*, **1**, pp. 237–58.

Boltzmann, L. [1871*c*]: 'Einige allgemeine Sätze über Wärmegleichgewicht', *Wissenschaftliche Abhandlungen*, **1**, pp. 259–87.

Boltzmann, L. [1872]: 'Weitere Studien über das Wärmegleichgewicht unter Gasmolekülen', *Wissenschaftliche Abhandlungen*, **1**, pp. 316–402.

Boltzmann, L. [1876]: 'Über die Natur der Gasmoleküle', *Wissenschaftliche Abhandlungen*, **2**, pp. 103–10.

Boltzmann, L. [1877*a*]: 'On the Relation of a General Mechanical Theorem to the Second Law of Thermodynamics', reprinted in S. G. Brush (ed.): *Kinetic Theory*, **2**, pp. 188–93.

Boltzmann, L. [1877*b*]: 'Über die Beziehung zwischen dem zweiten Hauptsatze der mechanischen Wärmetheorie und der Wahrscheinlichkeitsrechnung', *Wissenschaftliche Abhandlungen*, **2**, pp. 164–223.

[244] Smoluchowski in 1906, quite independently of Einstein, predicted the existence and magnitude of the Brownian motion from the kinetic theory of equilibrium. Smoluchowski's theory deals with a second aspect of the Brownian motion, i.e. the very small thermal inequalities which are produced irregularly and discontinuously by the Brownian agitation in spaces of the order of dimensions of one micron. The density of the fluid should, when it has reached equilibrium, vary from point to point. Smoluchowski calculated the mean density fluctuations for any fluid and predicted from this the phenomenon of *critical* opalescence. (Smoluchowski [1906], [1908].) The importance of Smoluchowski's theory is that it pinpoints exactly where the phenomenological second law is *false*, with the prediction of thermal fluctuations. Cf. Feyerabend [1966].

Boltzmann, L. [1884]: 'Ableitung des Stefanschen Gesetzes, betreffend die Abhängigkeit der Wärmestrahlung von der Temperatur aus elektromagnetischen Lichttheorie', *Annalen der Physik*, **22**, pp. 291–4.

Boltzmann, L. [1886]: 'The Second Law of Thermodynamics', reprinted in B. MacGuinness (ed.): *Ludwig Boltzmann: Theoretical Physics and Philosophical Problems*, pp. 13–32.

Boltzmann, L. [1895]: 'On Certain Questions of the Theory of Gases', *Nature*, **51**, pp. 413–15.

Boltzmann, L. [1896*a*]: *Lectures on Gas Theory*, in the translation of S. G. Bush, 1964.

Boltzmann, L. [1896*b*]: 'Reply to Zermelo's Remarks on the Theory of Heat', reprinted in S. G. Brush (ed.): *Kinetic Theory*, **2**, pp. 218–28.

Boltzmann, L. [1896*c*]: 'Ein Wort der Mathematik an der Energetik', *Annalen der Physik*, **57**, pp. 38–71.

Boltzmann, L. [1897*a*]: 'On the Necessity of Atomic Theories in Physics', *Monist*, **12**, pp. 65–79.

Boltzmann, L. [1897*b*]: 'On the Question of the Objective Existence of Processes in Inanimate Nature', reprinted in B. MacGuinness (ed.): *Ludwig Boltzmann: Theoretical Physics and Philosophical Problems*, pp. 57–75.

Boltzmann, L. [1897*c*]: 'On Zermelo's Paper "On the Mechanical Explanation of Irreversible Processes"', reprinted in S. G. Brush (ed.): *Kinetic Theory*, **2**, pp. 238–45.

Boltzmann, L. [1899]: 'Die Grundprinzipien und Grundgleichungen der Mechanik', *Populäre Schriften*, pp. 253–69.

Boltzmann, L. [1900]: 'The Recent Development of Method in Theoretical Physics', *Monist*, **11**, pp. 226–57.

Bridgeman, P. W. [1941]: *The Nature of Thermodynamics*.

Broda, E. [1955]: *Ludwig Boltzmann: Mensch, Physiker, Philosoph*.

Brown, R. [1827]: *A Brief Account of Microscopical Observations made in the Months of June, July, August 1827, on the Particles Contained in the Pollen of Plants; and on the General Existence of Active Molecules in Organic and Inorganic Bodies*.

Brush, S. G. [1958]: 'The Development of the Kinetic Theory of Gases, iv, Maxwell', *Annals of Science*, **14**, pp. 243–55.

Brush, S. G. [1962]: 'The Development of the Kinetic Theory of Gases, vi, Viscosity', *American Journal of Physics*, **30**, pp. 269–81.

Brush, S. G. [1965]: *Kinetic Theory*, **1**, *The Nature of Gases and Heat*.

Brush, S. G. [1966]: *Kinetic Theory*, **2**, *Irreversible Processes*.

Brush, S. G. [1967]: 'Foundations of Statistical Mechanics 1845–1915', *Archive for the History of Exact Sciences*, **4**, pp. 145–83.

Brush, S. G. [1968]: 'Mach and Atomism', *Syntheses*, **18**, pp. 192–215.

Brush, S. G. [1969]: 'A History of Random Processes', *Archive for the History of Exact Sciences*, **5**, pp. 1–36.

Brush, S. G. [1974]: 'Should the History of Science be Rated X?', *Science*, **183**, pp. 1164–72.

Brush, S. G. and Everitt, C. W. F. [1969]: 'Maxwell, Osbourne Reynolds, and the Radiometer', in R. McCormmach (ed.): *Historical Studies in the Physical Sciences*, **1**, pp. 104–25.

Bryan, G. H. [1891]: 'Researches Relating to the Connection of the Second Law with Dynamical Principles', *Reports of the British Association for the Advancement of Science*, **61**, pp. 85–122.

Bryan, G. H. [1894]: 'The Laws of the Distribution of Energy and Their Limitations', *Reports of the British Association for the Advancement of Science*, **64**, pp. 64–106.

Buijs-Ballot, C. H. D. [1858]: 'Über die Art von Bewegung, welche wir Wärme und Electricität nennen', *Annalen der Physik, 2nd series*, **103**, pp. 240–8.
Carnot, S. [1824*a*]: 'Reflections on the Motive Power of Fire and on Machines Fitted to Develop that Power', reprinted in E. Mendoza (ed.): *Reflections on the Motive Power of Fire*, pp. 3–59.
Carnot, S. [1824*b*]: 'Selections from the Posthumous Manuscripts of Carnot', in E. Mendoza (ed.): *Reflections on the Motive Power of Fire*, pp. 60–9.
Chaudesaigues, M. [1908]: 'Le mouvement brownien et le formule d'Einstein', *Comptes Rendus*, **147**, pp. 1044–6.
Clapeyron, E. [1834]: 'Memoir on the Motive Power of Heat', reprinted in E. Mendoza (ed.): *Reflections on the Motive Power of Fire*, pp. 73–105.
Clausius, R. [1850]: 'On the Motive Power of Heat, and on the Laws which can be deduced from it for the Theory of Heat', reprinted in E. Mendoza (ed.): *Reflections on the Motive Power of Fire*, pp. 109–52.
Clausius, R. [1854]: 'On a Modified Form of the Second Fundamental Theorem in the Mechanical Theory of Heat', reprinted in T. A. Hirst (ed.): *The Mechanical Theory of Heat*, pp. 111–35.
Clausius, R. [1857]: 'The Nature of the Motion which we call Heat', in S. G. Brush (ed.): *The Kinetic Theory*, **1**, pp. 111–134.
Clausius, R. [1858]: 'On the Mean Lengths of Paths Described by the Separate Molecules of Gaseous Bodies, reprinted in S. G. Brush (ed.): *The Kinetic Theory*, **1**, pp. 135–47.
Clausius, R. [1862*a*]: 'On the Conduction of Heat by Gases', *Philosophical Magazine*, **23**, pp. 417–35, 512–34.
Clausius, R. [1862*b*]: 'On the Application of the Theorem of the Equivalence of Transformations to Interior Work', in T. A. Hirst (ed.): *The Mechanical Theory of Heat*, pp. 215–50.
Clausius, R. [1863]: 'On an Axiom in the Mechanical Theory of Heat', reprinted in T. A. Hirst (ed.): *The Mechanical Theory of Heat*, pp. 267–89.
Clausius, R. [1865]: 'On Several Convenient Forms of the Fundamental Equations of the Mechanical Theory of Heat', reprinted in T. A. Hirst (ed.): *The Mechanical Theory of Heat*, pp. 327–65.
Clausius, R. [1870]: 'On a Mechanical Theorem Applicable to Heat', reprinted in S. G. Brush (ed.): *Kinetic Theory*, **1**, pp. 172–8.
D'Abro, A. [1939]: *The Rise of the New Physics*, **1**.
Dalton, J. [1808]: *A New System of Chemistry*.
Daub, E. E. [1967]: 'Atomism and Thermodynamics', *Isis*, **58**, pp. 293–303.
Daub, E. E. [1969]: 'Probability and Thermodynamics: The Reduction of the Second Law', *Isis*, **60**, p. 318–30.
Delsaulx, J. [1877]: 'Thermodynamic Origin of the Brownian Motion', *Monthly Microscopical Journal*, **18**, pp. 1–7.
Dugas, R. [1959]: *La théorie physique au sens de Boltzmann*.
Duhem, P. [1886]: *Le Potential Thermodynamique et ses applications à la mécanique chimique et à la théorie des phénomènes électriques*.
Duhem, P. [1895]: *Les Théories de la Chaleur*.
Duhem, P. [1900]: 'On the General Problem of Chemical Statics', *Journal of Physical Chemistry*, **2**, pp. 1–42.
Duhem, P. [1902]: *Le Mixte et la Combinaison Chimique*.
Duhem, P. [1906]: *La Théorie Physique: Son Objet; Sa Structure*.
Ehrenfest, P. [1911]: *The Conceptual Foundations of the Statistical Approach in Mechanics*.
Einstein, A. [1905*a*]: 'On the Movement of Small Particles Suspended in a Stationary Fluid Demanded by the Molecular–Kinetic Theory of Heat',

reprinted in R. Fürth (ed.): *Investigations on the Theory of the Brownian Movement*, pp. 1–18.

Einstein, A. [1905*b*]: 'Über einen die Erzaugung und Verwandlung des Lichtes betreffenden heuristischen Gesichtspunkt', *Annalen der Physik*, Series 4, **17**, pp. 132–48.

Einstein, A. [1907*a*]: 'Theoretical Observations on the Brownian Motion', reprinted in R. Fürth (ed.): *Investigations on the Theory of the Brownian Movement*, pp. 63–7.

Einstein, A. [1907*b*]: 'Plancksche Theorie der Strahlung und die Theorie der spezifischen Wärme', *Annalen der Physik*, Series 4, **22**, pp. 180–90.

Einstein, A. [1949]: *Albert Einstein Philosopher Scientist, The Library of Living Philosophers*, **7**.

Elkana, Y. [1970]: 'The Conservation of Energy: a case of simultaneous discovery?' *Archives Internationales d'Histoire des Sciences*, **90**, pp. 31–60.

Elkana, Y. [1973]: 'Boltzmann's Scientific Research Programme and its Alternatives' in Y. Elkana (ed.): *Some Aspects of the Interaction between Science and Philosophy*, 243–79.

Elkana, Y. [1974]: *The Discovery of the Conservation of Energy*.

Exner, F. [1900]: 'Notiz zu Browns Molekularbewegung', *Annalen der Physik*, **2**, pp. 843–7.

Feyerabend, P. K. [1966]: 'On the Possibility of a Perpetuum Mobile of the Second Kind' in P. K. Feyerabend and G. Maxwell (eds.): *Mind, Matter and Method, Essays in Honour of Herbert Feigl*, pp. 409–12.

Feyerabend, P. K. [1967]: 'Ludwig Boltzmann', in *Encyclopedia of Philosophy*, pp. 334–7.

Fitzgerald, G. F. [1895]: 'The Kinetic Theory of Gases', *Nature*, **51**, pp. 221–2, 452–3.

Garber, E. W. [1970]: 'Clausius and Maxwell's Kinetic Theory of Gases', in R. McCormmach (ed.): *Historical Studies in the Physical Sciences*, **2**, pp. 299–319.

Gibbs, J. W. [1875]: 'On the Equilibrium of Heterogeneous Substances', reprinted in *The Scientific Papers of J. Willard Gibbs*, **1**, pp. 55–371.

Gibbs, J. W. [1879]: 'On Vapour Densities of Peroxide of Nitrogen, Formic Acid, Acetic Acid, and Perchloride of Phosphorus', reprinted in *The Scientific Papers of J. Willard Gibbs*, **1**, pp. 372–403.

Gibbs, J. W. [1889]: 'Rudolf Julius Emanuel Clausius', reprinted in *The Scientific Papers of J. Willand Gibbs*, **2**, pp. 261–7.

Gibbs, J. W. [1902]: *Elementary Principles in Statistical Mechanics*.

Gouy, L. [1888]: 'Note sur le mouvement brownien', *Journal de Physique*, **7**, pp. 561–4.

Gouy, L. [1889]: 'Sur le mouvement brownien', *Comptes rendus*, **109**, pp. 102–5.

Gouy, L. [1895]: *Le Mouvement brownien et les mouvements moléculaires*.

Grünbaum, A. [1963]: 'The Bearing of Philosophy on the History of Science', *Science*, **143**, pp. 1406–12.

Helm, G. [1896]: 'Zur Energetik', *Annalen der Physik*, **57**, pp. 646–59.

Helm, G. [1898]: *Die Energetik nach ihrer geschichtlichen Entwicklung*.

Helmholtz, H. von [1882]: 'The Thermodynamics of Chemical Processes', *Wissenschaftliche Abhandlungen*, **2**, pp. 958–76.

Henri, V. [1908]: 'Etude cinématographique du mouvement brownien', *Comptes Rendus*, **146**, pp. 1024–6.

Hess, G. H. [1846]: 'Thermodynamical Investigations', *Annalen der Physik*, **50**, pp. 385–95.

Hibben, O. [1903]: 'The Theory of Energetics and its Philosophical Bearings', *Monist*, **13**, pp. 321–30.

Hiebert, E. N. [1962*a*]: *Historical Roots of the Principle of Conservation of Energy*.
Hiebert, E. N. [1962*b*]: 'The Concept of Chemical Affinity in Thermodynamics', *Ithaca*, **26**, pp. 871–3.
Hiebert, E. N. [1963]: 'Historical Remarks on the Discovery of Argon: The First Noble Gas', in R. Hyman (ed.): *Noble Gas Compounds*, pp. 3–20.
Hiebert, E. N. [1966]: 'The Genesis of Mach's Early Views on Atomism', in R. S. Cohen and R. J. Seeger (eds.): *Ernst Mach Physicist and Philosopher, Boston Studies in the Philosophy of Science*, **6**, pp. 79–106.
Hiebert, E. N. [1970]: 'The Conception of Thermodynamics in the Scientific Thought of Mach and Planck', *Ernst Mach Institut, Wissenschaftlicher Bericht*, **5**.
Hiebert, E. N. [1972]: 'The Energetics Controversy and the New Thermodynamics', in D. H. D. Roller (ed.): *Perspectives in the History of Science and Technology*, pp. 67–86.
Hopley, I. B. [1956]: 'Clerk Maxwell's Contribution to Physics', **2**, Ph.D. Dissertation, University of London.
Jaynes, E. T. [1967]: 'Foundations of Probability Theory and Statistical Mechanics', *Delaware Seminar in the Foundations of Physics*, **1**, pp. 76–101.
Jeans, J. H. [1925]: *The Dynamical Theory of Gases*.
Joule, J. P. [1845*a*]: 'On the Existence of an Equivalent Relation between heat and the ordinary forms of mechanical Power', *Philosophical Magazine*, **27**, p. 205.
Joule, J. P. [1845*b*]: 'On the Changes of Temperature produced by the rarefaction and condensation of air', *Philosophical Magazine*, **26**, p. 369.
Kelvin, Lord [1849]: 'An Account of Carnot's Theory of the Motive Power of Heat', *Mathematical and Physical Papers*, **1**, pp. 113–38.
Kelvin, Lord [1852]: 'On a Universal Tendency in Nature to the Dissipation of Mechanical Energy', *Mathematical and Physical Papers*, **1**, pp. 511–14.
Kelvin, Lord [1856]: 'On the Origin and Transformations of Motive Power', *Mathematical and Physical Papers*, **2**, pp. 182–8.
Kelvin, Lord [1880]: 'Vibrations of a Columnar Vortex', *Mathematical and Physical Papers*, **4**, pp. 152–65.
Kelvin, Lord [1886]: 'Steps Towards a Kinetic Theory of Matter', *Reports of The British Association*, **56**, pp. 613–23.
Kelvin, Lord [1888]: 'On the Average Pressure due to Impulse of Vortex Rings on a Solid', *Mathematical and Physical Papers*, **4**, pp. 188–92.
Kelvin, Lord [1901]: 'Nineteenth Century Clouds over the Dynamical Theory of Heat and Light', *Baltimore Lectures on Molecular Dynamics and the Wave theory of Light*, pp. 486–527.
Klein, M. J. [1962]: 'Max Planck and the Beginnings of the Quantum Theory', *Archive for the History of Exact Sciences*, **1**, pp. 459–79.
Klein, M. J. [1969*a*]: *Paul Ehrenfest, 1, The Making of a Theoretical Physicist*.
Klein, M. J. [1969*b*]: 'Gibbs on Clausius', in R. McCormmach (ed.): *Historical Studies in the Physical Sciences*, **1**, pp. 127–49.
Klein, M. J. [1972]: 'The Development of Boltzmann's Statistical Ideas', in E. G. D. Cohen and W. Thirring (eds.): *The Boltzmann Equation*, pp. 53–95.
Klein, M. J. [1974]: 'The Historical Origins of the Van der Waals Equation', *Physica*, **73**, pp. 28–47.
Krönig, A. K. [1856]: 'Grundzüge einer Theorie der Gase', *Annalen der Physik*, **99**, pp. 315–22.
Kuhn, T. S. [1957]: 'Energy Conservation as an Example of Simultaneous Discovery', in M. Clagett (ed.): *Critical Problems in the History of Science*, pp. 321–56.
Kundt, A. and Warburg, E. [1876]: 'Über die specifische Wärme des Quecksilbergases', *Annalen der Physik*, **157**, pp. 355–69.

Lakatos, I. [1970]: 'Falsification and the Methodology of Scientific Research Programmes', in I. Lakatos and A. Musgrave (eds.): *Criticism and the Growth of Knowledge*, pp. 91–195.

Lakatos, I. [1971]: 'History of Science and its Rational Reconstructions', in R. Buck and R. S. Cohen (eds.): *Boston Studies in the Philosophy of Science*, **8**, pp. 91–136. This volume, pp. 1–39.

Lakatos, I. [1974]: 'The Role of Crucial Experiments in Science', *Studies in the History and Philosophy of Science*, **4**, pp. 357–73.

Lakatos, I. and Zahar, E. G. [1976]: 'Why did Copernicus's Research Programme Supersede Ptolemy's?', in R. Westman (ed.): *The Copernican Achievement.*

Loeb, L. B. [1961]: *The Kinetic Theory of Gases.*

Loschmidt, J. [1865]: 'Zur Gröse der Luftmoleküle', *Sitzungsberichte der Kaiserlichen Akademie, der Wissenschaften in Wien*, **52**, pp. 395–406.

Loschmidt, J. [1870]: 'Experimental-Untersuchungen über die Diffusion von Gasen ohne pörose Scheidewände', *Sitzungsberichte der Kaiserlichen Akademie, der Wissenschaften in Wien*, **61**, pp. 367–77; **62**, pp. 468–76.

Loschmidt, J. [1876]: 'Über den Zustand des Wärmegleichgewichtes eines Systems von Körpern, mit Rücksicht auf die Schwerkraft', *Sitzungsberichte der Kaiserlichen Akademie, der Wissenschaften in Wien*, **73**, pp. 135, 366.

Loschmidt, J. [1886]: 'Schwingungen einer elastischen Hohlkugel', *Sitzungsberichte der Akademie der Wissenschaften in Wien*, **93**, pp. 434–6.

Mach, E. [1863]: *Compendium der Physik für Mediziner.*

Mach, E. [1872]: *History and Root of the Principle of Conservation of Energy*, in the translation of P. E. B. Jourdain, 1911.

Mach, E. [1882]: 'The Economical Nature of Physical Enquiry', reprinted in *Popular Scientific Lectures*, pp. 186–213.

Maxwell, J. C. [1860]: 'Illustrations of the Dynamical Theory of Gases', *Scientific Papers*, **1**, pp. 377–409.

Maxwell, J. C. [1865]: 'On the Viscosity or Internal Friction of Air and other Gases', *Scientific Papers*, **2**, pp. 1–25.

Maxwell, J. C. [1866]: 'On the Dynamical Theory of Gases', *Scientific Papers*, **2**, pp. 26–78.

Maxwell, J. C. [1872]: *The Theory of Heat.*

Maxwell, J. C. [1873]: 'On the Final State of a System of Molecules in Motion Subject to Forces of Any Kind', *Scientific Papers*, **2**, pp. 351–4.

Maxwell, J. C. [1875]: 'On the Dynamical Evidence of the Molecular Constitution of Bodies, *Scientific Papers*, **2**, pp. 418–38.

Maxwell, J. C. [1876]: 'Atom', *Scientific Papers*, **2**, pp. 445–84.

Maxwell, J. C. [1877]: 'The Kinetic Theory of Gases', *Nature*, **16**, pp. 242–6.

Maxwell, J. C. [1878]: 'Tait's Thermodynamics', *Scientific Papers*, **2**, pp. 660–71.

Maxwell, J. C. [1879]: 'On Stresses in Rarefied Gases Arising from Inequalities of Temperature', *Scientific Papers*, **2**, pp. 681–712.

Meyer, O. E. [1877]: *Die Kinetische Theorie der Gase, in elementarer Darstellung mit mathematischen Zusätzen.*

Nernst, W. H. [1906]: 'On the Calculation of Chemical Equilibrium from Thermal Measurements', reprinted in H. M. Leicester (ed.): *Source Book in Chemistry*, pp. 167–73.

Nye, M. J. [1972]: *Molecular Reality.*

Ostwald, W. [1891]: 'Studien zur Energetik', *Leipzig Maths. Phys. Ber.*, **42**, pp. 271–88.

Ostwald, W. [1892]: 'Studien zur Energetik', *Leipzig Maths. Phys. Ber.*, **44**, pp. 211–37.

Ostwald, W. [1895]: 'Emancipation From Scientific Materialism', *Science Progress*, **4**, pp. 430–6.
Ostwald, W. [1896]: 'Zur Energetik', *Annalen der Physik und Chemie*, **58**, pp. 154–67.
Ostwald, W. [1907]: 'The Modern Theory of Energetics', *Monist*, **17**, pp. 481–515.
Ostwald, W. [1927]: *Lebenslinien*, **2**.
Perrin, J. [1908]: 'La Loi de Stokes et le mouvement brownien', *Comptes Rendus*, **147**, pp. 475–6.
Perrin, J. [1910]: 'Mouvement brownien et réalité moléculaire', *Annales de Chimie et de Physique*, **18**, pp. 1–114.
Perrin, J. [1916]: *Atoms* (in the translation of D. Hammick of 'Les Atomes'), 1913.
Planck, M. [1879]: 'Über den zweiten Hauptsatz der mechanischen Wärmetheorie', *Physikalische Abhandlungen und Vorträge*, **1**, pp. 1–61.
Planck, M. [1887]: 'Über das Prinzip der Vermehrung der Entropie', *Physikalische Abhandlungen und Vorträge*, **1**, pp. 196–273, 382–425.
Planck, M. [1891]: 'Allgemeines zur neueren Entwicklung der Wärmetheorie', *Zeitschrift für Physikalische Chemie*, **8**, pp. 647–56.
Planck, M. [1893]: 'Der Kern des zweiten Hauptsatzes der Wärmetheorie, *Physikalische Abhandlungen und Vorträge*, **1**, pp. 437–41.
Planck, M. [1896]: 'Gegen die Neuere Energetik', *Annalen der Physik und Chemie*, **57**, pp. 72–8.
Planck, M. [1897*a*]: *Treatise on Thermodynamics* (third English edition, Dover, 1945).
Planck, M. [1897*b*]: 'Über irreversible Strahlungsvorgänge. Erste Mitteilung', *Physikalische Abhandlungen und Vorträge*, **1**, pp. 493–504.
Planck, M. [1900]: 'Zur Theorie des Gesetzes der Energieverteilung im Normalspektrum', *Physikalische Abhandlungen und Vorträge*, **1**, pp. 698–706.
Planck, M. [1909]: 'The Unity of the Physical Universe', reprinted in *A Survey of Physical Theory*, pp. 1–26.
Planck, M. [1910]: *Vorlesungen über Thermodynamik* (third German edition).
Planck, M. [1914]: 'Dynamical Laws and Statistical Laws', reprinted in *A Survey of Physical Theory*, pp. 56–68.
Planck, M. [1931]: 'James Clerk Maxwell in seiner Bedeutung für die theoretische Physik in Deutschland', *Physikalische Abhandlungen und Vorträge*, **3**, pp. 352–7.
Poincaré, H. [1890]: 'On the Three-body problem and the Equations of Dynamics', reprinted in S. G. Brush (ed.): *Kinetic Theory*, **2**, pp. 194–202.
Poincaré, H. [1893]: 'Mechanism and Experience', reprinted in S. G. Brush (ed.): *Kinetic Theory*, **2**, pp. 203–7.
Poisson, S. O. [1823]: 'Sur la chaleur des gaz et des vapeurs', *Annales de Chimie*, **23**, pp. 337.
Popper, K. R. [1934]: *The Logic of Scientific Discovery* (English edition, 1958).
Popper, K. R. [1963]: *Conjectures and Refutations*.
Popper, K. R. [1972]: *Objective Knowledge*.
Popper, K. R. [1974]: *The Philosophy of Karl Popper: The Library of Living Philosophers*, **14**.
Post, H. [1968]: 'Atomism 1900', *Physics Education*, **3**, pp. 1–13.
Ramsay, W. [1892]: 'Pedetic Motion in Relation to Colloidal Solutions', *Chemical News*, **65**, p. 90.
Rayleigh, Lord [1900]: 'The Law of the Partition of Kinetic Energy', *Philosophical Magazine*, **49**, pp. 98–118.
Schrödinger, E. [1944]: *Statistical Thermodynamics*.
Simon, W. [1956]: 'The Third Law of Thermodynamics: An Historical Survey', *Year Book of the Physical Society*, pp. 1–22.

Smoluchowski, M. V. [1906]: 'Zur Kinetischen Theorie der Brownschen Molekularbewegung und der Suspensionen', *Annalen der Physik*, **21**, pp. 756–80.
Smoluchowski, M. V. [1908]: *Abhandlungen über die Brownsche Bewegung.*
Suvorov, S. G. [1966]: 'Einstein's Philosophical Views and their Relation to his Physical Opinions', *Soviet Physics Ospekhi*, **8**, p. 578.
Svedberg, T. [1907]: *Studien zur Lehre von den kolloiden Lösungen.*
Tait, P. G. [1876]: *Lectures on Some Recent Advances in Physical Science.*
Ter Haar, D. [1954]: 'Foundations of Statistical Mechanics', *Reviews of Modern Physics*, **27**, pp. 289–338.
Thomson, J. [1849]: Theoretical Considerations on the Effect of Pressure in Lowering the Freezing Point of Water', *Mathematical and Physical Papers of Baron Kelvin*, **1**, pp. 156–64.
Truesdell, C. [1971]: *The Tragi-Comedy of Classical Thermodynamics.*
Urbach, P. M. [1974]: 'Progress and Degeneration in the "IQ Debate"', *British Journal for the Philosophy of Science*, **25**, pp. 99–135, 235–59.
van der Waals, J. D. [1873]: *Over de Continuitet van den gas – en vloeistoftoestand.*
van der Waals, J. D. [1910]: 'The Equation of State for Gases and Liquids', in *Nobel Lectures in Physics* (1901–1924), pp. 254–65.
van't Hoff, J. H. [1903]: *Physical Chemistry in the Service of the Sciences*, in the translation of A. Smith.
van't Hoff, J. H. [1904]: 'The Relation of Physical Chemistry to Physics and Chemistry', *Journal of Physical Chemistry*, **9**, pp. 81–9.
Watkins, J. W. N. [1958]: 'Influential and Confirmable Metaphysics', *Mind*, N.S. **67**, pp. 344–65.
Watkins, J. W. N. [1974]: 'Metaphysics and the Advancement of Science', *British Journal for the Philosophy of Science*, **26**, pp. 91–121.
Watson, H. W. [1876]: *The Kinetic Theory of Gases.*
Watson, W. H. [1894]: 'The Maxwell–Boltzmann Theorem', *Nature*, **51**, p. 222.
Zahar, E. G. [1973]: 'Why did Einstein's Research Programme Supersede Lorentz's?', *The British Journal for the Philosophy of Science*, **24**, pp. 95–123, 223–262. This volume, pp. 211–75.
Zermelo, E. [1896*a*]: 'On a Theorem of Dynamics and the Mechanical Theory of Heat', reprinted in S. G. Brush (ed.): *Kinetic Theory*, **2**, pp. 208–17.
Zermelo, E. [1896*b*]: 'On the Mechanical Explanation of Irreversible Processes', reprinted in S. G. Brush (ed.): *Kinetic Theory*, **2**, pp. 229–37.

Thomas Young and the 'refutation' of Newtonian optics: a case-study in the interaction of philosophy of science and history of science*

JOHN WORRALL

LONDON SCHOOL OF ECONOMICS AND POLITICAL SCIENCE

Introduction

Introduction

A great scientific revolution occurred in optics in the early part of the nineteenth century when the Newtonian corpuscular theory of light was rejected in favour of the wave theory. The majority of scientists became convinced of the superiority of the wave theory in the late 1820s to early 1830s, yet many historians and commentators have claimed that the superiority of the wave theory had already been established by the work of Thomas Young published in the first years of the 1800s. Hence there arises a famous and much discussed historical problem – why was Young's work ignored for so long?

* This is an extensively revised version of a paper first read before the British Society for the Philosophy of Science in May 1973. The paper also formed the basis of several seminar talks at the LSE and one at Chelsea College, London. A shortened version was presented at the conference on *Research Programmes in Physics and Economics* in Nafplion, Greece in September 1974. I should like to thank all those who offered helpful criticisms on these various occasions. I should particularly like to mention Paul Feyerabend, M. L. G. Redhead and my colleagues Peter Urbach, John Watkins and Elie Zahar. Only the first version was read by my late teacher and friend, Imre Lakatos. But he made some very helpful comments and, as will be clear from the paper, I owe him an enormous intellectual debt.

My paper consists of five parts. In the *first* part, I discuss briefly the reasons usually given for regarding Young as having established the wave theory's superiority. In the *second*, I argue that the usual explanations given by historians and commentators on science of why Young's work was ignored for a time are false. In the *third* (and by far the longest) part, I produce some arguments which challenge the fundamental assumption that Young did establish the superiority of the wave theory. In the *fourth* part, I consider how the (descriptive) historical problems are shifted by my (normative) methodological reappraisal of the merits of Young's theory and of the import of his experimental results. In the *fifth* part, I try to draw some of the general morals about the interaction of history of science and philosophy of science which seem to be illustrated by my case study.

1 *Young's alleged achievement*

The most widespread (and least sophisticated) account of Young's achievement is that his work decided the issue between the corpuscular and wave theories of light. By a series of experiments, of which perhaps the most conclusive was the two-slit experiment, Young, according to this account, established that beams of light are capable of interfering and, in particular, of interfering destructively. In other words, Young experimentally established the remarkable fact that light added to light may produce darkness. This possibility, as Young himself had pointed out, is a consequence of the wave theory of light: if at a given point the trough of one wave meets the crest of another equal and parallel wave, the net effect on the medium at that point will be zero. (Indeed, this is the central consequence of Young's famous 'principle of interference'.) Destructive interference is, however, not a possibility according to the Newtonian corpuscular theory – for how can the effect of two parallel streams of particles be less than the effect of either stream separately? Thus Young's experiments were crucial in deciding between the two rival theories of light.

This account is often given. R. W. Wood, for example, in his famous optics textbook reports that 'true interference was first observed by Dr. Young...whose justly celebrated experiments established almost beyond question the validity of the wave theory'.[1] Magie writes: 'Young's great importance for physics rests upon his discovery of the interference of light by which he established the undulatory theory.'[2] The story of the crucial nature of Young's experiments goes back to the mid-nineteenth century. Young's first biographer, George Peacock, Dean of Ely and a celebrated Cambridge mathematician, writes of one of Young's experiments: 'This was a *crucial* experiment and may be considered as having constituted an important epoch in the history of the undulatory theory.'[3]

[1] Wood [1905], p. 2. Similar statements are made in almost all optics textbooks.
[2] Magie [1935], p. 59. [3] Peacock [1855], p. 162.

And Peacock's friend Whewell suggests that 'the man whose name must occupy the most distinguished place in the history of Physical Optics, in consequence of what he did in reviving and establishing the undulatory theory, is Dr. Thomas Young'.[4]

Now even if we weaken the claim that Young established the truth of the wave theory to the less obviously false claim that Young at least established the superiority of the wave theory by refuting its corpuscularist rival, it is strange that, both in Britain and in France, most physicists interested in optics adhered to the emission theory well into the 1820s. If Young's experimental results (all of which had been published by 1807) were crucial, why this irrational delay?[5]

2 *Young's work allegedly ignored: the 'Newton-worship', 'poor presentation' and 'character assassination' explanations of the 'neglect' of Young's work*

One natural answer would be that Young's work was not known to his scientific contemporaries. This answer is, however, far from being true. His three major papers on the wave theory all appeared not in some obscure and unread journal but in the *Philosophical Transactions* – two of these papers had originally been delivered as Bakerian lectures at the invitation of the Council of the Royal Society. The fact that both Young and his work were known to major scientists of the time is well documented. He corresponded with Brewster, Herschel, and Wollaston[6] in England. In France, Laplace knew of, and refers to, Young's work; as did Biot and Arago. Indeed, Fresnel in his work consistently refers to 'the celebrated Dr. Young'.[7]

But if the corpuscular theory was not only refuted, but even *known* to be refuted (since scientists knew of Young's work), why did scientists continue to accept it?

Whewell in his *History of the Inductive Sciences* provided not just one answer to this question but three.[8] These answers were repeated and

[4] Whewell [1837], p. 21.

[5] Whewell puts the problem in almost as many words when he writes that Young's experiments, together with the theoretical results in Young's [1802*a*], 'certainly *ought* to have convinced all scientific men of the truth of the doctrine thus urged [i.e. the wave theory]'. (Whewell [1837], p. 324; my italics.) Cf. Jenkins and White [1957], p. 203: '[Young's] early work proved the wave nature of light but was not taken seriously by others until it was corroborated by Fresnel.'

[6] Wollaston actually wrote to Young: 'I like your Bakerian very much, but I cannot say that I have yet inserted the undulatory doctrine into my creed and it may be some time before I repeat it with fluency.' See Young [1855], p. 261.

[7] See, for example, Laplace [1813], volume 4 of Biot's enormous [1816], and Fresnel [1822], p. 1.

[8] Two of Whewell's three explanations (those which I call below the 'obscure style' and the 'character assassination' explanations) are to be found still earlier in Arago's 1832 '*Éloge*' on Young (see his [1859]). The actual source of the 'Newton-worship' and 'character assassination' explanations seems to have been Young's own ([1804*b*]) bitter reply to Brougham's reviews of his papers.

elaborated by Peacock (who admits that he had 'nothing to correct and very little to add' to Whewell's 'judicious treatment' of Young[9]) and eventually became almost as much a part of the scientific folklore as the crucial nature of Young's experiments. I shall call the three answers, in turn, the '*Newton-worship*' answer, the '*poor presentation*' answer, and the '*character assassination*' answer.

(i) According to Whewell's first explanation Young's theory was ignored because of the general hero-worship of Newton in the early nineteenth century. Whewell writes:

When Young, in 1800, published his assertion of the Principle of Interference, as the true theory of optical phenomena, the condition of England was not very favourable to a fair appreciation of the value of the new opinion. The men of science were strongly preoccupied in favour of the doctrine of emission, not only from a national interest in Newton's glory, and a natural reverence for his authority, but also from a deference towards the geometers of France, who were looked upon as our masters in the application of mathematics to physics, and who were understood to be Newtonians in this as in other subjects.[10]

Peacock copies Whewell:

The reverence...attached in this country to whatever was sanctioned by the authority of Newton, operated not a little to retard...the acceptance of any conclusion or theory which he had repudiated [as he had, of course, the wave theory of light[11]].[12]

The Newton-worship claim is repeated by recent historians. I. B. Cohen for example, in his preface to the Dover edition of Newton's *Opticks*, writes (p. xiii):

Had not Fresnel and Arago in France become interested in the work of Young, it seems probable that the great name of Newton would effectively have blocked any pursuit of Young's ideas...and any further development of the wave theory of light.

(ii) Whewell adds a mitigating circumstance which, in turn, becomes his second explanation for the lack of recognition of Young's achievements: namely that Young presented his theoretical and empirical results poorly so that their content and importance was obscured.

Whewell writes:

We may add, however, that Young's mode of presenting his opinions was [because of his 'style of writing'] not the most likely to win them favour.[13]

[9] Peacock [1855], p. 138.

[10] Whewell [1837], p. 346.

[11] Cf. below, p. 131.

[12] Peacock [1855], p. 182.

[13] Whewell [1837], p. 148. In 1832 Arago had written 'I should be wanting in frankness, I should be the panegyrist not the historian, if I did not avow, that in general Young did not sufficiently accommodate himself to the capacity of his readers; that the greater part of the writings for which the sciences are indebted to him, are justly chargeable with a certain obscurity.' (Quoted from the reprint in Arago [1859], p. 512.)

Peacock copies Whewell:

It is but an act of justice, however, both to those who neglected as well as to those who opposed the conclusions of these Memoirs[14] to admit that there is much in the form which they assumed which made it very difficult to appreciate their value. Like all Young's early scientific writings they were extremely obscure.[15]

So scientists from 1802 to the 1820s refused to allow that Newton might have been wrong. And they were unable to comprehend Young's papers. But, as if this were not enough, they were also, as we shall now see, taken in by a nasty character assassination.

(iii) Young's 1801 Bakerian lecture and his 1802 paper[16] were attacked by an anonymous reviewer in the *Edinburgh Review* of January 1803. The author was known to be Henry Brougham – later to be Lord Chancellor. Brougham also attacked Young's [1804*a*] paper, which had again been delivered as a Bakerian lecture (in 1803). This second attack appeared in the October 1804 issue of the *Edinburgh Review*. Brougham's comments cannot be construed as favourable. Brougham writes, for example, that Young's first lecture 'contains nothing which deserves the name either of experiment or of discovery and...is, in fact, destitute of every species of merit'.[17] Brougham finds that Young shifts his ground so often as to force him to ask if 'the world of science is to be as changeable in its modes, as the world of taste, which is directed by the nod of a silly woman, or a pampered fop'.[18] Brougham finds Young's law of interference 'one of the most incomprehensible suppositions we remember to have met with in the history of human hypotheses',[19] and he admits his final feelings to be ones of 'regret at the abuse of that time and opportunity which no greater share of talents than Dr Young's are sufficient to render fruitful, by mere diligence and moderation'.[20]

According to Whewell, these two reviews amount to an assassination of Young's scientific character and they retarded the growth of knowledge: 'We can hardly doubt that these Edinburgh reviews had their effect in confirming the general disposition to reject the undulatory theory.'

[14] These are Young's [1802*a*], [1802*b*] and [1804*a*].

[15] Peacock [1855], p. 185. Peacock obviously added the qualification 'early' (in 'Young's early scientific writings') so as not to have a problem of why people eventually came to understand Young. That Young's style became clearer as he got older is an interesting hypothesis. If true it would make Young a notable exception to the general rule. Unfortunately I can find no evidential basis for the hypothesis in Young's writings.

Pettigrew in his [1840], p. 23, also favours the 'poor presentation' hypothesis: 'Perhaps one reason for this apparent neglect is to be found in the object and style of his writings. In his mathematical...researches, he has presumed upon his readers being more acquainted with science than is really the case...' Crowther in his [1968] makes the poor presentation of his results the central reason for Young's neglect.

[16] Published as Young's [1802*a*] and [1802*b*].

[17] Brougham [1803], p. 450.

[18] *Ibid.*, p. 452.

[19] Brougham [1804], p. 97.

[20] *Ibid.*, p. 103.

Peacock elevates Brougham's reviews into the major cause of Young's neglect. He writes rather melodramatically:

The effect which these powerful and repeated attacks produced upon the estimate of Dr. Young's scientific character was remarkable. The poison sank deep into the public mind, and found no antidote in reclamations of other journals of co-ordinate influence and authority. We consequently [*sic*] find that the subject of Dr. Young's researches remained absolutely unnoticed by men of science for many years.[21]

Young's later biographer Wood copies Peacock – according to Wood, Brougham's 'bitter and virulent attacks on Young's work certainly did much to delay the acceptance of the wave theory.'[22]

Ernst Mach joins in:

In spite of his great work, Young was not to rejoice in the fruits of his labours immediately, for a scientific reactionary who felt himself called to uphold Newton's theory even to an iota, commenced an unparalleled attack upon him...Young's reputation was actually marred by it for many years...[23]

According to Whittaker, although Young's papers showed 'the superior power of the wave theory' their publication 'occasioned a fierce attack... from the pen of Henry Brougham...[T]here can be no doubt that Brougham for the time being achieved his object of discrediting the wave theory.'[24]

Let us examine each of these three explanations.

The claim that there existed in the late eighteenth and early nineteenth centuries a Newton-worshipping establishment ready to pounce on any unfortunate who was foolhardy enough to suggest the falsity of some theory proposed by Newton seems to be refuted by the historical facts.[25] Several scientists were able to state quite openly that a Newtonian theory is false without either forfeiting their good name, or preventing serious consideration of their subsequent work. Let me list a few of these scientists.

[21] Peacock [1855], p. 173. Peacock writes elsewhere (footnote on p. 192 of Young [1855]) that Brougham's reviews 'Not only seriously damaged, for a time, the estimation of the scientific character of Dr. Young, but diverted public examination of the truth of his theories...for nearly twenty years'.

[22] Wood [1954], p. 157. Young produced and had published a reply to Brougham's attacks (reprinted in Young [1855], pp. 192–215). It is alleged that only one single copy of this reply was sold. If so it was a well-thumbed copy, for there is documentary evidence that several people read it.

[23] Mach [1926], p. 156.

[24] Whittaker [1910], p. 108.

[25] Indeed, almost the only evidence in favour of the claim is to be found in Young's own [1802*a*], where he tries to defend his own theory by showing it to be not so very different from Newton's. Brougham did not let Young get away with this exercise in public relations (which, if the rest of my evidence is anything to go by, would anyway have had little effect on his fellow scientists). Brougham writes (it is an aspect of his attack which historians have tended to overlook): 'We hold the highest authority to be of no weight whatever in the court of Reason; and we view the attempt to shelter this puny theory under the sanction of great names, as a desperate effort in its defence, and a most unwarrantable appeal to popular prejudice.' (Brougham [1803], p. 454.)

Priestley wrote in 1772 of Newton's researches into the colours of thin plates that 'in no subject to which he gave his attention does [Newton] seem to have overlooked more important circumstances in the appearances he observed, or to have been more mistaken with respect to their causes.'[26]

G. W. Jordan produced, in 1799 and 1800, two tracts[27] on the 'inflection' of light, in which he is especially critical of Newton's hypotheses of 'fits of easy reflection and transmission', which, Jordan claims, 'are inconsistent with the actual condition of things, and the general phenomena of light and of nature...they obstruct all discovery concerning them; they interrupt the general progress of philosophy'.[28] Far from being dismissed as obvious rubbish since inconsistent with Newtonian doctrine, this work seems to have been given serious attention, and was certainly favourably received at least by the reviewer for the *Philosophical Magazine*. This reviewer records that Jordan

> has carefully repeated those experiments by which Sir Isaac Newton effected his analysis of light. The experiments have produced to his observation, phenomena materially different from those which appeared to Newton...this author infers from his observations...that...the Newtonian theory of light and colours is not fundamentally true.[29]

And the reviewer gives the following scrupulously fair and balanced summary:

> The apparent accuracy of these observations; the logical fairness of the induction; the literary composition of the essay, deserve every praise. Without having ourselves repeated the experiments, and without knowing them to have been repeated, with similar results, by others, we would not presume to decide concerning the truth of the doctrine.[30]

No sign here of any irrational desire to preserve Newton at all costs against the charge of having uttered a falsehood. (And this was written in 1800, only two years before Young's first major paper.)

In 1802 itself, Wollaston published the results of some experiments in crystal optics.[31] These results are stated to be inconsistent with Newton's law of the position of the extraordinary ray in extraordinary refraction; they are explicitly stated to confirm the law of Huygens. Wollaston maintained a high reputation amongst his fellow scientists (even to the extent of being nicknamed 'The Pope' because of his alleged infallibility) and his results (independently arrived at by Malus) were accepted as correct, were frequently quoted, and formed the basis of Laplace's theoretical work on the subject.

Young's senior contemporary William Herschel, in a series of papers on

[26] Priestley [1772], p. 498. This book, by the way, has a chapter entitled 'The Opposition which Newton's Doctrine of Light Met With'.

[27] Jordan [1799*b*] and [1800].

[28] Jordan [1799*b*], p. 126.

[29] *Philosophical Magazine*, **7**, 1800, p. 365.

[30] *Loc. cit.*, p. 366.

[31] Wollaston [1802].

'Newton's rings',[32] was able, without incurring the wrath of his contemporaries, to state explicitly that Newton's theory of fits 'cannot account for the phenomena'. Whilst even Brougham, Mach's 'scientific reactionary who felt himself called to uphold Newton's theory even to an iota', developed a theory of light which, while based on corpuscles, was openly different from Newton's theory. Newton regarded his theory of fits as experimentally proven. Brougham calls it, in print, a 'degrading' speculation, which hides the simple facts behind 'the smoke of unintelligible theory'.[33]

Moreover, as later commentators demonstrated, it is not too difficult to reconcile the view that the corpuscular theory is false with the views that it is a scientific sin to espouse a theory which turns out to be false, but that Newton was free from (scientific) sin. This is because it is not difficult to defend Newton against the 'charge' of having espoused the corpuscular theory. Although it is clear, I think, that Newton basically was a corpuscularist (he saw the problems in optics that a corpuscularist would see and tried to solve them in a corpuscularist way[34]) he did not explicitly adopt the corpuscular hypothesis. Indeed there are passages in his writings in which he is explicitly non-committal;[35] moreover, Newton endowed rays of light with periodic properties ('fits'), properties which he was inclined to attribute to periodic disturbances in the ether overtaking the rays of light.[36] Anyone wanting to espouse the wave theory could easily have argued that Newton was not *really* a proponent of the opposing theory. Indeed, this became standard practice amongst commentators once the wave theory had been accepted. One example of this occurs in the translators' footnotes to Arago's '*Éloge*' on Young: 'Upon the whole it appears that the name of Newton can in no way be legitimately claimed as a partisan of either theory.'[37]

Apart from all this, one wonders why, if criticism of Newton was not allowed during the period from 1800 to 1820, the Newton-worshipping scientific establishment crumbled so quickly afterwards. For by about 1830 all major scientists in Britain and France accepted the wave theory as superior to its corpuscular rival.[38]

[32] William Herschel [1807], [1809*a*], [1809*b*].

[33] Brougham [1796], p. 272.

[34] See below, p. 159.

[35] For example, [1672], p. 5086 (reprinted in Cohen [1958], p. 116): 'Tis true that from theory I argue the corporeity of light: but I do it without any absolute positiveness, as the word 'perhaps' intimates; and make it at most but a very plausible consequence of the doctrine, and not a fundamental supposition, nor so much as any part of it.' See also [1675], p. 249 (reprinted in Cohen [1958], p. 179).

[36] See, e.g., Newton [1730], Query 17, and [1675], p. 251 (reprinted in Cohen [1958], p. 181).

[37] Arago [1859], p. 515.

[38] This does not mean that all major scientists in 1830 *believed* the wave theory. Many, in part put off by the all-pervading but invisible ether required by the wave theory, felt that the wave theory could not constitute the truth about the universe. Biot, Brewster, Herschel, Potter and even Arago seem to have put themselves into this category with

Next, the 'poor presentation' explanation of the neglect of Young's theory.

No doubt some parts of Young's work are rather obscure. This is particularly true of the mathematical sections. This is seized upon by Whewell, Peacock and some other commentators. Whewell for example writes:

> [Young's] mathematical reasonings placed them out of reach of popular readers, while the want of symmetry and system in his symbolic calculations, deprived them of attractiveness for the mathematician.[39]

This will, however, hardly do as an explanation of the neglect of Young's theory of light, for of the three early papers which are alleged to constitute Young's major contribution to the wave theory (and which amount to some 50 pages of the *Philosophical Transactions*), precisely *one paragraph*[40] contains any 'symbolical calculations'. The vocabulary of the rest is restricted to that of ordinary English, plus the natural, rational and real numbers, the equality sign and a few variables which are always given an explicit physical interpretation.

In fact, those aspects of his work which are generally alleged to constitute his primary achievement and to have established the wave theory's superiority – namely the qualitative aspects of his 'crucial' experimental results and his principle of interference – are on the whole presented with admirable clarity.[41] And in fact Young's accounts of the qualitative aspects of his experimental results (namely the appearance of interference fringes in certain circumstances and their disappearance in others) seem to have been clearly understood by corpuscularists like Biot and Brougham, both of whom report them accurately.[42]

varying degrees of firmness. Some scientists (like Potter; see for example his [1833]) continued to work in an effort to discomfit the wave theory. Yet no serious scientist of 1830 denied that, *as things stood*, the wave theory was superior to the corpuscular theory. This is nicely exemplified by the attitude of Sir David Brewster. It is clear from his work (though he never makes it explicit) that he was very attached to the corpuscular theory, and Moon (amongst others) classified Brewster as an 'anti-undulatist' (see Moon [1849], p. 58). Yet Brewster admits that 'the theory of undulations has made great progress in modern times, and derives such powerful support from an extensive class of phenomena, that it has been received by many of our most distinguished philosophers' (Brewster [1831], p. 135), and that 'The inability of the undulatory theory to explain the phenomena of inequal refrangibility, is almost the only exception to its universal application in accounting for the most complicated phenomena of light.' (Brewster [1832], p. 317.) All of which illustrates the importance of sharply distinguishing between scientists' beliefs about the universe, their choice of which of the various competing theories to try to develop, and their appraisals of the merits and demerits of these theories. See below, §5, pp. 161–3.

[39] Whewell [1837], p. 348.

[40] This occurs on p. 165 of Young [1802*a*]. (All page references to Young's work – except for his [1807] lectures – are to the reprints in the [1855] Collected Works.)

[41] This applies to all the different versions of the interference principle; see below, pp. 138–42. (For an important qualification in the case of the two slit experiment, however, see below, pp. 152–6.) But it does not mean that these clear statements were clearly problem-free.

[42] See Biot and Pouillet's appendix to volume 4 of Biot [1816] and Brougham [1803] and [1804]. Admittedly Brougham makes one or two minor mistakes but these clearly arise from simple misreading and not from fundamental misunderstanding.

> This ought then to have been enough. As long as these allegedly crucial aspects of his work were clearly presented, any obscurity in the rest of Young's work is unimportant and certainly cannot be used to explain other scientists' failure to accept the wave theory of light as superior to its rival.[43]

Finally the 'character assassination' explanation of the neglect of Young's theory. The claim here, remember, is that Young's reputation was so marred by Brougham's reviews that no one took him seriously.[44]

This claim is – at least *prima facie* – rather implausible. First of all, even if the claim were true it would only explain Young's theory's lack of success in England, and not in France where the *Edinburgh Review* was not at all widely read. But is the claim true? If it is, there must have been very few counter-suggestible people around at the beginning of the nineteenth century. For one would have thought that Brougham, by attacking Young so loudly and aggressively, gave Young so much (adverse) publicity as to make it extremely *unlikely* that his researches would remain 'absolutely unnoticed by men of science for many years'.

Peacock himself in the end admits that the story is implausible, although he still expects us to believe it. He tries to make it more credible by referring to the great authority of the *Edinburgh Review*. Peacock writes that the influence of Brougham's reviews 'upon public opinion was *more remarkable than could reasonably be expected*, even from *the great authority of the publication in which they appeared*...'[45] Young's second biographer Wood also tries to make the 'character assassination' claim more plausible by telling us how authoritative the *Edinburgh Review* was.[46]

[43] This already indicates one way in which methodological considerations can shift historical problems. If a methodology were to pronounce Young's work as having established the wave theory's superiority, but having established it for reasons different from those normally alleged, then different aspects of Young's work may have to be looked at and these may indeed turn out to be obscure. (See below, p. 120.)

[44] Some of my arguments for the untenability of this claim were anticipated by David Hargreave of the University of Wisconsin in an excellent short paper entitled 'A New Look at the Young–Brougham Controversy', which was read before the Mid-West Junto meeting of the History of Science Society (University of Oklahoma, Norman, Oklahoma, April 4, 1969), and of which Professor Hargreave kindly allowed me to read a copy. (Hargreave's paper unfortunately remains unpublished.)

[45] Footnote in Young [1855], p. 192 (my italics). As we have seen above, pp. 111–12, most commentators report the hypothesis of Brougham's influence as a fact. Larmor (in his [1934] essay on Young in *Nature*) is one of the few who admit the implausibility of the hypothesis, though even he does so in a half-hearted way. Larmor writes that Brougham's 'satire [!] is commonly held to have diverted men from any attentive consideration of the new discoveries [of interference effects], by discrediting their author, and so as is said managed to postpone the progress of optical science for twenty years. *But that is possibly ascribing to him too much credit*...' (p. 277, my italics). Larmor goes on to mention that Young's personal reputation does not seem to have suffered at Brougham's hands – cf. below p. 118.

[46] Wood [1954], pp. 168–9. This Peacockian explanation which Wood supports is, however poor, an improvement on Arago's earlier attempt to make the story more

But Brougham's first two articles appeared in the *very first* volume (second number) of the *Edinburgh Review*.[47] It seems unlikely that the *Review* achieved its authority instantly; unless for some reason the British educated world was awaiting the launching of the new literary enterprise with bated breath.[48] This *may* have been the case, but it is certain that those who thus awaited its inception did not include its editor-writers – at least if the attitude of one of them, Francis (later Lord) Jeffrey, is representative.[49] Five months before the first issue actually appeared Jeffrey wrote to his friend Morehead:

> Our Review has been postponed till September, and I am afraid will not go on with much spirit even then. Perhaps we have omitted the tide that was in our favour. We are bound for a year to the booksellers, and shall drag through that I suppose for our indemnification.[50]

And a month later, writing to his brother John, Jeffrey was even gloomier:

> Our review is still at a stand. However, I have completely abandoned the idea of taking any permanent share in it, and shall probably desert after fulfilling my engagements, which only extend to a certain contribution for the first four numbers. I suspect that the work itself will not have a much longer life. I believe we shall come out in October, and have no sort of doubt of making a respectable appearance, though we may not perhaps either obtain popularity, or deserve it.[51]

Of course, Jeffrey's gloom turned out to be unjustified and the *Review* was a success.[52] However, (*a*) its success was almost entirely based on its social and political content and in particular on its great reforming campaigns (e.g. against slavery) – campaigns which had scarcely started even in 1804 when Brougham's second and final attack on Young appeared; and (*b*) as one might have expected, in the course of these campaigns it

plausible by referring not to the eminence of the *Edinburgh Review*, but to the eminence of Henry Brougham. He claims (see his [1859], p. 516) that we cannot blame the public for adopting in this case the views of the journalist Brougham for 'The journalist, in fact, was not one of those unfledged critics whose mission is not justified by any previous study of the subject. Several good papers, received by the Royal Society, had attested his mathematical knowledge...the profession of the bar in London had acknowledged him one of its shining luminaries: the Whig section of the House of Commons saw in him an efficient orator, who in parliamentary struggles was often the happy antagonist of Canning: this was the future President of the House of Peers – the present Lord Chancellor. How could opposition be offered to unjust criticisms proceeding from so high a quarter?' Arago conveniently forgets that, although this was all true when he wrote in 1832, when Brougham wrote against Young he was not an experienced and revered Lord Chancellor but only a young man of 23.

[47] Hargreave made this same point before me (in the paper quoted above, note 44).

[48] Paul Feyerabend drew my attention to this possibility.

[49] The instigators and writers-cum-editors of the early editions of the *Edinburgh Review* were Sidney Smith, Jeffrey, Horner and Brougham. Each of these was to achieve national fame, but in 1802 (the first number of the *Review* appeared in October of that year) they were relatively unknown young men. (Their respective ages in 1802 were 31, 29, 24 and 23.) Jeffrey soon became its first full-time editor. Brougham, in particular, seems then to have had only a limited reputation (as a rather unpredictable up-and-coming young man).

[50] Quoted in Cockburn [1852], p. 129.

[51] *Ibid.*

[52] For the full story of this success, see Clive [1957].

built up a substantial band of devoted enemies – these people would if anything have been predisposed *in favour of* anyone who had been attacked in the *Review*.[53]

Before leaving this 'character assassination' claim, I should like to point out that none of the historians who have written on the Young–Brougham controversy have considered it worthwhile to inquire in detail into the intellectual merits and demerits of Brougham's arguments. Did Brougham have any good arguments, for example, for his claim that Young's wave theory 'teaches no truth, reconciles no contradictions, arranges no anomalous facts, suggests no new experiments, and leads to no new inquiries'?[54] All these historians wrote after the wave theory had triumphed, and so perhaps they felt that the fact that Brougham supported the corpuscular theory meant that his arguments could not be worth considering.[55] I shall not follow this tradition. Indeed, we shall see as we go along that although Brougham certainly made mistakes in these attacks (which must have been very hurriedly written), some of his arguments are quite perceptive.

Apart from all this, the three socio-psychological explanations of the lack of recognition of Young's work would, if true, explain too much. We should expect someone who had contradicted the holy word of Newton, who wrote so obscurely that no one could understand him, and whose reputation had been savaged by a highly reputable journal to meet with little personal success. And yet Young was highly successful.[56] Well before the acceptance of the wave theory (an event which Young lived to see and profit by[57]) Young had become part of the scientific establishment. He was

[53] I should add that Brougham is attributed with attacks on other scientists in the *Review* almost as vituperative as the attacks on Young. In particular he attacked Count Rumford and W. H. Wollaston, neither of whom can exactly be said to have had his scientific career or scientific credibility destroyed!

[54] Brougham [1803].

[55] G. N. Cantor is a partial exception to this. In his scholarly and interesting [1971] (one of a series of interesting papers by Cantor connected with Young's optical researches) he does mention some of Brougham's arguments and even mentions that Young in his 1804 *Reply* 'failed to appreciate many of Brougham's more pertinent remarks, when they did not coincide with his conceptual scheme' (*op. cit.*, p. 88). (Unfortunately Dr Cantor does not go on to substantiate this remark.) But even Cantor largely concentrates on the question of the *motives* behind Brougham's attack and (usually) fails to go into the question of the scientific merits and demerits of the criticisms Brougham puts forward. Cantor mentions the usual explanation (based on personal animosity) of the ferocity of Brougham's attack, but instead traces the differences between Young and Brougham to differences of a '*methodological* character'. By this he means that Brougham had certain views about science which might have led him to find any hypothesis unacceptable if it involved the 'unobservable' ether. I don't think that this will do. Positivistic attitudes, like the one attributed to Brougham, tend to disappear if the theories involving the disputed entity are good enough. After all, Brougham certainly accepted 'unobservable' forces. Anyway why may not Brougham's main motive have been scientific rather than methodological: perhaps he just wanted to get at the truth; perhaps he was genuinely convinced that Young's theory was hopelessly false?

[56] Hargreave too (see note 44, above) points out that 'there is little evidence that Young's reputation within the scientific community was dimmed by his encounter with Brougham'.

[57] Young died in 1829.

on three separate occasions invited to give the Bakerian lecture to the Royal Society.[58] He was appointed Foreign Secretary of the Royal Society in 1802 and he also became (at the invitation of Rumford) Davy's professorial colleague at the Royal Institution. In 1812 he was offered the secretaryship of the Royal Society. What more could a full-time physician expect? Moreover, Young during his career had no less than 16 papers published in the *Philosophical Transactions* (11 of them after Brougham's attacks, which had allegedly destroyed his scientific reputation and in which the Council of the Royal Society had been rebuked for admitting his and others' 'paltry and unsubstantial papers into its Transactions'). Young was also invited to contribute well over 50 articles on a variety of subjects to the *Encyclopaedia Brittanica*, as well as a supplement to an article by the great Arago.

Something then is wrong with each of the three time-honoured explanations of why Young's fellow scientists failed to recognise the superiority of the wave theory and pushed on with Newtonian optics, despite all Young's counter-arguments and crucial experiments. Are we then to look for other non-intellectual, 'external' factors to explain the action of Young's contemporaries? This is certainly a possible course of action (and may possibly be successful). It is not, however, the only such possibility. But these other possibilities will not be visible until it is recognised that the reception of Young's work appears noteworthy and in need of explanation only in view of the assumption that his work *ought* to have been taken seriously because it established the superiority of the wave theory. The problem of what features a theory must possess in order to be superior to a rival theory is, however, a (normative) methodological problem – and different methodologies may specify different features. Our historical problem thus depends on the application of methodological considerations. Once this is realised new avenues of approach to the problem are opened.

The highly favourable appraisal of Young's work is, no doubt, usually taken for granted by historians, or, perhaps, argued for (as I indicated earlier) in rather uncritical falsificationist terms. It would be interesting to sharpen up these implicit methodological assumptions, and indeed to look at Young's work through the eyes of *various* explicit methodological criteria. We could then see, in each case, how much of this historical episode the methodology could explain 'internally' (i.e. in terms of the intellectual merits of Young's work, as it sees them) and how much was left to be explained with the help of 'external' factors.[59] This would,

[58] The information that these were probably three separate invitations was kindly supplied by the Librarian to the Royal Society.

[59] I use the terms 'internal' and 'external' in the redefined (and methodology dependent) sense given them in Lakatos [1971], reprinted above. Note in particular that in Lakatos's sense not all external factors need be social, though most methodologies will characterise social factors as external.

however, be a very protracted exercise.[60] Instead I intend to apply to Young's work the explicit criteria of appraisal supplied by one methodology – the methodology of scientific research programmes.[61]

This restricted exercise will in any case highlight the two main ways in which historical problems like that of Young's 'neglect' may be shifted via methodological considerations (and it will provide an illustration of one of the ways).

The first main way is this. Suppose that explicit methodological analysis of Young's work reveals that it did indeed contain the vital intellectual turning-point in the wave–corpuscle debate, but that this analysis locates the turning point in a place other than those generally pointed to. In this case the problem of Young's neglect will still exist and will still be in need of an external solution. Now, however, the search for external factors will have been redirected, historians will have to explain why Young's fellow scientists failed to spot the importance of this *new* aspect of Young's work.

The second way in which this historical problem (and others like it) may be shifted by methodological analysis is as follows. Suppose that the methodological analysis of Young's work reveals that it did *not* establish the wave theory's superiority. In that case there will be no reason to invoke external factors to explain the failure of Young's contemporaries to see the superiority of the theory. If, moreover, the methodology can also point out the factor which did eventually turn the debate in favour of the wave theory, then the whole episode of the nineteenth-century revolution in optics (both its failure to occur in the 1800s and its occurrence in the 1820s) will be capable of an internal explanation in terms of the differing objective merits of the (developing) theories involved.[62]

3 *What did Young really achieve? A reappraisal of Young's work*

My intention is to use the methodology of research programmes to provide a new answer to the question of what scientific conclusions Young's contemporaries ought to have drawn from his work. This reappraisal of Young's work turns out to provide a solution to many of the problems associated with its immediate reception, by showing the reactions of

[60] The exercise would hardly be complete unless it included a look at the whole episode of the early nineteenth century revolution in optics. For a methodology may pronounce Young's work non-crucial, thus avoiding the need to invoke external factors to explain its 'neglect'; but it may also be unable to point to anyone else's work as establishing the wave theory's superiority – thus incurring the need to explain the eventual *acceptance* of the wave theory in external terms.

[61] For the methodology of scientific research programmes, see especially Lakatos [1970], and Zahar [1973].

[62] This, I shall argue, is precisely what does happen when Young's work is reappraised using the criteria supplied by the methodology of scientific research programmes. The question of how far this ought to be regarded as a success for that methodology is a difficult one which I shall leave until §5.

Young's contemporaries to be allied much more closely than is usually believed to the intellectual situation in early nineteenth-century optics.

3(*a*) *Two expectations based on the methodology of scientific research programmes*

Some experiments come out in favour of one theory and against another. As Duhem showed,[63] however, these are not crucial experiments in any conclusive sense, for, in drawing from the two theories the consequences which are compared with experiment, extra assumptions must be made, and it may be these assumptions which are false. The proponent of a theory may therefore defend it against an apparent refutation, an apparent 'crucial experiment', by modifying not the theory itself, but one of the requisite auxiliary assumptions. According to Lakatos, this is typically what happens in the development of science. In other words, the rivalry between two theories (or rather programmes) is not considered settled at a single blow by 'crucial experiments'. Rather the title of crucial experiment is an honorific one (mistakenly) conferred by *later* commentators on one of the many 'anomalies' which the eventually defeated theory never managed to solve except in an *ad hoc* way. Indeed, on the basis of the methodology of research programmes one would expect[64] that, for any famous crucial experiment in the history of science, the protagonists of the theory allegedly defeated by the crucial result could, and did, develop their theory so as to explain this result – the trouble was that such explanations, though certainly constructible, were unacceptable because in various senses *ad hoc*. I shall argue that this first expectation is satisfied in the case of Young's allegedly crucial experiments in §3(*a*1) below. The methodology of scientific research programmes also leads one to expect that in the case of any crucial experiment there were, at the time, other experiments whose results were just as anomalous[65] for the eventually successful theory as the 'crucial' result was for the eventually defeated theory, and that it was only in view of later developments (i.e. the 'progressive' explanation of some of these results by the successful theory) that the title 'crucial' was not conferred on *these* experiments. I shall show that this expectation too is satisfied in the special case of Young's 'crucial' experiments (§3(*a*2)).

[63] See Duhem [1906].

[64] The status of these 'expectations' is considered in §5, pp. 173–6.

[65] I talk here as though there are *degrees* of anomalousness. It surely is the case that either a theory conjoined with accepted statements of initial conditions is inconsistent with an experimental result, or it is not. This is a straightforward black and white situation. The grey enters when we consider the real historical facts rather than methodologists' fairytales. In fact theories are not normally made sufficiently precise for them to make predictions in certain areas. In order for these theories to be capable of being inconsistent with results in these areas certain auxiliary assumptions have to be added. The range of auxiliary assumptions which would render the basic theory inconsistent with these results and their 'naturalness' within the programme of which the theory is a part are then questions of degree (which moreover may be characterisable only intuitively).

3(*a*1) *Corpuscularist explanations of Young's 'crucial refutations'*

The results of two of Young's experiments are alleged by various authors to have provided crucial evidence in favour of the wave theory of light and to have crucially refuted the corpuscular theory. These experiments are the 'two slit' experiment, reported in Young's *Lectures in Natural Philosophy*, delivered in 1802–3 but published only in 1807, and a special kind of diffraction experiment, described in a paper delivered to the Royal Society in 1803 (and published as Young [1804*a*]).

The two slit experiment consists in allowing homogeneous light, diverging from a single slit in a first screen, to fall on a second opaque screen in which there are two narrow and closely adjoining slits (symmetrically placed with respect to the first slit); the effects are observed on a third screen some distance from the second. The result of the experiment is that the observation screen is covered with a series of bands alternately light and dark, the centre of the pattern being light. (If the experiment is performed with white light a series of coloured bands will be observed, only the central band being white.[66]) I shall consider below what positive role this experiment plays in supporting the wave theory.[67] Let us for the moment assume that Young's wave theoretic explanation of these bands, as due to the alternately constructive and destructive interference of light coming from the two slits, is acceptable and unproblematic. Did the experimental result refute the corpuscular theory?

Certainly no one, so far as I know, produced a corpuscular theory of light which could explain all the details of the fringes and their positions on the observation screen. Also, precise versions of the corpuscular theory no doubt could be constructed which, together with accepted initial conditions, would be inconsistent with the result of the two slit experiment. Thus there is no denying that the two slit result presented a problem for the corpuscular theory. It was, however, not Young who first presented corpuscularists with this problem, for it was of a kind that they had known about since 1665.

From the point of view of the corpuscular theory, Young's two slit experiment was merely a more complicated form of the diffraction experiments first recorded by Grimaldi in 1665. Grimaldi had reported that the real shadow of a narrow opaque object in sunlight is wider than its 'geometrical' shadow,[68] and that coloured fringes are observable both inside and outside the shadow.

[66] Young discusses only the case of homogeneous light. (See below p. 153.)

[67] See pp. 143–6.

[68] This whole situation is somewhat confused. What one counts as the actual, as opposed to the geometrical, shadow of an opaque object was later shown (by Fresnel) to be arbitrary. However, for all but the smallest opaque objects, Fresnel's theory predicts that while the intensity on the observation screen will tend to zero as one goes into the object's geometrical shadow on either side, the intensity will nevertheless be quite high

The results of Grimaldi's experiments and of similar ones performed by Hooke, Newton[69] and others had convinced corpuscularists that the assumption that light is propagated in straight lines had to be amended. The diffraction effects[70] had to be explained, however, as small-scale *exceptions* to the general rule of rectilinear propagation, for facts like those of eclipse phenomena and our inability to see through a bent tube were taken on all sides as conclusive evidence that at least most of the light is propagated rectilinearly most of the time.[71]

some distance *into* the geometrical shadow – i.e. that what one might naturally take as the 'apparent real shadow' will be *smaller* than the geometrical one (altough there will of course be dark bands *outside* the geometrical shadow). Grimaldi, however, categorically states that the 'shadow...appears considerably larger in fact that it ought to be, if the whole thing is supposed to act by straight lines...' (Grimaldi [1665]; see Magie [1935]). Most modern textbooks nevertheless inform us that Grimaldi discovered that real shadows are smaller than geometrical ones. This is true even of the 'popular textbook' by Holton and Roller which has at least one eye on history. Holton and Roller write: 'Grimaldi put a tiny obstacle in the path of the beam and found its shadow...to be smaller than that to be expected...' (Holton and Roller [1958], p. 545). Young, like Newton, accepted that Grimaldi had shown that real shadows are enlarged relative to their geometrical counterparts.

[69] Some interesting problems arise from Newton's various accounts of fringes inside and outside shadows. In his 1675 'Second Paper on Light and Colours', Newton has a diagram (p. 199 of the reprint in Cohen [1958]) in which he drew three fringes *inside* the shadow of a *large* wedge. Now obviously Newton could not have performed this experiment, since it is a 'diffraction by a straight edge' experiment, and as is now well known, no fringes are observed inside the geometrical shadow in such a situation. (Historians have failed to spot this point, so far as I can see.) In Part I of the third book of the *Opticks*, however, no mention is made of fringes inside the geometrical shadow, and all the diagrams, even of the shadows of hairs, show only external fringes. This fact has been much commented upon. Mach, in his [1926], for example, writes of this investigation in the *Opticks*: 'Strange to say, Newton completely ignored the inner fringes observed by Grimaldi; he even explicitly asserted that diffraction only took place in an outward direction from the edge of the shadow-forming body, and all his diagrams give effect to this opinion.' (p. 143). This mystery was cleared up by Professor Stuewer who, in what seems to me an exemplary piece of work (Stuewer [1970]), repeated Newton's experiments, using Newton's own descriptions and other historical evidence to produce experimental equipment which he conjectured was very similar to the equipment Newton used. The result of the experiment was that no fringes were observed inside the shadow. Since Newton does not claim to have performed the experiment he describes in his [1675] but only to be reporting Grimaldi, Stuewer's conjecture is that Newton at first accepted Grimaldi's results (though what I say above about Newton's diagram suggests that he did not understand Grimaldi) but then repeated the experiments himself and decided on the basis of his own results that only external fringes exist.

[70] I use '*diffraction*' (a term introduced by Grimaldi) in a theory-neutral sense. As we shall immediately see, Hooke, and following him Newton, explained diffraction effects as due to the '*inflection*' of light by material bodies. Young (and at first Fresnel) described them, rather confusingly, as due to the interference of inflected rays. Fresnel eventually satisfactorily explained them on the basis simply of his theory of the propagation of light.

[71] This was Young's position too. His theory faced two problems here, however (those of explaining both rectilinear propagation and the exceptions to it), while the corpuscular theory faced only one (rectilinear propagation being taken care of, as we shall immediately see, by Newton's first law of motion). Young's rather unsuccessful attempt to solve these two problems of the wave theory will be discussed in detail below pp. 144–6. *Fresnel completely transformed this situation by progressively explaining (near) rectilinear propagation as a special case of diffraction.*

Corpuscularists solved this problem by assuming that light, in passing by a material body, is acted upon by short-range forces which emanate from the body and which deflect the light corpuscles from their naturally rectilinear path. A fundamental assumption of the corpuscular theory (it is better described as part of the hard core of the corpuscular programme[72]) was that light corpuscles obey the ordinary laws of particle mechanics. Any deviation, therefore, from a straight path indicates the presence of external forces acting upon the light corpuscles.[73]

Thus the protagonists of the corpuscular theory had a clearly defined way of explaining Grimaldi's diffraction fringes. The existence of fringes both inside and outside the geometrical shadow showed that both attractive and repulsive forces were active, which had anyway already been 'established' by the fact that transparent bodies could both reflect and refract light. The fact that in sunlight the fringes are variously coloured showed that the same forces have different effects on the particles corresponding to the different colours. But this had already been 'established' by the fact of prismatic dispersion.[74] When homogenous light is used, the corpuscular theory could explain the light fringes inside the geometrical shadow as due to the short-range forces deviating the corpuscles towards places they would not otherwise reach; and it could explain the dark fringes outside the geometrical shadow as due to the forces deviating the corpuscles away from places they *would* otherwise reach.

By assuming that the action on the light is alternately attractive and repulsive at different distances from the body, any given set of fringes could, given sufficient ingenuity, be explained. Thus, so could Young's two slit 'interference' pattern. But given that corpuscularists had not so far dealt (in quantitative fashion) with the simpler cases of diffraction, they were unlikely to bother much with Young's two slit case, which seems from this point of view a complicated affair involving three sets of fringes – those produced by the narrow section of screen between the two slits and those produced by larger sections of the screen on either side.[75]

[72] I shall characterise the corpuscular and wave *programmes* more explicitly below, pp. 156–7 and 136.

[73] The aim of the corpuscular programme was thus to explain all such deviations, whether by reflection, refraction or diffraction as due to one single set of forces emanating from matter.

[74] The assignment to the corpuscles of a mechanical parameter, the different values of which would explain this different effect, constituted a major problem for Newton and the whole corpuscular programme. For Newton's difficulties on this point, see Bechler [1974]. Bechler does not remark that Laplace provided a solution of the problem within the corpuscular programme; see below, p. 128.

[75] The complexity of the two slit situation (the apparent simplicity of the account in most modern textbooks is, of course, only achieved by forgetting about the diffraction pattern produced by each slit separately) was pointed out even after the wave theory had been accepted. For example, a translators' footnote in the 1859 English edition of Arago's *Biographies of Distinguished Scientific Men* (p. 493) tells us that 'some of the earliest of Young's researches were complicated by unnecessary conditions. Thus, to exhibit the

It may be objected that although corpuscularists like Brougham,[76] Jordan[77] (and later Biot[78]) might succeed through a suitable assignment of short range forces in explaining the two slit interference pattern, they could not, at the same time, possibly succeed in explaining – even in an *ad hoc* way – the change in pattern when one of the slits is closed. For, as we now know, points on the observation screen which had been dark when both slits were open, are illuminated when one slit is closed. Surely, this establishes that light can be subtracted from darkness to produce light and hence establishes Young's fundamental claim that light added to light can produce darkness. Is not *this* the final crucial blow to the corpuscular theory?

It may seem strange that Young never records having performed this simple and apparently vital modification of his experiment.[79] It is, however,

effect of *two* rays interfering, he at first not unnaturally transmitted the narrow beam of light through *two* small apertures near together. In point of fact, though the real effect is here seen, it is mixed up with others of a more complex kind.'

By the way, the possibility of corpuscularists appealing to the effect of the screen on the light in order to account for the result of the two slit experiment, is often pointed out by textbook writers. (I am indebted to M. L. G. Redhead for pointing out to me just how widespread this comment on Young's result is.) For example, Hardy and Perrin write ([1932], p. 569): 'Young's explanation of the phenomenon was not immediately accepted because of the possibility that light undergoes some modification in passing through the pin-holes.' And Jenkins and White write ([1957], p. 239): 'Soon after the double-slit experiment was performed by Young, the objection was raised that the bright fringes he observed were probably due to some complicated modification of the light by the edges of the slits and not to true interference. Thus the wave theory of light was still questioned.' Unfortunately these remarks are usually introduced as an afterthought to highlight the importance of experiments like Fresnel's biprism and two mirror experiments in which the interfering beams are created from one beam without the use of slits. The remarks are usually inconsistent with others made about Young's experiments. This is certainly the case, to take an example, with Jenkins and White. A mere eight pages before their remark just quoted they write: 'The first man successfully to demonstrate the interference of light, and thus establish its wave character, was Thomas Young. In order to understand his crucial experiment performed in 1801, we must first...' (*op. cit.*, p. 225). They go on to describe the two slit experiment. And 30 pages earlier than that they state '[Young's] early work proved the wave nature of light...' (p. 203). Of course one can rescue them from the charge of inconsistency by assuming that they meant that *given what we now know* Young established the wave theory of light, but people at the time did not know enough about the nature of diffracted light to realise it. This position, however, is surely untenable. One may as well say that the wave theory of light was established by the first person to observe what we now know to be diffraction fringes. This was probably the first creature that evolution endowed with eyelashes, who would see diffraction bands when looking at the sun through half closed eyes.

[76] See Brougham [1796]. Some of Brougham's theories were criticised and developed in Prévost [1798]. [77] See Jordan [1799*a*], [1799*b*] and [1800].

[78] See especially Biot [1816], vol. 4 and appendix.

[79] Holton and Roller decided that this is *so* strange that it cannot be true and wrote 'Young, the first to demonstrate the "double-slit experiment", also pointed out that this interference pattern is immediately destroyed when either of the two slits is covered up.' (Holton and Roller [1958], p. 557.) Humphrey Lloyd had, much earlier, made the same mistake: 'That these alternations of light and darkness [in the two slit experiment] are produced by the mutual actions of the two pencils, Young proved by the fact that when one of the beams is intercepted, the whole system of fringes instantly disappears.' (Lloyd [1833], p. 303.) Mach ascribes the true situation to luck, reporting that Young 'did not happen to chance' on this modification of his two slit experiment (Mach [1926], p. 147).

not so strange. For suppose Young *had* closed one of the two slits and observed the one slit diffraction pattern, how could he possibly have explained it himself? For where now are the two portions of light which are to interfere to produce the fringes? Light 'diffracted in all directions' by the now *single* slit is of no use, since two 'portions' of light have to coincide in direction in order to interfere. This difficulty is symptomatic of a much more fundamental defect in Young's theory which I shall discuss later;[80] for the moment I return to Young's experiments. Although Young does not record having performed this modification of the two slit experiment, he does record having performed what we should now regard as an analogous experiment.[81]

This is the 1803 diffraction experiment – the second of the experiments alleged to be crucial by commentators on Young's work. Young himself, his early commentators Whewell and Peacock and some later writers like Whittaker all record this as *the* experiment which established Young's principle of interference and hence refuted the corpuscular theory.[82]

Young's experiment consists, like Grimaldi's,[83] in observing the fringes inside and outside the 'shadow' of a narrow opaque object. Young then modified Grimaldi's experiment by screening off the light passing one side only of the diffracting object. The result is that the fringes previously visible inside the shadow disappeared.[84] Young at the time claimed this to be a 'simple and...demonstrative proof' that the two portions passing on either side of the object are concerned in the production of these inner fringes. Peacock, following Whewell, writes of this result: 'This was a *crucial* experiment, and may be considered as having constituted an important epoch in the history of the undulatory theory.'[85]

Now it would be strange if this were the case, for then the 'epoch' ought

[80] See below, pp. 144–7. I do not, of course, deny that Young could have developed an *ad hoc* explanation of the single slit diffraction pattern on the basis of his theories. In fact this pattern would no doubt be (for Young) a combination of the 'external fringes' produced by the two sections of screen making up the slit. These external fringes Young variously ascribed to interference between direct (rectilinearly propagated) rays and light inflected by the side of the screen, and to interference between direct rays and light reflected by the side of the screen. Neither explanation works; either explanation involves four portions of light – illustrating the arbitrariness pointed to below, p. 140.

[81] Which pairs of experiments are 'analogous' depends on theory. These two experiments are analogous according to the wave theory as later developed by Fresnel. Interestingly enough the two experiments are not analogous on the 'natural' interpretation of Young's theory. In the single slit case there is, apparently, only one 'portion' of light and one would expect, on Young's theory, no interference. In the narrow diffracting object case Young *can* see two portions of light; one passing either side of the object. (But cf. footnote 80 above.)

[82] See Whewell [1837], p. 326; Peacock [1855], p. 162; Whittaker [1910], p. 108.

[83] See above, p. 122.

[84] It is this qualitative aspect of the experimental outcome which normally is considered important. There are, nevertheless, other aspects of this result which turn out to be important in the light of the methodology of research programmes. See below, pp. 147–56.

[85] Peacock [1855], p. 162.

already to have been 'constituted' in 1665 when Grimaldi recorded that, while fringes are visible inside the shadows of narrow objects, there are no fringes visible inside the shadows of large opaque objects. Young's narrow object plus screen combination amounts simply to a large opaque object.[86] But let us forget this problem, and again assume for the time being that Young's wave theory provides an acceptable and unproblematic explanation of this experimental result, and merely ask whether the result refutes the corpuscular theory.

As in the case of the two slit experiment there doubtless are precise versions of the corpuscular theory which contradict this 1803 experimental result. This is not, however, to deny the possibility of constructing assumptions consistent with the fundamental tenets of the corpuscular theory which, together with these tenets, explain Young's result. This possibility was, moreover, actually explored.

There were, in fact, two major ways of developing the corpuscular theory so that it explained Young's results. The first involves admitting that the joint effect of two beams of light may be darkness; this, however, is given a physiological rather than a straightforwardly physical explanation. This first explanation (which was only *in fact* resorted to when many more results concerning what we now call interference effects had been established) shows that the folklore is quite trivially wrong in claiming that Young's experiments crucially refuted the corpuscular theory. Some corpuscularists were willing to admit that particles of light *can*, in a sense, destructively interfere. This corpuscular explanation makes the apparent lack of illumination the result of two sets of particles arriving at the eye in a fixed 'phase' relation. There are two possibilities. It could be claimed that in the case of destructive interference the two sets of particles arrive in such a way that particles of the first set having stimulated vibrations in the eye, particles of the second set produce vibrations which interfere destructively with the first at the retina.[87] Alternatively it could be claimed that no light is experienced when the elements of the two streams of particles arrive in *unison* so that they, as for example Potter put it, 'cease by their combined bulks being too large to create that sensation which we call light.'[88]

These explanations cannot be dismissed *automatically* as cheap *ad hoc*

[86] This simple point must have been well known. To my knowledge the point was first made in print by Brewster (see Brewster [1831], p. 130). Later commentators seem to have missed it altogether; Mach is an exception, but he simply asserts that despite Young's result being well-known it was important. Mach writes: 'This experiment would, as Brewster rightly notes, have been really unnecessary, if Young had considered that wide shadow-forming objects show no fringes in the shadow. Nevertheless, the result is an important one.' (Mach [1926], p. 145.)

[87] See e.g. Young [1817], p. 328 and Fresnel [1822], p. 25. (It is often a sign that a programme is getting ahead of its rival, as the wave programme was beginning to in 1817 and definitely was in 1822, when its protagonists start to give their rivals ideas for new auxiliary assumptions!)

[88] Potter [1833], p. 33.

manoeuvres simply because they invoke physiological considerations.[89] Fresnel's wave theory also used physiological factors to explain certain phenomena. For example, the non-appearance of interference patterns in some circumstances is explained by Fresnel as due, not to the *non-occurrence* of interference, but rather to the *invisibility* of the interference fringes caused by limitations in our visual apparatus.

Nor was this use of physiological considerations an isolated event in the development of the corpuscular programme. For example, Laplace held that various experiments had shown that 'the relative velocity of a homogeneous light ray is constantly the same',[90] that is, that the observed velocity of light is always the same no matter what the relative velocity of the observer and the light source. He conjectured that this constancy was 'determined by the nature of the fluid which it [the light particle] puts in movement in our organs, to produce the sensation of light'.[91] More exactly, Laplace's hypothesis was that 'luminous bodies launch an infinity of rays subject to different velocities' but that the eye is capable of registering only those rays whose speed, relative to it, is comprised between certain very narrow limits.[92] This conjecture was regarded as confirmed by the discovery, made by Ritter and Herschel, of the 'invisible rays' emitted by the sun. And indeed Laplace's theory is independently testable (it was in fact eventually independently refuted), for it predicts that ultra-violet and infra-red rays have velocities different from that of light.

The proponents of the corpuscular programme, however, also had a second, rather less drastic, method of accounting for Young's 1803 diffraction experiments; a method which did not invoke any physiology at all. This method of explanation (already available to corpuscularists *before* Young produced his result) involved assuming that the forces, which were 'known' to emanate from the sides of bodies and to act on light, were capable of 'interfering'. This method was never (as far as I know) actually used in any published account to explain Young's result, but that it could be so used is clear from corpuscularists' explanations of other experiments.

Brougham, for example, published a long paper principally on diffraction effects in 1796 in which he explains the result of a kind of one slit experiment by assuming that the light-bending forces emanating from the sides of material bodies 'interfere'. He speaks of light being 'inflected' by

[89] Although as a matter of fact these particular explanations are *ad hoc* at least in the empirical sense, i.e. they predict no novel facts. (One can, as usual, envisage ways in which they might have been developed so as to become independently testable.)

[90] Laplace [1813], p. 327. Since, by the way, Laplace regarded this as empirically established and came to a theory very different from the special theory of relativity, this seems to provide a 'constructive' disproof of the (once popular) idea that Einstein induced his theory from the 'facts' about the propagation of light.

[91] *Ibid.*

[92] *Ibid.*

one side of the slit and then, when the slit is narrow enough, being 'deflected' by the other side.[93]

A second example of a corpuscularian who used the interference of the light-bending forces to explain diffraction effects is G. W. Jordan. Jordan published in 1799 and 1800 a series of three monographs which record carefully the results of a great number of painstaking experiments on diffraction.[94] These experiments include the observation of the shadows of pieces of lead of steadily decreasing width. Jordan observed that 'By reduction of the breadth of the piece of lead... thin slender streaks of white light, with dark intervals began to make their appearance [within the shadow of the piece of lead]...By further reduction... these white streaks became broader and more distinct and approached the central dividing line of the shadow...'[95] This is substantially Young's 1803 diffraction experiment (at least in its qualitative aspects[96]) only in reverse. Young first observed the shadow of a narrow object and then saw the internal fringes disappear as the object was widened; whereas Jordan observed the internal fringes appear as the diffracting object was slimmed down. Jordan does not follow the above description of his experiment with an explicit explanation of its outcome. He does, however, explain why the fringe pattern in the one slit experiment changes as one narrows the slit. The changed pattern, according to him, is caused by the fact that 'by their [i.e. the sides of the slit] mutual approach *their various actions upon different portions of the intervening light begin to intermingle with each other*...'[97]

The change in diffraction pattern in the Grimaldi–Young experiment when light passing one side of the narrow object is screened off could thus be given a corpuscular–theoretic explanation without allowing that light beams mutually interfere and without invoking physiology. One has only to assume that in the absence of the screen, light passing one side of a narrow opaque object is acted on, not just by forces emanating from that side, but also by forces emanating from the other. However, screening off the light passing one side also screens off the forces emanating

[93] The experiment is the one first performed by Newton in which a beam of a light passes through a small gap between two knife blades which meet at a very small angle (see Newton [1730], p. 329). Brougham writes of the fringes produced in this experiment that they were 'formed by the *inflexion* of one knife and were moved into its shadow and separated and dilated by the *deflection* of the other' ([1796], p. 235).

[94] These monographs are Jordan [1799*a*], [1799*b*] and [1800]. Jordan's experiments were, in fact, too painstaking. He had no clear theories about what conditions were important for the production of diffraction effects and so he modified *all* the experimental conditions he could think of. The result is that Jordan's work provides a beautiful illustration of the anti-inductivist thesis that experimentation undirected by theory is worse than useless and leads nowhere.

[95] Jordan [1799*b*], p. 78.

[96] I consider below, pp. 147–56, the role of the quantitative aspects of Young's experiment in the wave-corpuscle debate.

[97] Jordan [1799*a*], p. 81; emphasis supplied.

from that side, and hence the path of the light not screened off is also affected.[98]

I do not claim that either of these two corpuscular explanations of Young's experimental results was unproblematic. The second explanation in terms of the 'inference of forces', for example, seems to require the light-bending forces to be sensible at small but nonetheless sensible distances, whereas other phenomena seem to require these forces to be insensible at any sensible distance. Furthermore, no one succeeded, in anything but an *ad hoc* way which required modification with each new experimental result, in assigning to opaque bodies specific light-bending forces obeying specific laws.[99] But no matter what difficulties these explanations run into elsewhere, they were possible corpuscularist explanations of Young's results. Hence neither of these now famous results crucially refuted the corpuscular theory in the usual sense.[100]

[98] My account here is oversimplified in at least one respect. The explanation in the text had, I think, struck Young himself as possibility. This is suggested by a further modification of his experiment, which he reports, but whose importance has not been commented on in the subsequent literature. Young, apart from placing the screen adjoining one side of the narrow diffracting object, also places it some distance behind the object, still screening off the light passing by one side only. In this case too the fringes inside the shadow disappear, despite the fact that, as Young significantly adds, the light passing the unobstructed edge of the object '*must have undergone any modification that the proximity of the other edge of the* [*object*] *might have been capable of occasioning*' (Young [1804*a*], p. 180; my italics). I did not include this modification in the text since it only complicates matters without affecting my claim that a corpuscular explanation of the experiment is possible. For the corpuscularist will now say that when the screen is held some distance from the diffracting object, the light, in passing the object, is again acted on by forces emanating from both sides, but the fringes thus produced are shifted and modified by the subsequent effect of the forces emanating from the screen.

[99] In previous drafts of this paper, written before I had found Jordan's little-known work, I had said that my account was simplified in a respect other than the one mentioned in footnote 98, viz. that no corpuscularist ever actually published the explanation I give in the text. Jordan's explanation certainly does not seem to have been applied to Young's 1803 diffraction result by other corpuscularists. Biot, for example, records Young's result in his [1816] (appendix to volume 4) as a so far unresolved anomaly. Brougham, in his second review, challenges the accuracy of the result itself (see Brougham [1804], p. 99). Nevertheless, Jordan's explanation came as no surprise for it is (in some sense) suggested by the corpuscular theory (or is within the spirit of the corpuscular programme). Of course, we should not expect the claim that it is generally easy for a programme to produce *ad hoc* explanations even of refutations of its theories always to be supported by the actual production and publication of such explanations by scientists at the time. Indeed, scientists will often fail to publish such explanations precisely because they intuitively recognize them to be *ad hoc*.

[100] Still a third corpuscular explanation might be possible in terms of the light-molecules' fits of easy transmission and reflection, since these were interpreted as dispositions of the molecule to be acted on by the attractive and repulsive forces respectively. (This might also indicate how the coloured fringes were formed in diffracted light since the molecules corresponding to different colours have different intervals of fit.) See below, p. 157. No one seems, however, even to have started to develop this as an explanation of the diffraction fringes. Brougham, as I pointed out, regarded Newton's fits hypothesis as 'unintelligible' (see above, p. 114). Biot, who developed the theory of fits, steered clear of any explicit attempt to *explain* diffraction effects (although he recorded lots of diffraction experiments).

I now turn to the second expectation based on the methodology of research programmes: that the successful theory in any allegedly crucial experiment (in this case the wave theory) will have faced (at the time of the discovery of the 'crucial' result) anomalies which posed for it problems just as big as those posed for the defeated theory by the 'crucial' result. This expectation too is satisfied.

3(*a*2) *'Crucial refutations' of the wave theory*

Newton had long ago claimed that there was 'both Demonstration and Experiment' against the wave theory. The demonstration concerns the rectilinear propagation of light. I shall argue later that Young provided no acceptable reply to Newton's objection on this point.[101] But as well as this 'Demonstration' there were two distinct experiments which, in Newton's view, crucially falsified the wave theory of light.

The first of these experiments concerns the phenomenon of dispersion. Briefly, Newton held that his famous experiments on prismatic dispersion established that the splitting of a ray of sunlight into a spectrum of coloured rays was the result not of a modification of the ray of sunlight during its passage through the prism, but rather of the separation of the rays of different refrangibilities already present in the original ray.[102] Newton alleged that such a selection process would be impossible if light were a wave motion.[103] Newton thus regarded these researches (and especially his '*Experimentum Crucis*' in which the differently coloured rays of a spectrum are allowed to pass separately through a second prism) as especially damning for the wave theory.[104] Any wave theory, it seemed, would have to explain the creation of coloured light out of white light as due to a modification of the light by the prism – but then why did not the same modification occur when one of the coloured rays thus created crossed the second prism?[105]

These experimental results were not actually inconsistent with any specific version of the wave theory up to and including Young's. Huygens, for example, in his famous *Treatise on light* never mentioned, let alone explained, any colour phenomena, and thus *without the addition of extra auxiliary assumptions* Huygens's theory has no implications about dispersion. Similar considerations apply in Young's case. Although Young does make the odd suggestion about dispersion,[106] he makes no attempt to give a systematic account of it, and most of the time appears to be unaware of

[101] See below, pp. 144–7.

[102] 'Colours are not *Qualifications of light*, derived from Refractions, or Reflections of natural Bodies (as 'tis generally believed), but *Original* and *connate* properties...' (Newton [1671–2]; quoted from the reprint in Cohen [1958], p. 53).

[103] See, for example, Newton [1730], Queries 27 and 28.

[104] Because it hits at the whole wave *programme*. See below, pp. 136–8.

[105] The coloured rays in fact cross the second prism without any further modification of their colour.

[106] See, for example, his [1807], p. 463.

the problems it poses for his theory.[107] But this failure explicitly to predict the wrong results is to be regarded neither as a virtue of the wave theory nor as a criticism of Newton's argument. Indeed, scientists are not likely to oblige the methodologist–historian by developing their theories in an attempt to explain some group of phenomena if they guess that any specific theory to which this attempt will lead will be refuted by the facts. They are *not*, however, to get credit for simply failing to predict the wrong result (as was explained by Popper). What Newton was claiming was that any foreseeable addition of auxiliary assumptions to the basic wave assumptions which would produce a theory which made predictions about dispersion *would* be inconsistent with his experimental results.[108]

The second experiment which, Newton held, refuted the wave theory concerns double refraction and was first discussed by Huygens. Indeed, the full title of Huygens' famous book on light is: *Treatise on Light. In which are explained the cause of that which occurs in Reflexion, and Refraction. And particularly in the strange refraction of Iceland Crystal.* This title is, however, a fraud. The causes of 'the strange refraction of Iceland Crystal' are not explained in this work.

Bartholinus, in 1669, became the first to record that when a small object is viewed through two opposite faces of a crystal of calcite ('Iceland spar') it appears double. When a ray is incident on such a crystal it is, in general, split into *two* refracted rays. One of these rays, called the *ordinary ray*, obeys the usual law of refraction, the other does not and hence is called the *extraordinary ray*. Huygens's attempted wave theoretic explanation of this

[107] For example, Young often cites the fact that wave theories predict uniform velocity of propagation of light in a given homogeneous medium (since homogeneous elastic substances transmit all disturbances no matter what their size and frequency with the same velocity) as a great advantage of such theories over their corpuscularist rivals (see, e.g., his [1810], p. 253: 'Among the facts which appear favourable to the Huygenian theory we must first enumerate the uniformity of the velocity of light, in any one medium...'). But in both theories different refrangibilities are associated with different velocities, and so it ought to follow on the wave theory from the fact of prismatic dispersion that one disturbance (emanating from the sun, say) can be split into many different disturbances each crossing a single homogeneous medium (glass, say) with a velocity *different* from the others. This was indeed regarded as a major difficulty for the wave theory even after the revolution had occurred.

[108] This is a further argument for speaking of the wave *programme*. The situation can then be characterised more precisely as follows: no explanation of dispersion in line with the heuristic of the wave programme was possible either in Newton's or in Young's time. It also constitutes a further argument against what Kuhn and Lakatos call naive falsificationism. Kuhn, Feyerabend and Lakatos all point out that scientists often disregard inconsistences between accepted theories and accepted observational results in the hope that the theory concerned will eventually be so modified that this inconsistency is removed. But, as this example shows, 'problems' or 'anomalies' for a programme in actual scientific practice need not be clear cut inconsistencies with fully articulated theories. (No account of both prismatic dispersion and the propagation of white light in a vacuum was developed within the wave theory, until Gouy, in his [1886]. Even this account was far from being uncontroversial. For the controversy over Gouy's solution see Wood [1905], chapter VI. The account of the controversy was dropped from later editions of Wood's book after the new photon theory of light became known.)

phenomenon was that when a wave front impinges on such a crystal surface it generates two waves which cross the crystal simultaneously. The first yields the 'ordinary' ray and travels through the ether alone, the 'wave fronts' corresponding to it being spherical. The other (which yields the extraordinary ray) travels through both the particles of the ether *and those of the crystal itself.* It has different velocities in different directions, being transmitted by spheroidal surfaces. Huygens's sphere/spheroid construction gives the correct paths of the rays for various angles of incidence, as was later confirmed by Malus and Wollaston. As an explanation of the phenomenon, however, it is refuted by a simple experiment performed and reported by Huygens himself.[109]

There is one direction through a calcite crystal in which no double refraction occurs. This direction is called the optic axis. If two crystals placed one behind the other are similarly oriented with respect to their optic axes, a ray doubly refracted by the first will not be refracted doubly by the second – the ordinary ray produced by the first crystal will in fact be propagated as an ordinary ray by the second, and the extraordinary ray produced by the first as an extraordinary ray in the second.

Undeterred, Huygens explained the results of this experiment by assuming that a ray incident upon a calcite crystal is modified in such a way that the emergent ordinary ray consists of a disturbance which no longer has the capability of travelling through matter. (He has explained, remember, the extraordinary ray as yielded by a disturbance which travels through the particles both of the ether and of the crystal.) This explanation is immediately refuted by altering the inclination of the second crystal. If the principal sections (planes passing through the optic axis normal to a crystal surface) of the two crystals are perpendicular, the ordinary ray emergent from the first crystal is refracted extraordinarily by the second, and *vice versa.* At all relative inclinations of the principal sections apart from parallel and perpendicular, both rays emerging from the first crystal are doubly refracted by the second – i.e. four rays are produced. At this stage Huygens admits himself at a loss: 'But to tell how this occurs, I have hitherto found nothing which satisfies me.'[110] And so 'leaving then to others this research',[111] Huygens dropped his central problem. He should have also dropped the title!

Young was unable to offer any precise way out of Huygens's *cul-de-sac.* The only mention he makes of double refraction in his early papers is when he states that Huygens's *experimental* law, giving the position of the extraordinary ray as a function of the angle of incidence and the inclination of the optic axis, is more accurate than the one suggested by Newton.[112] In

[109] See Huygens [1690], pp. 92–4.

[110] Huygens [1690], p. 94.

[111] *Ibid.*, p. 95.

[112] See his [1802*a*], pp. 166–7. Young was, of course, correct, but Huygens's experimental law was not dependent on the wave theory and could, as Laplace showed (see, e.g., Laplace [1809]), be explained within the corpuscular theory.

his later work,[113] he does present one or two suggestions on how one might construct a wave theoretical account of double refraction. He presents in fact the conjecture that double refraction might be explained by assuming that the ether inside birefringent crystals has different elasticities in different directions. This is, of course, as one would expect from Young, a highly intelligent suggestion, but as in other cases, Young never developed his idea into anything more than a suggestion. Suggestions have no influence on the appraisal of the existing merits of two theories.[114]

On the other hand, Newtonian optics could at least offer a *post hoc* explanation of these double refraction effects. It explained how the action of the second crystal on the rays transmitted by the first crystal changes as we turn the second crystal around the rays. Newton endowed his light corpuscles with sides – two opposite sides which have the property on which double refraction depends and two other sides which do not have this property. Whether any particle is transmitted as part of an ordinary or of an extraordinary ray was then made to depend on the inclination of its sides relative to the optic axis of the crystal. No such explanation of the results of the two crystal experiments could be given by either Huygens's or Young's wave theory. For the disturbance in a longitudinal wave is, by definition, *in* the direction of its propagation, and thus such a wave motion, unlike a Newtonian different-sided corpuscle, could not have different properties in different planes through the direction of its propagation.[115] The situation for the wave theory became, in this respect, much worse in 1810 when Malus recorded that light beams could be made to show this 'sidedness' (or could be 'polarised'), not just by passage through birefringent crystals, but by simple reflection at certain angles.[116]

The only major optical discovery made in the period between Newton and Young also turned out to be anomalous for the wave theory. This was the phenomenon of aberration, an apparent periodic movement of the fixed stars discovered by Bradley in 1728. Aberration received a simple explanation by the corpuscular theory in terms of the addition of the velocity of light particles and that of the earth. No such explanation could be given within the wave theory unless, as Young was to suggest, the earth

[113] See especially Young [1809] and [1817].

[114] Commentators generally, in fact, give Young great credit for this *aperçu*, regarding it as a remarkable anticipation of the 'Truth'. But since Young never turned it into a real explanation he should presumably be given credit only if his suggestion influenced the man who eventually produced the explanation – in this case, Fresnel. Young's work in this respect (as in others) seems to have had precisely no effect on Fresnel.

[115] Thus Newton had written: 'Are not all Hypotheses erroneous, in which light is supposed to consist in Pression or Motion, propagated through a fluid Medium? For Pressions or Motions propagated...through a uniform Medium, must be on all sides alike; whereas by those [double refraction] Experiments it appears that the rays of light have different properties in their different sides.' ([1730], Query 28.) Newton never fills in the sketch of his own theory.

[116] See below, p. 141.

moves freely through the ether.[117] This was, however, inconsistent with wave theoretic explanations of other phenomena,[118] and it even predicts that *all* bodies are transparent![119]

3(*b*) *The wave optics research programme and its eventual degeneration at Young's hands*

The standard account of the situation in optics in the first years of the nineteenth century is then inaccurate in both the ways the methodology of scientific research programmes would have led us to expect. Corpuscularists could deal with Young's alleged crucial refutations; and the wave theory faced empirical difficulties which seemed, at the time, just as intransigent as those faced by the corpuscular theory.[120] This shows that if we take into account the historical facts about the state of the two theories and the problems they faced, the decision on which one was better is not as straightforward as it is usually made to appear. The attitude of those scientists like Laplace, who preferred the corpuscular theory, cannot be dismissed as entirely intellectually baseless. Indeed, given the two theories as they stood at the time and the experimental evidence, Laplace's attitude seems rather reasonable. He wrote:

> The phenomena of double refraction and of the aberration of the stars, seem to me to give to the system of the emission of light, if not complete certainty, at least an extreme probability. These phenomena are inexplicable by the hypothesis of the undulations of a fluid ether. The singular property of a ray polarised by a crystal, of no longer dividing in passing into a second crystal parallel to the first, evidently indicates different actions of a single crystal on the different faces of a molecule of light, whose movements are...subject to the general laws of the movements of projectiles. [121]

Of course, as I stressed at the beginning, the investigation of the relative merits of two theories involves the application of some set of *norms*. I now intend to apply the set of explicit norms provided by the methodology of scientific research programmes to the wave and corpuscular theories as

[117] See Young [1804*a*], p. 188. Young, as usual (cf. above, p. 134), did not work out the consequences of his suggestion.

[118] Specifically in Young's case with his original explanation of the coloured diffraction fringes – a phenomenon he made depend on the presence of an 'inflecting atmosphere' of higher density ether around opaque bodies (see Young [1802*a*]). See also Cantor [1970–1] for an account of the various hypotheses Young made about the ether during his career.

[119] This rather striking refutation of the wave theory with its hypothesis of an all-pervading ether was apparently first pointed out by Halley.

[120] Of course the fact that a fact can be dealt with by a theory (i.e. that assumptions can be introduced which together with the theory imply a statement describing the fact) does not mean that it poses the theory no problem. The problem is not simply to produce any old explanation of a fact, but to produce a specific kind of explanation: a non-*ad hoc* one which is testable independently of the fact concerned.

[121] Laplace [1813], I IV, chapter XVIII; my translation. See also below, p. 160. The English translation by H. H. Harte has 'seemed' instead of (the correct) 'seem' – perhaps Harte, writing after the wave theory had been accepted, thought that Laplace was too great a scientist not to have come to realise that his appraisal was 'mistaken'?

they had been developed up to and including the first few years of the nineteenth century. (They are, in fact, better described as the wave and corpuscular programmes, as we shall immediately see.)

Various aspects of my account so far have suggested that there was no such thing as *the* wave theory. There was rather a *series* of such theories. Each theory was built around (and therefore logically implies) the fundamental assumption that light is some sort of disturbance in an all-pervading elastic medium. But this assumption alone does not account for specific optical phenomena – for this, extra assumptions have to be invoked. A sequence of different such additions to the fundamental assumption were made. But the shift from one set of assumptions to the next was not made in a random trial-and-error fashion. These shifts were occasionally spurred by empirical refutations (although refutations always remained). But more importantly, these shifts were guided by the aim of the whole wave optics enterprise – the union of optics with the mechanics of elastic media. This aim both guided the extension of the programme (via the accretion of suitable additional assumptions so that the whole theory made predictions about fields it had not hitherto covered), and it ruled out as unacceptable (and suggested replacements for) certain auxiliary assumptions which were actually made – and it ruled them out *independently of whether or not they were part of an empirically successful theoretical system.*

There was therefore a wave optics *research programme* complete with 'hard core' ('Light is a disturbance in an all-pervading elastic medium') and 'positive heuristic' ('Reduce all optical phenomena to the mechanics of elastic media, without invoking any force not already made available by theoretical mechanics'). In this programme mechanics, as we shall see more clearly as we go along, played a dual role. Newton's laws of mechanics were used (often implicitly) in deducing consequences from testable versions of the wave programme and thus played a part in explaining the phenomena. But mechanics also played a second, prescriptive role – all optical phenomena were to be explained as consequences of ethereal disturbances obeying the ordinary laws of mechanics without invoking 'special' or 'abnormal' forces not amongst those already made available by mechanics. There were several testable versions of the wave programme which, independently of any particular empirical difficulties, were pronounced unacceptable by this heuristic principle.[122]

[122] This heuristic principle plays a largely negative role as far as Young's work is concerned, pointing to certain inadequacies in his theories. It really begins to show its fertility in a positive way at the hands of Fresnel. Let me cite one very clear example. The law of reflection (that the angle of incidence equals the angle of reflection) was certainly amongst the best corroborated empirical laws in optics. Huygens's wave theoretical account of reflection has this law as a consequence (hardly surprisingly, since Huygens constructed the theory *ad hoc* precisely so as to account for the law!). Huygens's theoretical account can hardly then be said to have been in empirical difficulties. Yet the account is unacceptable from the point of view of mechanics since it requires certain point centres of disturbance in the ether to produce an *actual* disturbance only in one direction (the

The wave optics programme existed long before Young. Indeed it existed long before Huygens. As Verdet remarks in his preface to Fresnel's collected works: 'Neither Huygens, nor any of the other authors who, in the seventeenth century, considered light as a motion, present this idea as a personal invention. They treat it as one of those current hypotheses which do not belong to anyone, but which everyone is required to discuss.'[123]

The major contributor to the wave optics programme prior to the nineteenth century was undoubtedly Newton. It was Newton who had 'discovered' the periodicity of light,[124] and he who had suggested that Huygens's theory, though false, would have had greater verisimilitude and would have dealt with a greater range of phenomena had it characterised light as a *periodic* motion and the light of different colours as having different periods.[125] Newton also was the first to provide an analysis of harmonic motion;[126] and the first to base an explanation of a physical phenomenon (the peculiar tides at the East Indian port of Batsha) on a principle of interference or superposition.[127]

We see already from this the heuristic role of mechanics in Young's major 'theoretical discovery' – the principle of interference. All wave theorists prior to the latter half of the nineteenth century (Huygens included) assumed that the medium in which light is propagated obeys the laws of mechanics. Newton had produced an explanation of certain wave effects on the basis of an interference principle (which principle, it is interesting to note, Newton presents not as a new discovery but rather as a simple consequence of the mechanics of fluid media).[128] It was therefore not too much of a step to 'invent' the hypothesis that light waves interfere, given mechanics and given Newton's work both on periodicity and on interference. The only difference between Newton's principle of interference and Young's is that in Newton's tidal case the restoring force, which brings the particles back towards the equilibrium position from

disturbance being 'infinitely feeble' in all other directions). Fresnel set out to replace this account with a new one in line with mechanics, in which the disturbance was real in more than one direction. He produced in this way a theory which both explains the success of the previous theory and *corrects* it in certain cases (where the reflecting surface is very narrow).

[123] Verdet [1866], p. xv; my translation.

[124] That is, he conjectured that light was periodic and he supported his conjecture with his experiments on the colours of thin plates ('Newton's rings').

[125] It is not always realised that Huygens, usually regarded as the founder of the modern wave theory, made light consist not of nice smooth periodic wave-like disturbances, but rather of irregular pulses. Newton calls the assumption that the disturbances alleged to make up light are periodic, 'the most free and natural application of this [light is motion] hypothesis to the solution of the phenomena'. (Newton [1672].) Verdet remarks ([1866], p. xvii) that 'The first one who (after Newton) would dare to return to the wave theory, could not fail to consider the light waves succeeding each other periodically at regular intervals, depending on their colour...'

[126] Newton [1729], Book II.

[127] Newton [1729], System of the World, p. 586.

[128] *Ibid.*

which they have been disturbed, is supplied by gravity, whereas in Young's optical case the restoring force is supplied by the elasticity of the ether.[129]

Thus Young was not conjuring theories out of the blue, but was pursuing an already existing research programme. The question is how Young developed the programme – whether it progressed or degenerated at his hands.

There are several significantly different formulations of the law of interference to be found in Young's work. Together they form a classic case of a 'degenerating problemshift'.[130] I shall first chart this degeneration and then investigate the (independent) question of whether or not the original principle of interference constituted progress.

Young's first version of the law of interference states:

> Whenever two Undulations, from different origins, coincide either perfectly or very nearly in Direction, their joint effect is a Combination of the Motions belonging to each.[131]

This proposition is very different from the interference or superposition principle which appears in modern textbooks, which is suggested by mechanics and which states that, whenever several waves are propagated simultaneously, the disturbance at a given point is the sum of the disturbances originating from the individual waves. This modern principle explains both the interference of light in certain circumstances (if the two light beams act at the same point their effect is the vector sum of the disturbances belonging to each) *and* the failure of light beams to 'interfere' in the sense that a light beam carries with it beyond the point of crossing no sign of having crossed another beam. According to this modern principle (developed by Fresnel), light disturbances emitted from any number of sources and acting at the same point always interfere or superpose there. The circumstances in which observable interference fringes are produced are, however, limited by a variety of factors.

Young's claim differs from this modern principle in claiming only that the disturbances (positive or negative) sum when the two waves 'coincide either perfectly or very nearly in Direction'. This restriction of his principle to only parallel (or near parallel) light beams constitutes, I believe, an immediate indication of Young's lack of that tenacity which is generally necessary if important scientific discoveries are to be made. It marks the beginning of Young's policy of (too readily) modifying his wave theoretic assumptions in the face of empirical difficulties (even when these assumptions were dictated by the heuristic of the wave optics programme). Moreover, the modifications Young makes are always *ad hoc* – they are always made precisely so as to remove the difficulty in question without making any extra predictions about other situations.

No doubt Young produced this first version of his interference principle

[129] This is not said in an attempt to belittle Young's achievement, but to point to the power (and autonomy) of the programme.

[130] See Lakatos [1970], p. 118.

[131] Young [1802*a*], p. 157.

to take account of the admitted experimental fact that interference fringes are not in general *observable* unless the interfering beams are close to parallelism. This does not mean, however, that two non-parallel beams do not interfere or superpose. Indeed Fresnel gave a wave theoretic explanation of the non-appearance of interference fringes in these cases which presupposes that the two beams do superpose, but that the interference pattern produced changes so rapidly that the illusion of uniform illumination is created.[132] But rather than attempt to explain the refutation of the general superposition principle in the case of non-parallel beams as only an 'apparent refutation', Young restricts the principle to apply to only parallel or near parallel beams.[133]

In his second major paper on light published later in 1802 Young writes:

> The law [of interference] is, that 'Wherever two portions of the same light arrive at the eye by different routes, either exactly or very nearly in the same direction, the light becomes most intense when the difference of the routes is any multiple of a certain length, and least intense in the intermediate state of the interfering portions; and this length is different for light of different colours.'[134]

Young clearly wants to create the impression that he is quoting himself. This is, however, far from being the case. Apart from the rather problema-

[132] See below, footnote 137.

[133] This is the method which Lakatos in his [1963–4] calls 'exception-barring'. There may well, by the way, have been a further difficulty (this time of a conceptual nature) behind Young's presentation of the principle in this restricted form. (That this, and the difficulty mentioned in the text, did play a role in Young's thought is, of course, a conjecture on my part – Young gives no explicit indication of what lay behind this formulation.) The conceptual difficulty is as follows. As we shall see, Young wants light to be propagated rectilinearly (with just a little fuzziness around the edges of beams). He also describes light as a disturbance in a fluid medium – the disturbances as in the case of sound being in the direction of propagation. Simple mechanical considerations suggest that he cannot then have *general* interference of light. Consider for example the following situation. Two rays of light (from coherent sources) are made to cross one another at an angle of 90°. If, as Young assumes, the disturbances are in the direction of the ray's propagation, then the resultant disturbance at the point of the medium where the rays cross will produce an *elliptical motion*, the exact form depending on the phase relation between light from the two sources (assumed constant) and on the relative intensities. If the two disturbances are of equal intensity and are either in phase or 180° out of phase, then the resultant disturbance will be in a straight line at an angle of 45° with both rays. Thus the disturbance here is in *neither* direction of propagation. (I should add that Fresnel too had some difficulty early on with the question of the interference of non-parallel beams, though this conceptual difficulty was completely solved when he adopted – and provided independent evidence for – the transverse wave hypothesis.)

The associated problem of the apparent failure of light rays to 'interfere' when they cross was by the way a standard problem for all theories of light. It is now often presented as an objection to the corpuscular theory: when two streams of light corpuscles cross, should not at least some of them collide and fly off in directions other than those of the two rays? But given that the speed of light is about 200000 miles per second, then even if one particle in a ray were separated from its predecessor by 1000 miles, around 200 particles would arrive at the eye in a second, easily creating what would on this view be the *illusion* of continuous emission. With particles so far apart the probability of a collision is of course negligibly small. This explanation of 'non-interference' seems to have been well known in the nineteenth century. See, for example, Herschel [1827].

[134] Young [1802*b*], p. 170.

tical extension of the principle to light of different colours,[135] whereas previously light from any two origins could interfere, now all that is claimed is that two portions of the 'same' light (by which he presumably means light from the same origin) will interfere. In this passage Young's requirement that the two portions be of the 'same' light appears as a sufficient condition, but in the published version of his Royal Institution lectures it appears as a necessary condition. There Young writes:

In order that the effects of the two portions of light may be thus combined [to produce 'the alternate union and extinction of colours'] it is *necessary* that they be derived from the same origin...[136]

So the 1807 version of Young's principle states that two 'portions' of light interfere if and only if they have a joint origin (and if and only if they are parallel or nearly so). By 1807 Young had clearly become aware of an obvious objection to his 'law of interference': by allowing any two sources of light (for example, two candles) to illuminate a screen we should cover the screen with interference fringes, which is contrary to experience.[137]

This 1807 version of the law is, however, open to a new and devastating objection of a conceptual kind already hinted at in Brougham's second attack. What is to count as two 'portions' of the 'same' light? If we screen off an arbitrary part of the light coming from a single source, ought we, according to Young's law, to create interference fringes in the light reflected by some convenient wall simply by removing the screen? For the light now coming from the two arbitrarily selected parts of the source

[135] Why should light of a specific wavelength be preserved as a separate entity when mixed with light of other wavelengths? This problem was not solved until Fourier analysis was applied to it in the late nineteenth century.

[136] Young [1807], p. 464; my italics.

[137] It is not, however, according to Fresnel's theory, contrary to fact! Fresnel, sticking to the heuristic of the wave programme, upheld the principle that any two optical disturbances acting at the same point superpose or interfere, and he explained the *invisibility* of the interference pattern in circumstances like the 'two candle' case as due to the fact that the pattern changes position so rapidly (because there is no constant relation between the phases of the various disturbances from the various parts of each source) as to create the *illusion* of uniform illumination.

This is in fact a typical example of Young's failure to stick to his (theoretical) guns when faced by empirical difficulties. In all his later published material (with one exception) he restricts the domain of application of the law of interference instead of trying to explain the refutation as only apparent. The exception is in his reply to Brougham (who makes something like this objection). There Young does half-heartedly try to produce such an explanation. He writes: 'Let us suppose the assertion [that his theory predicts interference fringes whenever two sources of light illuminate an area] true – what will be the consequence? In all common cases the fringes will demonstrably be invisible...' (Young [1804*a*], p. 203). His explanation of their invisibility is however, not very good: 'if we calculate the length and breadth of each fringe, we shall find that a hundred such fringes would not cover the point of a needle...' (*ibid.*). But it is a consequence of Young's theory (together with a few simplifying assumptions) that the breadth of the fringes is proportional to D/d where d is the length of the line joining the two sources, and D is the distance from the mid-point of that line to the point of observation. For a given separation of the sources, the size of the fringes can thus in principle be increased at will by increasing the distance between the sources and the observation screen.

travels different distances before it reaches the wall. We can in fact regard the light emitted from, say, a single candle flame, as coming from any number of sources. If *we* decide to regard it as from two sources, since it is two sources of the 'same' light, Young's law of interference tells us it will produce interference fringes. If we decide to regard it as from one source, Young's law of interference tells us it will not produce fringes. Brougham rightly objects to this degree of 'observer interference':

> It must follow [from Young's theory] that, by doubling the quantity of light on any place, we can cover it with coloured fringes; *or, which is the same thing, that coloured fringes are nothing absolute, but a mere relative idea*...[138]

The next version of the interference principle appeared in Young's 1817 article on 'Chromatics'. By then Young's theory had run into new empirical difficulties. Apart from the fact that half an undulation had to be assumed lost in various and rather ill-defined circumstances,[139] the discovery of polarised light by Malus in 1810 finally convinced Young that there was something fundamentally wrong with the wave theory.[140] Young in fact wrote of the state of the debate between the corpuscular and wave theories in the year of the publication of Malus's discovery (1810):

> All the satisfaction we have derived from an attentive consideration of the accumulated evidence, which has been brought forward...on both sides of the question, is that of being convinced that much more evidence is still wanting before it can be positively decided.[141]

Young was even less happy with his theory in 1815 when he wrote to Brewster:

> With respect to my own fundamental hypotheses respecting the nature of light, I become less and less fond of dwelling on them as I learn more and more facts like those which Mr. Malus discovered...[142]

In view of all this, it is not surprising that Young's 1817 formulation of his law is full of cautious qualifications. The central part of it reads:

> ...When two equal portions of light, in circumstances exactly similar, have been separated and coincide again, in nearly the same direction, they will either

[138] Brougham [1804], p. 99; my italics. (This, by the way, is far from being the only correct argument against Young's theory that Brougham had.) This objection to the wave theory was again dealt with by Fresnel, who showed that light coming from a single light source (a candle say) could indeed be regarded as coming from any number of sources – no interference fringes are observed, however, because there is no constant phase relation between these different sources.

[139] This lost half wavelength, although Young gave it the beginnings of dynamical explanation (see e.g. his [1817], p. 330), remained a bugbear for the wave theory for some time. See also below, p. 149, footnote 173.

[140] Young was not here being a good falsificationist (as might at first appear). Malus discovered only that there are ways of 'polarising' light other than by passing it through a birefringent crystal. (The ordinary and extraordinary rays in double refraction are, of course, each polarised in different planes.) But polarised light was as old as Bartholinus's experiments of 1665. Cf. above, p. 132.

[141] Young [1810], p. 249.

[142] Letter to Brewster of 13 April 1815, published in Young [1855], pp. 360–4.

co-operate, or destroy each other, accordingly as the difference of the times, occupied in their separate paths, is an even or an odd multiple of a certain half interval...[143]

The two 'portions' of light have now for some reason to be 'equal' (though admittedly only if they are of equal intensity is *complete* destruction possible). They also have to be in 'circumstances exactly similar'.[144] Young also states that

in reflections at the surface of a rarer medium, and of some metals, in all very oblique reflections, in diffractions, and in some extraordinary refractions, a half interval appears to be lost.[145]

Moreover,

it is said that according to some late observations of Mr. Arago, two portions of light, polarised in transverse directions, do not interfere with each other.[146]

By 1817, in fact, Young had been reduced to defending 'the general law of interference' for which he had found in 1803 'so simple and so demonstrative a proof' as nothing more than 'a temporary expedient for assisting the memory and the judgement' to be employed only 'hypothetically'.[147]

Thus at each stage Young modified his claims precisely so as to reconcile his principle of interference with known refutations, without providing the principle with extra content which would be exposed to possible refutation. Some modifications were actual logical weakenings;[148] others consisted essentially of weakening the previous version so that it no longer said anything about those cases in which it had been refuted and then adding to it *correct* descriptions of the previously refuting instances.[149] The various versions of Young's principle do then form a classic case of a 'degenerating problemshift'.

[143] Young [1817], p. 287.

[144] This seems to amount to a declaration by Young that he is no longer going to bother rewriting the law in the event of any further refutations, but will simply assert that the law was only apparently refuted because the circumstances of the two beams were not *exactly* similar.

[145] *Ibid.*

[146] It was of course 'said' correctly. As Fresnel was to explain, however, the beams do superpose. They fail to interfere only in the sense that no fringes are produced. (Not this time through lack of coherence, but because the disturbances in the two beams are in different planes.)

[147] Young [1810], p. 249. Young seems then to have been an instance of the psychological law (to which no doubt many counter-instances also exist) that degree of belief in a programme decreases as the programme degenerates. (This psychological law is not, of course, a consequence of the methodology of scientific research programmes.)

[148] This applies to the switch from the [1802*a*] version to the [1802*b*] version (ignoring the addition about coloured light). Such moves are, in the jargon, '*ad hoc*$_1$': see Lakatos [1970], p. 125, footnote 1.

[149] This applies to the switch from the [1807] to the [1817] version. Such moves produce theories whose content is *not* a subset of the content of their predecessor theory, but such extra content as they have consists merely of a substitution for the consequences of their predecessor known to be false of the corresponding 'correct' results yielded by experiment.

3(*c*) *Did the wave optics programme ever progress at Young's hands?*

But, given what we now know about the power and verisimilitude of the wave theory, it would be very surprising if Young's development of it produced no novel predictions whatsoever. I now turn to the question of whether the wave optics programme always degenerated at Young's hands.

According to the standards laid down by the methodology of scientific research programmes, the shift from one theory to the next constitutes scientific progress if the new theory not only explains the facts it was introduced to explain but makes extra predictions as well, some of which are empirically confirmed.[150] The positive weight such confirmed predictions lend to the programme is not at all diminished if some other programme eventually succeeds in explaining in an *ad hoc* way the facts originally predicted by its rival. Thus the fact that, as we saw earlier,[151] the corpuscularists could explain the two results, now usually alleged to have crucially refuted their theory, does not, according to this methodology, preclude the possibility of these results constituting vital positive evidence for the rival wave theory.

According to the methodology of research programmes, a given factual proposition 'supports' a programme (by showing that it is progressive) if

(i) it was predicted by the latest theory produced by the programme (in conjunction with appropriate and experimentally accepted initial conditions);

(ii) the factual proposition is accepted as empirically accurate according to available experimental techniques;

(iii) the factual proposition was not used in conjunction with some previous theory in the programme to *construct* the theory which entails it.

Did either of the two experimental results cited by Young's commentators as crucial contributions to the wave–corpuscle debate satisfy these three requirements?

There is in fact some doubt whether the results as usually described satisfy even the basic requirement (i) – i.e. there is some doubt whether they could be predicted on the basis of Young's theory. This doubt stems from some vagueness in Young's account of the basic propagation of light. In the case of the two slit experiment, for example, although Young's theory predicts what would happen *if* two nearly parallel beams of light affect the same portion of the observation screen, it is not clear whether the theory predicts that any portion of the observation screen *is* affected by two beams of light. Whether it does or not depends on the account Young's theory gives of how light is propagated after being admitted through a single slit or aperture. Let us look into this.

[150] See Lakatos [1970], p. 118. This formulation actually incorporates a modification of Lakatos's original proposal due to Zahar; see Zahar [1973]; reprinted below pp. 211–75.
[151] Above, pp. 127–30.

As I mentioned earlier,[152] one of the fundamental objections against the wave theory urged by Newton was its difficulty in explaining the rectilinear propagation of light. Newton wrote in 1672:

To me the fundamental supposition itself seems impossible, namely, that the waves or vibrations of any fluid can like the rays of light, be propagated in straight lines, without a continual and very extravagant spreading and bending every way into the quiescent medium where they are terminated by it.[153]

And in the *Opticks* he asked:

Are not all Hypotheses erroneous, in which Light is supposed to consist in Pression or Motion, propagated through a fluid Medium?...if it consisted in Pression or Motion...it would bend into the Shadow. For Pression or Motion cannot be propagated in a Fluid in right Lines, beyond an Obstacle which stops part of the Motion, but will bend and spread every way into the quiescent Medium which lies beyond the Obstacle.[154]

Huygens had tried to produce a wave theoretical explanation of rectilinear propagation,[155] but this was completely *ad hoc* (it was in fact completely circular) and involved assumptions about the ether which were mechanically totally unacceptable.[156] Newton justifiably ignored it. Young did almost nothing to improve the wave theory's plight in this respect.

Young shared the basic position of the corpuscularists on the propagation of light. Light was basically rectilinearly propagated, and any deviation from a rectilinear path had to be explained as an exception to the general rule. Thus, for example, when Young needed two 'portions' of light whose interference would explain the fringes outside the shadows of opaque objects, he plumped for one 'portion' propagated rectilinearly past the object and another 'portion' passing nearer the object and 'inflected' from its naturally rectilinear path.[157] In fact, Young's basic method in explaining interference and diffraction effects (this was also the heuristic principle followed by Fresnel in his very early work) was to forget about the wave character of light except where two light beams crossed. Light coming from two sources was treated as consisting of rectilinear rays (left theoretically uninterpreted, as it were) but then, when the 'rays' crossed, their wave character was remembered and thus their interference explained.[158] Nevertheless Young did in at least one place in his three

[152] Above, p. 131. [153] Newton [1672]; see the reprint in Cohen [1958], p. 121.

[154] Newton [1730], p. 362. [155] See Huygens [1690], pp. 19–21.

[156] This is a rather controversial assertion, for which I intend to argue elsewhere.

[157] See for example Young [1804*a*], p. 180.

[158] This heuristic principle has lived on as a pedagogical device. In most modern textbooks on optics, the chapter on interference precedes the chapter(s) on diffraction. In the interference chapter, light is considered as consisting of rectilinearly propagated rays up till the point where two or more rays interfere, when the ray suddenly becomes wavelike. The justification for this procedure then comes in the diffraction chapter where it is shown that, although all light is diffracted, in most circumstances the assumption that light is rectilinearly propagated is a good enough approximation. This justification was of course not provided by Young, but only later by Fresnel.

major early papers seem to be attempting a wave theoretical justification of this procedure, i.e. to be attempting a wave theoretical explanation of the rectilinear propagation of light. He argues there that for an undulation in an elastic medium to

> continue its progress to any considerable distance, there must be in each part of it a tendency to preserve its own motion in a right line from the centre; for if the excess of force at any part were communicated to the neighbouring particles, there can be no reason why it should not very soon be equalized throughout, or, in other words, become wholly extinct, since the motions in contrary directions would naturally destroy each other.[159]

As far as I can understand this passage, the explanation it provides of rectilinear propagation is unacceptable (it is *ad hoc* in the main sense specified by the methodology of research programmes);[160] for it argues *back from* the 'fact' of rectilinear propagation to the (mechanically unrealistic) assumption that the particles of an elastic medium, when agitated, do not pass on some of their momentum to all particles in contact with them, but only (or in great part only) to those particles on the extension of the straight line joining the agitated particle to the original source of the disturbance.[161]

Notice that Young talks only of a 'tendency' for the light disturbance to maintain a rectilinear course. Young decided that a slight divergence from rectilinear propagation could be a quite useful explanatory instrument.[162] Thus he gives in his [1802*a*] paper the following account of what happens when light is admitted through an aperture. (This is in fact the only detailed statement on this part of the theory in his three major papers on light. It occurs as 'Proposition III' immediately after the passage just quoted.)

> *A Portion of a spherical Undulation, admitted through an Aperture into a quiescent Medium, will proceed to be further propagated rectilinearly in concentric Superficies, terminated laterally by weak and irregular Portions of newly diverging Undulations.*[163]

There is no danger of this divergence ruling out a wave theoretic explanation of rectilinear propagation because the divergence is 'weak

[159] Young [1802*a*], p. 149.

[160] See particularly Zahar [1973], p. 102.

[161] Young in effect immediately admits that this is a mechanically (or 'mathematically') unsound assumption, but simply states that we can hardly expect any more, given the contemporary state of hydrodynamics: 'It may be difficult to show mathematically the mode in which this inequality of force is preserved, *but the inference from the matter of fact* appears to be unavoidable; and while the science of hydrodynamics is so imperfect that we cannot even solve the simple problem of the time required to empty a vessel by a given aperture, it cannot be expected that we should be able to account perfectly for so complicated a series of phenomena as those of elastic fluids.' (Young [1802*a*], pp. 149–50; emphasis supplied.) Perhaps in view of this Young would have been better not to insist so loudly that he had solved the problem of reconciling the wave theory and light's (apparent) rectilinear propagation.

[162] This divergence is, however, to be in addition to, and *not* a replacement for, the 'inflection' light undergoes in the neighbourhood of opaque bodies.

[163] Young [1802*a*], p. 151.

and irregular'. But how does this divergence operate? In the text to the above 'Proposition', Young uses the 'weak and irregular...diverging Undulations' to explain the apparent narrowing of the beam of light, and hence the apparent enlargement of the shadows of the two sides of the aperture, reported by Newton, following Grimaldi.[164] This is alleged to depend on the 'divergence...diminish[ing] the force' of the 'principal' (rectilinearly propagated) rays, particularly near the edges of the beam. The divergent disturbance does not create sensible light (Young says it does not 'add materially' to the 'force' of the 'dissipated light') and thus the beam is narrowed.[165] Young immediately admits, however, that 'in other circumstances the lateral divergence might appear to increase, instead of diminishing, the breadth of the beam'.[166] Thus in these 'other circumstances' – the diverging light *is* 'sensible', i.e. sensible light does get into the geometrical shadow. But there is not a word about what these other 'circumstances' are.

Young is therefore in the happy position of being able to account, without further effort, for *two quite different* possible results of the two slit experiment – *which the actual result will be, he can tell only by performing the experiment.*

(i) He can assume that the divergent disturbance produces no sensible light. In this case his theory predicts that the two slit pattern will consist of exactly two illuminated bands, adjoined and separated by unilluminated bands, one illuminated band being due to light emerging from one slit, the other illuminated band being due to light coming through the other slit. The light from both slits is propagated more or less rectilinearly, the two illuminated bands being in fact slightly narrower than they would be were light propagated in *strictly* rectilinear fashion. No point of the observation screen will be illuminated by sensible light from both slits and so no interference will occur.

(ii) Young can assume that the two slit arrangement constitutes one of those 'circumstances' in which light diverging into the geometrical shadow is sensible. In this case various points of the observation screen may be illuminated by light from both slits. Thus, according to the interference principle, a series of interference fringes may be produced.[167]

Thus Young's theory explains the existence of fringes in the two slit experiment only in an *ad hoc* way – Young reads off from the result of the experiment that this is an instance of case (ii). In other words, Young's theory does not predict the result of the two slit experiment. All Young can justifiably claim of the result is that it establishes which of his two kinds of exception to rectilinear propagation operates in these experimental circumstances.

[164] See above, p. 122, footnote 68. [165] Young [1802*a*], p. 152. [166] *Ibid.*

[167] I say 'may be produced' because, as we have seen (above pp. 138–42), Young very soon gave up the claim that *any two* overlapping beams interfere.

The case of the 1803 diffraction experiment is similar. Again it is a question of first observing fringes and then deciding which two 'portions' of light interfered to produce them, rather than of predicting the existence of fringes. For example, the well-known fringes outside the geometrical shadow of an opaque object are alleged by Young (after the event, of course) to be produced by the interference of two 'portions' of light. One of these is the 'portion' passing rectilinearly by the object. The other 'portion' Young had different conjectures about in different papers – he sometimes alleged it was 'inflected' by the object and sometimes that it was reflected from its sides. As for the fringes *inside* the shadow of a narrow object held in a diverging cone of light, neither of the two required portions of light could, in this case, be rectilinearly propagated. Both portions would, on Young's assumptions, somehow have to be bent. Young in fact assumes (and I stress again that this is only *after* observing the fringes) that this is one of those 'circumstances' in which the divergence from rectilinear propagation (on both sides of the object) produces 'sensible' light. It is these two 'divergent' portions which interfere to create the internal fringes.

It cannot then be the frequently cited qualitative features of either of Young's two famous experimental results which provide the 'novel facts' which the methodology of research programmes requires Young's theory to predict if it is to constitute progress. For not only were these features already known (as I stressed earlier), they were not predicted by Young's original theory and subsequent versions only dealt with them in an *ad hoc* way (i.e. in a way which violates requirement (iii), above, p. 143).

A further possibility remains open, however.[168] While I have shown that Young's theory did not characterise the circumstances in which two 'portions' of light would overlap so as to produce fringes, and hence that that theory did not predict the existence of fringes, the theory may still predict what happens *if there is overlap*.[169]

There is, of course, a quantitative aspect to Young's principle of interference. Two 'portions' of the 'same' homogeneous light (i.e. light from a single source and of a given wavelength, λ) will destructively interfere if their path difference is an odd number of half wavelengths. So, while Young must, in various circumstances, wait to observe the fringes before being able to say that interference is occurring, once he knows that there is interference, and once he has identified the two 'portions' of light which are interfering, he ought to be able to predict some of the quantitative features of the fringe pattern, such as the spacing between fringes. For example, the first dark fringe on either side of the central fringe ought to

[168] I should like to thank Paul Feyerabend who, in criticising an earlier draft of this paper, pointed out that I had omitted to consider this possibility.

[169] This would be very much in line with the heuristic procedure outlined above, p. 144, of considering beams of light as consisting of rectilinearly propagated rays until they cross, at which point they assume wave characteristics.

occur at points where the difference of the paths of the two 'portions' of light which produce it is $\frac{1}{2}\lambda$. Geometry will then give the distance between the centre of the pattern and the centre of the first dark fringe as a function of λ. Provided that Young had an independent method of arriving at values of λ for various kinds of homogeneous light, it seems that this fringe spacing forms a prediction of Young's theory.[170] It is, moreover, a prediction which, at least in the case of the two slit experiment, is not matched either by the corpuscular theory or by 'background knowledge'.

In fact Young did arrive at values of λ for the various spectral colours independently of these diffraction results through his reinterpretation of Newton's results on the colours of thin plates. [I should perhaps digress a little here and make a few remarks on Young's theory of thin plates, for this is sometimes considered one of Young's important achievements. Newton's theory of this phenomenon implied that, in thin plates of varying thickness, the first dark band occurs where the distance from the first surface of the plate to the second is just sufficient for the light particle to change from a fit of easy transmission (in which it must have been at the first surface in order to cross it) into a fit of easy reflection and back into a fit of easy transmission so that it is transmitted rather than reflected at the second surface. Thus no light is reflected back to the eye where the thickness of the plate is D, where D is the 'interval' between two successive fits of the same nature. (Similarly the first bright band occurs where the thickness of the plate is $D/2$.) Young, on the other hand, ascribed the dark bands to the destructive interference of the light reflected at the first surface and the light which was transmitted at the first surface, reflected at the second and then re-transmitted at the first.[171] This meant that for Young the first dark band on a plate of varying thickness occurs where the thickness is $\lambda/4$ – for then the second beam will have travelled a total distance of $\lambda/2$ further than the first beam, and beams originally in phase but with a path difference of $\lambda/2$ will destructively interfere. I would say that the scores which these two theories chalked up for their respective programmes were about equal. On the one hand, Newton had produced the *first* decent explanation of the phenomenon (and his theory can predict, for example, all the radii of the successive rings in the 'Newton's rings' experiment[172] having measured the radius of just one), whereas

[170] Alternatively, one could regard this as a prediction of the value of λ (which can then be checked against values arrived at by other means) on the basis of the measured distances between the fringes. (It is, in fact, very often the case that the values of the free parameters of a theory can be 'read off' from any one of a group of n experimental results and then the theory predicts the other $n-1$ results.)

[171] See e.g. his [1802*a*], pp. 160–2.

[172] This consists of allowing light to fall at (approximately) normal incidence on a plano-convex lens lying convex side down on a glass plate. This provides a plate (of air) of varying thickness. It produces a series of concentric coloured rings – the thickness of the air plate corresponding to each ring can then easily be calculated (it clearly depends on the radius of curvature of the lens).

Young simply re-interpreted Newton's results and theory (his values of the various wavelengths are obtained from Newton's values for the various intervals of fits by a linear transformation). On the other hand, Young was able successfully to resolve the major problem which this re-interpretation presented to him.[173] Each theory also faced one rather obvious major conceptual difficulty. On the one hand, it looked as if Newton would have to make the interval of fits dependent on the obliquity of the incident rays (a very unnatural assumption) to account for the fact that the radii of the rings increase with increasing obliquity. On the other hand, it was clear that Young's theory could not account for the (more or less perfect) blackness of the dark rings, for the two beams whose interference he alleged caused them were of *very* unequal intensities; and it was anyway clear that a full wave theory of the phenomenon would have to take account of many more beams than two (beams 3 times, 5 times, etc. internally reflected and then transmitted at the first surface), but it was not clear how taking these extra beams into account would affect the prediction.]

Returning to the main problem, Young could obtain values of λ for various kinds of light from the thin plates experiments. If Young did make quantitative predictions, based on these values of λ, about the fringe positions in his diffraction and interference experiments, and if he experimentally confirmed them, then he would, according to the standards of the methodology of research programmes, have provided the wave programme with new and dramatic confirmation. In view of the apparent fact that the corpuscular programme had no 'novel facts' of its own with which to match this, it would then seem that the methodology of research programmes, if not exactly leaving us with the same historical problems as before,[174] would not have led to much significant historical advance. For even when we analyse the intellectual situation in optics in the early

[173] This problem was that the centre of the ring system viewed by reflection was, according to Newton, black. But there the plates were assumed to be in contact and one would expect no path difference between the two (?) beams and therefore *constructive* interference. Young was forced to assume *ad hoc* that the phase of one of the beams was altered by half a wavelength (Young actually speaks of one of them *losing* half a wavelength) in being reflected. He assumed that this was the beam reflected at the first surface, which underwent a dense-to-rare reflection, and that in fact a half wavelength is 'lost' in all dense-to-rare reflections, but not in rare-to-dense ones. (There is obviously some conceptual difficulty here, for the two surfaces are supposed *in contact* and so it is not clear that it makes sense to speak of *two* reflected beams. Also, it is hard to know how Newton and Young could be so certain about what happened at the centre of the pattern. For even in the best experimental conditions the two plates will not really be in contact, there will always be dust particles or something similar interposed between them.) Anyway, Young based on this explanation the conjecture that if the lens and the plate were of different optical densities, and if a substance of intermediate density were interposed, both reflections would be of the same nature and so there would be no loss of half a wavelength, and the centre of the pattern would thus be white. Young claimed to have confirmed this prediction.

[174] As I shall show later, the corpuscular programme had certain advantages (in terms of heuristic strength) over the wave programme. See also p. 152, footnote 181 below.

1800s in its terms, Young's contemporaries would still seem to be displaying an embarrassing lack of interest in his achievements.

I shall now reconsider Young's two famous experiments in an attempt to decide whether Young's theory did indeed make such novel predictions, and whether Young experimentally confirmed them.

First, a methodological preliminary. It is never difficult to dress up what are in fact *ad hoc* manoeuvres as genuine predictions of a theory. Indeed, one cannot tell merely from looking at the deductive structure of a test of a theory, whether or not one is dealing with a genuine prediction. For the scientist has only to disguise (or omit to mention) the fact that some initial condition or 'auxiliary hypothesis' assumed in the test, rather than having been independently ascertained, was 'read off' from the experimental result. In such cases the initial condition or auxiliary hypothesis concerned is arrived at, not on the basis of some *independent* experimental technique, but by first looking at the experimental result and then working backwards to find some assumption which when fed into the theory gives the already known result. This is a consistency proof rather than a prediction![175]

Let me give a very simple example. Suppose our theory is that all *A*s are *B*s. If we have an independent method of deciding whether some given object x is an A, then we can use the theory to *predict* that x is a B, which prediction we can then check against experiment. It is possible, however, that our only evidence for x being an A was the *observed* fact that x is a B (together with our theory that all *A*s are *B*s). In that case, although we could dress up the situation in predictive form 'All *A*s are *B*s; x is an A, therefore x is a B', we intuitively would not count the theory as having made a genuine prediction.

A second, slightly less simple example is, I claim, provided by Young's procedure in the case of his 1803 diffraction experiment. Young could have predicted the spacing of the fringes in this experiment had he had an independent method of identifying the 'portions' of light which do the interfering at the various points of the observation screen. But Young had no such method; rather he *worked back* from the observed fringe spacings (and the known value of the wavelength) to an identification of the 'portions' of light whose interference produced the fringes and the paths these 'portions' took. In other words Young was *not* in a position to say: 'These fringes are caused by the interference of these two portions of light; at this point the path difference of the two portions is an odd multiple of $\lambda/2$; therefore a dark fringe must be observed at precisely this point.' His method is rather: 'Here is a dark fringe. It must have been caused by the interference of two portions of light which have crossed

[175] That these are the important kinds of 'predictions' which ought not to count as giving genuine support to a theory was originally pointed out by Zahar in his [1973].

paths which differ by an odd multiple of $\lambda/2$. Which two portions and which paths might they be?'

It is clear that this *was* Young's method of proceeding, for he writes, in his account of this 1803 experiment:

If we now proceed to examine the dimensions of the fringes under different circumstances, we may calculate the differences of the lengths of the paths described by the portions of light...concerned in producing these fringes; ...[176]

Had Young been able independently to identify the sources and paths of the two interfering 'portions', then the path differences for various points on the observation screen would, of course, have been given simply by geometry. It seems, then, that the quantitative aspects of this 1803 diffraction experiment cannot provide the required 'novel fact'. We are left then with the two slit experiment.

Here there is a 'natural' conjecture about the sources and paths of the two beams interfering at any point. This conjecture is that the sources are the centres of the two slits and the paths are straight lines from the sources to the point of interference. There seems little doubt that Young does make this conjecture, for he writes that the two apertures 'may be considered as centres of divergence, from whence the light is diffracted in every direction'.[177] This 'natural' conjecture together with the interference principle yields, as we now know, predictions about the fringe spacing in the two slit experiment which are correct, at least as first approximations.[178]

Now had Young used this conjecture together with his interference principle to predict the quantitative features of the two slit result, and had he confirmed this prediction, it would still seem a debatable issue whether this prediction should count as a success for Young's fully fledged wave theory. For however 'natural' the conjecture about the paths of light may be, it is inconsistent with Young's wave theory. As we have just seen,[179] what Young says happens when light passes through a single aperture is that it is propagated essentially in a beam terminated by straight lines, but that there is a 'divergence' especially near the edges of the beam. Given that the two slit case is one in which interference fringes are produced, it follows from Young's theory that this is a 'circumstance' in which this 'divergence' produces sensible light. But this 'divergence' emanates from at least every point on the edge of the beam – presumably in a variety of directions. This means that the total light arriving at any

[176] Young [1804*a*], p. 181.

[177] Young [1807], p. 464.

[178] This differentiates the two slit case from that of the 1803 experiment where the analogous 'natural' conjecture – that the sources of the two 'portions' which are to interfere to produce the inner fringes are the two sides of the diffracting object – leads to predictions which are rather far from being correct. Fresnel later explained the narrow object diffraction pattern as caused by the 'interference' not just of waves emanating from the edges of the object but from all unobstructed parts of the original wave front.

[179] Above, p. 145.

point of the observation screen is a rather complicated compound of light rays which have taken a variety of paths in arriving there.

This raises rather a tricky methodological problem. Let us assume that the methodology of research programmes is correct in claiming that it is a theory's dramatic unexpected predictive successes which constitute the vital evidence in its favour. Are such predictions still to count as important successes for a theory when the derivation of these predictions from the theory requires auxiliary assumptions which are *inconsistent with other parts of the theory*? Fortunately we can decide whether or not Young demonstrated the superiority of the wave theory without answering this question. For there are, I claim, sufficiently many suspicious aspects of Young's account of the two slit case to support the belief that he never performed the experiment!

It is now known that, together with some simplifying assumptions, the interference principle yields, by some very elementary considerations, the prediction that the two slit interference pattern will consist of a series of equally spaced fringes – the centre of one bright fringe being a distance $\lambda D/d$ from the centre of the next bright fringe (where D is the distance of the screen with the two slits from the observation screen, d is the distance between the two slits and λ is, as usual, the wavelength of the light involved). This is more or less what is observed, at least near the centre of the pattern.

Now Young himself never makes any attempt to *derive* any prediction about the two slit case, but he does produce, out of the blue, some predictions which conform to the above approximations. He writes that the light on the observation screen is 'divided by dark stripes into portions nearly equal' and that the fringes become wider 'as the surface is more remote from the apertures (i.e. as D increases) and 'wider in the same proportion as the apertures are closer to each other'[180] (i.e. as d decreases). Assuming, as we are, that Young could independently support all the required premisses,[181] these predictions would, if confirmed, be good

[180] Young [1807], p. 464.

[181] If we further assume for a moment that Young did derive and confirm explicit predictions about the fringe spacing, then he would have shown the wave programme to be progressive. In that case we might again have to resort to 'external' considerations to explain Young's theory's lack of success; but the methodological appraisal in which we have indulged still would not have been historiographically pointless. For it would have shifted and more closely characterised the problem, which would now be the very specific one of why Young's fellow scientists failed to spot the importance of these *particular* aspects of this *particular* experimental result. Funnily enough this problem (which I am about to show does not in fact arise) has a rather plausible solution: Young's fellow scientists may not have known of the result. Unlike his 1803 diffraction experiment which was recorded in his [1804*a*] paper, published in the *Philosophical Transactions* and originally given as a Bakerian lecture, the account of the two slit experiment is rather hidden away in Young's [1807] *Lectures* delivered to an 'audience of fashionable ladies' (Brougham's phrase) in 1802–6. Although his [1807] book did eventually achieve a certain popularity, it seems hardly to have been read at the time. Furthermore, it was deliberately aimed at a popular audience (see Wood [1954], chapter vi) and so was perhaps unlikely to be read

enough to show the empirical superiority of the wave theory according to norms supplied by the methodology of scientific research programmes. However, as I said, there is evidence that Young never performed the experiment.

The first piece of evidence is that Young never explicitly claims to have performed the experiment. His usual practice is to head his accounts of experiments with the word '*Experiment*' or '*Observation*' and to give a full account of how he carried out the experiment.[182] He never describes the two slit case as an experiment, however, referring to it as merely 'the simplest case' of interference.[183]

The most worrying feature of Young's account of this 'simplest case' of interference is that he describes it as involving 'a screen in which there are *two very small holes or slits...*'[184] Assuming, as seems reasonable, and as do those few commentators who remark on this fact, that Young means by holes, round pinholes, then great doubt is thrown on whether Young ever performed the experiment. For the result he describes (dark and light *stripes*) is by no means obtained if the two apertures in the screen are round pinholes.[185] Furthermore it is more difficult to observe any interference effects when circular apertures are used.[186]

But perhaps still more worrying than this detail which Young includes are the many details he leaves out. First of all, and again contrary to his usual practice, he gives no numerical details whatsoever either of the experimental set-up or of the result. Also, he talks simply of allowing a 'beam of homogeneous light' to fall on the screen with the 'two slits or holes' without remarking at all on the source of this beam or the distance of the source from the double slit. Neither does he give any indication of

by scientists. Young himself never stressed the importance of the result: he never returned to it in his later writings (as he did to many of his other results) and even in his *Lectures* it is not given any emphasis. (This would be explained if my conjecture below about his performance of the experiment is correct.) None of his immediate commentators, Pettigrew, Whewell and Peacock, as much as mention the two slit experiment.

[182] See for example his [1804*a*], p. 179, where he heads the account of an experiment '*Exper. 1*' and begins his account 'I made a small hole in a window shutter, and covered it with a piece of paper which I perforated with a fine needle...'.

[183] Young [1807], p. 464.

[184] *Ibid.*

[185] Rather, what one gets is interference between the two circular diffraction patterns centred on the so called 'Airy disks' produced by each aperture separately. (This assumes that the original source of light was also circular.)

[186] Mach, in his [1926], pp. 147–8, pointed out the fact that the visibility of the interference effects is greatly enhanced when slits are used. He failed to see any difficulty in Young's account, however, by curiously misreading it. Mach refers to the account in Young's [1807] but somehow omitted to read there the 'or slits'. He in fact attributes the discovery of the two slit experiment to Fresnel. Mach writes: 'Young...illuminated two small apertures in close proximity by light from a single opening in a window shutter [in fact Young does *not* mention this window shutter]...Fresnel however replaced the two round apertures by two narrow vertical slits, which effected a great improvement.' Young had managed thoroughly to confuse Mach, by the way, for Young states (*ibid.*) that he is considering the case of a 'beam of homogeneous light' which he could hardly have obtained directly through a hole in a window shutter!

the separation and dimensions of the 'holes or slits', nor of the distance of the double slit from the observation screen.

Yet had he performed the experiment (*and still more had he wanted others to be able to repeat it*) he could reasonably be expected to have mentioned each of these factors. For instance, had he successfully performed it he must surely have remarked that for the experiment to succeed the original source of light has to be very small and its distance from the double slit very large. (We know in fact that if the difference of the paths from the two edges of the source to either of the slits is greater than about $\lambda/4$ the fringes will not in general be observable![187]) The width of the slits and the relation between this width and the width of the screen separating the two slits are also not specified by Young and yet are vital to the successful performance of the experiment. Only if the slits are very narrow and the distance between them very small will fringes be observable. Moreover, there are opposing tendencies here – as the slit separation increases, the *number* of dark fringes increases, but the *sharpness* of the fringes decreases.[188]

Furthermore, neither Young's theory of light nor the analogy between light and water waves, which Young always had very much in mind, gave him any reason to suspect that any of the features I just mentioned are important for the successful performance of the experiment. For example, as we saw earlier,[189] Young had no clear ideas on what was later called coherence which (had he had them) would have both pointed to, and explained the necessity for, a small source at a great distance from the double slit. Also his theory gave him no reason to suspect that narrowing the slits would lead to a greater 'spreading' of light from each of the slits and hence to greater definition of the fringes, for, according to Young's single slit theory, 'weak and irregular' undulations diverge from the sides of *any* beam.[190]

As for the interference of water waves, Young, we know, gave 'ripple tank' demonstrations of the pattern produced when a train of water waves meets a barrier in which there are two apertures. He always had water waves very much in mind when considering light, and he in fact prefaces his account of the optical two slit case with an account of the analogous water wave case. But, with water waves, interference is observable no

[187] This is explained in the 'classical' wave theory as due to the fact that each point of the slit source produces its own interference pattern. These patterns are displaced slightly relative to each other and hence their superposition may mask the individual interference effects.

[188] To give an example, if the double slit is separated from the observation screen by 70 cm, interference fringes will not normally be visible if the slits are separated by as much as ½mm. Without the aid of lenses the two slit interference effects are never very sharp since where the overlap occurs the light supplied by each slit separately is not very great. *Remember also that Young was observing the effects on a screen, a method which renders the effects much less easily observable than the later method (invented by Fresnel) of observing the fringes directly with the aid of an eye-glass.*

[189] See above, p. 140.

[190] See above, pp. 145–6.

matter what the form of the original disturbance or its distance from the double slit, and no matter what (within wide limits) the distance between the two slits.

Thus, there were no theoretical considerations which would have directed Young to the successful performance of this experiment. Moreover, the dearth of details in Young's account make it seem unlikely that Young ever did successfully perform it, and certain that he did not give sufficient information about the conditions of the experiment to ensure its repeatability by others.[191] My claim is supported by the otherwise rather strange (and certainly unargued) assertion by Houston in his text book on optics that 'the real obstacle [to the acceptance by Young's contemporaries of the crucial nature of the two slit experiment] was the difficulty of repeating the experiment'.[192]

Houston's remark suggests that other scientists actually attempted to 'repeat' the experiment and failed. Unfortunately (for it would obviously strengthen my case) I have not been able to find any evidence for this.[193] It is clear, however, that people were confused about Young's claims. Certainly anyone attempting to repeat it (or rather 'repeat' it, since it is likely Young never performed the experiment) following the account by Arago, probably the most famous optical scientist of the time, would scarcely have succeeded. For Arago writes that Young's experiment involves introducing 'sunlight into a dark room through two little holes not very far apart'.[194] But, of course, no interference fringes would be produced in these circumstances because of the lack of a constant phase relation between disturbances emanating from the two holes. For the experiment to stand any chance of success the light must be introduced first through a single slit (or small hole) before impinging on the double slit.

If my claim that Young never successfully performed the two slit experiment is correct, then it explains Young's reticence about the experiment. He never gave the two slit 'case' any prominence, hiding it away in his *Lectures* of 1807 and never returning to it in later works (unlike many of

[191] In a very similar context, Arago, in discussing Grimaldi's claim to have observed interference effects, makes it clear that we ought not to be at all surprised if scientists are not convinced that some experimental discovery has been made, if the result is not very clear and if the conditions for repeating the experiment are not given. He writes (in his [1859], p. 420): 'Grimaldi had long ago (before 1665) formed some notion of the action which one beam of light may exercise upon another: but in the experiment which he cites the action was but obscurely manifested; and besides this, the *conditions* which were essential to its production had not been pointed out, and thus no other experimenter followed up the inquiry.' What I am suggesting is that precisely similar considerations apply to Young and his two slit experiment.

[192] Houston [1938], p. 135.

[193] It seems unlikely that I shall find any such evidence. For only in the relatively short period between 1807 and the 1820s is an account of such a failure to obtain interference fringes likely to have been published. For after the acceptance of Fresnel's work, it was 'known' that interference fringes were the 'correct' result.

[194] Arago [1819]. The passage occurs on pp. 232–3 of the reprint in Fresnel's *Oeuvres Complètes*.

his other experiments, including the 1803 diffraction experiment). Indeed, it should be noted that, although it is probably the two slit experiment for which Young is now most famous, Young himself made so little fuss of it that none of his early commentators, Whewell, Peacock or Pettigrew, give it the slightest prominence.

4 *Was Young's work really ignored? The heuristic superiority of the corpuscular programme* circa *1810*

I think I have shown then that, when the intellectual background to the reception of Young's work is appraised using the criteria supplied by the methodology of research programmes, the reactions of Young's fellow scientists do not seem so strange. First of all, by this appraisal, instant rationality is exposed for the myth it is[195] – the corpuscular programme was certainly not instantly and once and for all made untenable by Young's experimental results (though the programme had so far failed to produce anything but *ad hoc* explanations of these and other, already well known problematical results). Secondly, the wave programme also faced some experimental difficulties which it had so far failed to resolve except in an *ad hoc* way.

Moreover, the reception of Young's work appears even less strange in view of the fact that from the point of view of heuristic strength (or perhaps heuristic definiteness) it was the corpuscular programme which seemed the stronger at the time Young wrote. It is this fact that I intend to establish next.

It is clear from what has been said that (as with 'the' wave theory) there was no such thing as the corpuscular theory of light, there was rather a series of such theories. Each of them implied that light consists of corpuscles emitted from luminous objects. The various corpuscular theories of light were related by a heuristic principle which was supplied by particle mechanics. A fundamental assumption behind all corpuscular theories was that the corpuscles of light obey the ordinary (*and already known*) laws of particle mechanics. This assumption played a wider programmatic role in addition to its descriptive one. For it both specified the problems the corpuscular programme faced and guided their solution. Thus, given that the particles of light are to obey the same laws as ordinary material particles, it follows from Newton's first law that any deviation from rectilinear propagation on the part of the light corpuscles was caused by some net external force (or forces) acting upon the light.[196] The fact

[195] See Lakatos [1970], pp. 154ff.

[196] The corpuscular programme thus seems to be one of those interesting cases (for other alleged examples, see Urbach [1974]) in which the heuristic is more important in characterising the programme than the 'hard core'. The natural 'hard core' assumption of the corpuscular programme would be that light consists of bits of *matter* thrown out by light sources. Yet many of the protagonists of the programme would have been

that the corpuscles producing different colours are refracted by different amounts meant, given the basic corpuscular assumptions, and given that only one set of forces could presumably be operative at the surface of a refracting body, that the corpuscles had to have distinguishing features which explained their different refractions.[197]

Moreover, the phenomenon of partial reflection showed that even particles incident on a transparent body at the same angle and velocity are acted upon differently by the same set of forces, some particles being refracted (by the attractive forces) and some reflected (by the repulsive forces). Newton explained partial reflection using his famous theory of fits of easy reflection and easy transmission, which he had previously used to explain the phenomenon of 'Newton's rings'.[198] Biot in turn explained these 'fits' as due to the light particles having repulsive and attractive 'poles', the particles being in uniform rotation (hence one could talk of a particle's 'phase').[199] The polarity of the corpuscles could also be used to explain polarisation effects.

Thus, by the early years of the nineteenth century, the corpuscularists had built up for themselves a quite complicated conceptual apparatus. Once, however, the relevant parameters had been assigned to the particles and the relevant forces to the opaque or transparent bodies, the path of any given ray of light was to be completely specified by known and simple particle mechanics.[200]

sceptical about this assumption. But Newton (and later corpuscularists) applied particle mechanics to optical phenomena (Newton did this, for example, for refraction). But, if mechanics is to apply, the corpuscles must have mass. Thus one can pursue a programme without believing in its hard core.

[197] The 'natural' candidate for this job was *mass*. Newton seems, however, to have been aware of the difficulties to which assigning the particles different masses leads. The principal difficulty is that, given that *in vacuo* the velocity of any kind of light is the same, the force which launches the light particles from a luminous source would have to be different according to the particle's mass: see Bechler [1974]. Bechler does not mention that this implausible assumption about a differential emissive force is made more plausible by Laplace's suggestion that lots of particles are thrown off by a luminous body at *different velocities*, but that our senses register only those having velocities within a certain range (cf. above, p. 128). Thus there may be several emissive forces each acting constantly on the particles, *no matter what their mass*, but only those particles to which a given force gives velocities within this range will be registered by the eye as light. Which kind of particles these are will differ with different emissive forces. Whatever may be the case, as late as 1827 Sir John Herschel records that it is part of the corpuscular theory that the 'particles differ from each other...in their actual masses, or inertia' (Herschel [1827], p. 439). Many attempts were made to detect light's mass (e.g. by Michell and by Bennett). Quite sophisticated attempts to work out the consequences of light's possessing inertia were also made by various corpuscularists (e.g. by Soldner who was concerned with the effect on light of the gravity of highly massive bodies, and by Laplace who worked out the conditions under which 'black holes' might exist). See also footnote 201.

[198] For the theory of fits see, e.g., Newton [1730], pp. 278–88.

[199] See Biot [1816], vols. 3 and 4.

[200] There is no doubt that it was Newton who almost singlehandedly invented the corpuscular programme, despite the fact that, as is well known, he deliberately refrained in the *Opticks* from explicitly asserting that light consists of corpuscles. It was Newton who

The problem with the corpuscular programme was that it lagged behind the facts: none of the modifications the corpuscularists made to their assumptions led to new predictions which were subsequently empirically confirmed. Each new diffraction effect, for example, far from being predicted by corpuscularists, required a rearrangement of the light-bending forces they assigned to matter. This, however, does not detract from the fact that, faced with any experimental result, the corpuscularist had a clear idea of how to go about explaining it (even, and indeed especially, if the result was anomalous). The presence or absence of light in any particular region, for instance, was to be explained as the result either of direct (rectilinear) propagation, or of some net force emanating from material bodies and acting on the light corpuscles.[201]

As I remarked earlier,[202] there was also, in 1810, such a thing as the wave optics programme. As in the case of the corpuscular programme, the wave programme's aim – the explanation of all optical phenomena in terms of the mechanics of an all-pervading elastic medium – also provided its heuristic. *Because of the comparatively less developed state of the mechanics of elastic media relative to the mechanics of rigid particles, the heuristic of the corpuscular programme was, however, rather more definite than that of the wave programme.*[203]

suggested that reflection, refraction and 'inflection' are all referrable to one set of forces emanating from matter and acting on light. Proposition IX of Book II, Part III of his [1730], for example, reads: '*Bodies reflect and refract light by one and the same power, variously exercised in various circumstances.*' (Notice, by the way, that this is a '*Proposition*' not a '*Query*'.) As mentioned above, Newton regarded '*inflection*' as but a special kind of refraction (this was also Hooke's position). It was also at Newton's hands that the corpuscular programme achieved one of its major heuristic successes, cf. below p. 159.

[201] See above, p. 156. Professor Erwin Hiebert, having heard a shortened version of this paper at the Nafplion conference, argued that I had not established the existence of a serious corpuscular optics research programme after Newton. Now a programme once created may exist (in Frege's and Popper's 'World 3' – see Popper [1972]) without anyone at all working on it. But in fact the corpuscular programme made many appearances in 'World 2' (the world of human consciousness) in the period between Newton and Young, as is evinced by its being quoted by a variety of authors. Let me give two examples. The Scots physicist John Robison writes that Newton's theory is 'that light may perhaps consist of small particles emitted by the shining body with prodigious velocity, which are afterwards acted upon by other bodies, with attracting or repelling forces like gravity, which deflect them from their rectilineal courses...' (Robison [1788], pp. 96–7). The programme is even more fully articulated in Herschel [1827] – the first optics textbook. Moreover, the scientists who worked on this programme in this period cannot all be regarded as lightweights. They include: Boscovich, the Scottish physicists Robison, Wilson and Playfair (they were particularly concerned with the optics of moving media), Michell and Bennett (they were particularly concerned with testing for light's inertia), Soldner (he was concerned with the question of massive bodies bending the rays of light-particles), Malus (who with Berthollet investigated, as many other scientists had done, the effect on the fringe patterns of varying the form and substance of the diffracting objects), Laplace, Biot and Brewster.

[202] See above, p. 136-8.

[203] This was admitted by Fresnel who agreed that the 'system of undulations' was 'more difficult to follow in its mechanical consequences than the emission hypothesis' (Fresnel [1822], p. 1; my translation). Lloyd in 1833 also admits that the corpuscular programme had an advantage since: 'The nature and laws of projectile movement are

Neither the corpuscular nor the wave programme had had anything in the way of *predictive empirical success* up to this time: both programmes were in the business of *post hoc* explanations. However, as regards giving explanations of optical phenomena which (whether or not *post hoc*) were in line with the heuristic of the programme, there is no doubt that the corpuscular programme was way ahead of its rival.

The first great heuristic boost to the corpuscular programme had been given by Newton in the *Principia*. He had there demonstrated that, if a moving particle were subject to no net external force, except during its passage through a narrow region bounded by two parallel planes, in which region the force on it satisfied certain conditions,[204] then (*a*) 'the sine of incidence [of the particle] upon either plane [bounding the 'active region'] will be to the sine of emergence from the other plane in a given ratio',[205] (*b*) the incident velocity of the particle will be to its emergent velocity as the sine of the angle of emergence from the 'active region' is to the sine of incidence,[206] and (*c*) if the particle is slower on emergence, then there is a critical angle at which it will be reflected from the 'active region' at an angle equal to the angle of incidence.[207]

Although Newton is at pains here to insist that he is 'not at all considering the nature of the rays of light, or inquiring whether they are bodies or not',[208] but rather merely remarking on 'the analogy there is between the propagation of the rays of light and the motion of bodies',[209] it is clear that both he and later scientists were very impressed by the fact that, as he had demonstrated, if a light ray *were* a stream of particles subject to no external force except in the neighbourhood of the interface between two different media, then the laws of mechanics imply that light would be refracted according to the known laws.

In fact this result of Newton's was the most quoted result in optics in the

far more familiar to every lover of mechanical philosophy than those of vibratory propagation; and the triumphant career of the former branch of this science, in its application to the movements of the heavenly bodies, is in itself sufficient to induce every one to lean to a theory which proposes to account for the phenomena of light on similar principles.' (Lloyd [1833], p. 296.) Some important results in continuum mechanics *were* available, thanks largely to Newton (see above p. 137) and Euler. (The wave programme is rather exceptional in that its heuristic shifted in the course of its development. The attempt to explain optical phenomena in terms of the mechanics of elastic *fluids* was, as is well known, replaced by the attempt to explain optical phenomena in terms of the mechanics of elastic *solids*.)

[204] These conditions are that the force be perpendicular to the two bounding planes and that the force is a function only of the distance from either of the planes.

[205] Newton [1729], p. 226. This is in fact the content of Proposition XCIV, Theorem XLVII of Section XIV.

[206] *Loc. cit.*, p. 228. This is the content of Proposition XCV, Theorem XLIX.

[207] *Ibid.* This is the content of Proposition XCVI, Theorem L.

[208] *Loc. cit.*, pp. 230–1. He also writes here: 'These attractions bear a great resemblance to the reflections and refractions of light...' (*loc. cit.*, p. 229).

[209] *Loc. cit.*, p. 230. Newton's *beliefs* are unimportant as far as methodological appraisal is concerned. The important fact is that his heuristic reflects the assumption that light consists of particles. Cf. above p. 156, footnote 196.

period from the *Principia* up to 1810. Thus, for example, Biot and Arago in a paper written prior to the latter's 'conversion' to the wave theory wrote: 'Newton *proved* that [the] change of direction [in refraction] was owing to an attraction which bodies exercise upon the elements of light...'[210] And Prévost wrote: 'nothing is better proved in optical theory than the proposition which establishes that refraction is produced by an attraction directed perpendicularly to the refringent surface (Newton. Princip. *1.* 1 *prop.* 94)'.[211]

A second example of a development within the corpuscular programme which was an advance from the heuristic point of view is Laplace's demonstration that Huygens's law for the position of the refracted rays in double refraction is reducible to the principle of least action. Laplace himself was clearly aware wherein the significance of this result lay, for he frequently remarks that the most important aspect of the result is that the principle of least action is one of the general principles of the action of 'intermolecular attractive and repulsive forces'.[212]

On the other hand, up until 1810, the protagonists of the wave programme had, at every sign of empirical difficulties, perverted their mechanical assumptions in the interests of explanatory expediency. Thus, for example, Huygens pointed out that it is a consequence of mechanics that each disturbed particle in an elastic medium may be regarded as a source of the disturbances beyond it.[213] But then faced with the 'fact' of rectilinear propagation, Huygens spoiled matters by assuming that each 'secondary' disturbance 'is infinitely feeble'.[214] As we have seen, Young hardly improved things in this respect. Indeed the story of the various modifications he made to the principle of interference is the story of how

[210] Biot and Arago [1806]; emphasis supplied.

[211] Prévost [1798].

[212] See e.g. Laplace [1813] pp. 320–2. Laplace was clear about the analogy between his demonstration and Newton's. Laplace writes (e.g. in his [1809]) that he had decided it 'would be extremely interesting to reduce [the law of double refraction], as Newton has reduced the law of ordinary refraction, to the action of attractive or repulsive forces, of which the effects are sensible only at insensible distances...'. This shows that Young was missing the point in his [1809] review of Laplace's paper when he complains that Laplace overestimated his achievement. Young points out that it was known that Huygens's theory assumed Fermat's principles of least time and that this is the same as Maupertuis's principle of least action 'supposing only the proportion of the velocities [in the two media, in this case a vacuum or air and the birefringent crystal], inverted'. But, Young goes on to point out, it was also well known that the proportion assigned to these two velocities by the corpuscular theory ($v_r = nv_i$ where v_r is the velocity of the refracted ray, v_i the velocity of the incident ray and n is the index of refraction) *is* inverted by the wave theory (which gives $v_r = v_i/n$). Young was obviously correct in all this, but the point he misses is that, while the principle of least time is assumed by Huygens, it was by no means clear that it is a fundamental law of disturbances in elastic media, whereas the principle of least action *was* known to be a fundamental law of particle mechanics. Laplace himself concluded that his demonstration 'leaves no doubt' that double refraction is due to such attractive and repulsive forces. What he ought, more modestly, to have concluded is that he had shown the heuristic power of the corpuscular programme.

[213] Huygens [1690], p. 19.

[214] *Ibid.*

he allowed empirical difficulties to bully him into describing optical disturbances as increasingly different from disturbances in an elastic medium obeying the ordinary laws of mechanics.

Comparing the achievements of the two programmes in, say, 1810, it thus seems that there was nothing to choose between them in terms of empirical progress, but that, as regards heuristic power, it was the corpuscular programme which had so far shown itself the better. This situation did change, but only at the hands of Fresnel.

The analysis provided by the methodology of research programmes thus yields, I claim, a completely 'internal' explanation of the reception of Young's work. Despite the impression most historians try to give, Young's work was received with some interest, but it did not 'persuade scientists of the truth' of the wave theory, as Whewell, for example, said it ought to have done.[215] This, if my analysis is correct, is hardly surprising, for although there is no denying that Young had some intelligent and suggestive ideas, his work neither established the truth of the wave theory nor its superiority over its rival. Thus there is, on this account, no need to invoke external factors like 'Newton worship' to explain Young's alleged neglect. What I should now like to do is to consider how far this should be regarded as a success for the methodology of scientific research programmes. This will also involve looking at the general lessons about the interaction of philosophy of science and history of science which are illustrated by my case study.

5 *The interaction between history of science and philosophy of science*

I shall show that there is an acceptable, general method by which methodologies may be tested against the history of science and shall then apply this method to see if my solution of the problem of Young's neglect counts as a confirmation of the methodology of scientific research programmes.

If there is a general way in which methodologies may be tested against history, then, since there is a well-known logical gap between normative and descriptive statements, there must be some principle which bridges this gap, and which methodologists are prepared to (and ought to) accept. I shall claim that there is such a principle. But the character of this principle depends on which aspects of the development of science methodologists are entitled to pass normative judgements; and so it is to this question that I turn first.

Methodologies, or philosophies of science, ought to provide, amongst other things, general criteria for appraising scientific theories.[216] Such

[215] Cf. above, p. 109, footnote 5.

[216] I shall use the term 'theory' to denote the unit of scientific discovery specified by a methodology, whether or not this is a theory in the intuitive sense.

a criterion will provide an ordering of the competing theories available at any time.[217] Assume that two theories, *A* and *B*, were rivals at some stage in the development of science, and that some methodology unequivocally pronounces theory *A* better than theory *B*.[218] This methodological pronouncement presumably has consequences about *some* of the decisions and actions of those scientists who were actually confronted with the two theories. But which decisions and actions?

First, the methodology certainly tells such a scientist that he ought to accept *A* as currently a better scientific theory than *B*, since, according to the methodology, *A is* the better theory.

Suppose that some of these scientists had previously been trying to develop theory *B*: does the methodology tell them that they ought to stop doing so and work instead on theory *A*?

Many methodologists assume that the answer to this question is 'yes'. Or at least, they seem to think that the methodology's judgement that *A* is better than *B* should carry with it *some* consequences about which theory a scientist should work on – perhaps the consequence that if working on *B* does not improve it relative to *A*, within a certain time period, then the scientist ought to switch to working on *A*. For example, Feyerabend argues that the standards provided by the methodology of research programmes are empty unless some time period is specified such that, if a programme has consistently degenerated throughout that period, then the methodology says that the programme should be abandoned, and that further work on it is 'irrational'.[219] Indeed, Feyerabend claims that if a methodology does not advise scientists on which theory to work, then its standards are mere 'verbal ornament[s]', 'a memorial to happier times when it was still thought possible to run a complex and often catastrophic business like science by following a few simple and "rational" rules'.[220]

[217] This ordering need not be a total ordering. The criterion of scientific merit provided by some methodology may, for example, have two parts, and there may be theories *A* and *B* such that *A* is better than *B* according to the first part, but *B* better than *A* according to the second part. Unless there is still a third part to the criterion for resolving such conflicts, *A* and *B* would then be incomparable according to this methodology.

[218] I am assuming throughout that the methodologies I deal with aim to provide *the* criterion of scientific merit; so that if *A* comes out better than *B* on this criterion, it would follow simply that *A* is a better scientific theory than *B*. (Some methodologists, e.g. some inductive logicians, make the more modest claim that the criterion they provide is only one possible way of judging theories, but that there may be others, equally good, which rank theories differently. This modesty is usually discovered only after the methodology concerned has run into difficulties.)

[219] Feyerabend writes: 'it is easy to see that [the] standards [of the methodology of scientific research programmes]...have practical force only if they are combined with a *time limit* (what looks like a degenerating problem shift may be the beginning of a much longer period of advance). But introduce the time limit and the argument against naive falsificationism reappears with only a minor modification (if you are permitted to wait, why not wait a little longer?) Thus the standards which Lakatos wants to defend are either *vacuous*...or they can be *criticized* on grounds very similar to those which led to them in the first place.' Feyerabend [1970], p. 215.

[220] *Ibid.*

But why should a methodology advise a scientist always to work on (i.e. to try to develop) only the theory it characterises as presently the best available? Say that A is better, according to some methodology, than B. It is possible that a scientist developing the ideas behind (the inferior) theory B will produce a theory C, which, *according to the same methodology*, is even better than A. Indeed most major innovations in science consist precisely of some great scientist developing an idea which had so far led to inferior scientific theories, and producing out of it a superior scientific theory. For example (and reverting to the terminology of the methodology of research programmes), Fresnel adopted the wave programme which before him had degenerated and was (at least in some senses) inferior to the corpuscular programme. He made the wave programme progress, and eventually made it much superior to its corpuscle-based rival. Hence, assuming that we do not want methodologies to yield advice which if followed would, according to its own canons, retard the development of science, it ought not to advise scientists to work on only the best available theory.[221]

There seems to be no difference on this score between Popperianism and the methodology of research programmes (which thus is not, as both Feyerabend and Kuhn imply, an intellectual retreat in this respect from the austere Popperian standards). Why should one not try to develop a refuted or even irrefutable theory in competition with a refutable and corroborated theory? Indeed, Popper says in many places that many scientific breakthroughs have been achieved through the development of originally irrefutable ideas.[222]

I reject Feyerabend's suggestion that if a methodology does not imply advice to scientists about which theories they should work on, then it is

[221] Some attempts have been made to weaken the advice which methodologies ought to yield so as to take account of this argument. Musgrave, for instance, in his [1976], wants to make methodological advice, advice to the scientific community rather than to individual scientists – the scientific community should invest *most of* its (manpower and financial) resources into the best available programme. Grünbaum had earlier also pointed out that the 'appraisals imply advice' thesis can be saved from the above difficulty by making the advice, advice to the majority of scientists – the 'gifted minority' is excused from following the advice (see Grünbaum [1975]). But to adopt this position seems to me really to give up the game. The main reason for adopting it seems to be the idea that methodologies without advice are empty, an idea I show to be false below. The other arguments for the position are dealt with more satisfactorily by my position. I argue that the methodology of research programmes can *explain* why most scientists work for most of the time in the most progressive programme available to them, but it does not *advise* scientists to do so (see below, p. 175).

[222] See, for example, Popper [1963], chapter 2. In fact the 'adviceless' nature of falsificationism is, if anything, clearer than that of the methodology of research programmes, which includes in its appraisal of the *present state* of a programme an appraisal of its heuristic power, and hence of the likelihood (in some intuitive sense) of its making *further* progress, *without the injection of major new ideas*. This does not, however, refute what I have said above. Some scientist may decide to work in a degenerating programme and, in doing so, endow it with powerful new heuristic principles. (This is what, for example, Fresnel did.) Such decisions must not, therefore, be advised against by the methodology.

empty. Such a methodology will still appraise and rank theories in terms of their present scientific merits, and it will tell a scientist what general features a new theory must have if it is to be even better than any of the existing theories. The fact that these requirements do not indicate how to go about constructing a theory which satisfies them does *not* mean that these requirements are empty (just as the notion of a proof in a formal system is not empty despite the fact that it does not – in general – indicate how to go about constructing a proof of any particular proposition).

Returning to my main concern – the problem of whether or not methodologies are testable against the history of science – our considerations suggest that a methodology pronounces directly only on the 'rationality' or 'irrationality' ('correctness' or 'incorrectness' would be better) of scientists' appraisals of existing theories, and not on the 'rationality' or 'irrationality' of working on one theory rather than another.[223] Thus, in the special case of Young, the directly embarrassing aspect of the affair for those who claim that Young established the wave theory's superiority is not that scientists did not stop working on corpuscular ideas, but rather that most seem to have continued to regard the corpuscular theory as superior.

But, the naturalistic fallacy being genuine, even a methodology's pronouncements on scientists' intuitive judgements of the theories which confronted them cannot be *directly* tested against history. A methodology that tells us what scientists *ought* to prefer will not be directly refuted if scientists' actual preferences are different. Nevertheless, a methodology which has to admit that scientists have often preferred the 'wrong' theory would certainly be embarrassed by the enormous increase (at least since Newton) in science's predictive power and practical, technological success. Also, the recent upsurge in the use by philosophers of science of historical examples to illustrate their own position and to undermine the positions of other philosophers, would be inexplicable unless these philosophers suppose that their appraisals have to reflect accurately the intuitive appraisals made by those scientists who were actually confronted with these theories.

My claim is, in other words, that we should regard the following proposition, which connects the methodologist's normative claims with descriptive ones, as an implicit adjunct to any methodology, M:

* Other things being equal, working scientists have accepted theory A as better than theory B if, and only if, A was better than B; moreover,

[223] But, of course, a methodology may indirectly say something on this point, as I show below, pp. 174–5. The point is simply that in order to make predictions about these aspects of the development of science a methodology will have to invoke extra premisses on whose truth the methodology is not obliged to insist. Thus if these predictions turn out to be historically false, there is no need to regard the methodology as disconfirmed – it may be these extra premisses which are false.

we can tell whether *A* was better than *B* by applying the criterion of scientific merit supplied by the methodology *M*.[224]

And indeed some contemporary philosophers do explicitly subscribe to something like *. Popper, for example, writes:

[My] view of scientific method is corroborated by the history of science, which shows that scientific theories are often overthrown by experiments, and that the overthrow of theories is indeed the vehicle of scientific progress.[225]

The conjunction of a particular methodology and its version of * (which is a propositional schema) does have descriptive consequences. If theories *A* and *B* were once rivals and the methodology makes *A* better than *B*, then this appraisal, plus *, entails that, unless things were unequal, scientists of the time did regard *A* as better than *B*. In order to make consequences of this kind historically testable, we must further require that any attempt to save such a prediction by an appeal to the inequality of things must result in a *specific* conjecture about why things were unequal. But methodologists will, in fact, when faced with a case in which scientists' intuitive appraisals differed from the appraisals provided by their methodology, feel obliged to attempt a specific explanation. We have seen this, for instance, in the case of Whewell's account of the reception of Young's work.

Given all this, there is one way in which a historical case study may disconfirm a methodology, and there are two ways in which a historical case study may confirm a methodology. I shall illustrate these three possibilities using the example of (naive) falsificationism.

According to naive falsificationism, theory *A* is better than theory *B* if *A* and *B* explain a certain range of phenomena equally well except that, in one instance, they give conflicting predictions and only *A*'s prediction survives experimental test: *A*'s superiority has been established by a 'crucial experiment' between the two. In the case of the early nineteenth-century revolution in optics, two of the most popular contenders for the

[224] Peter Urbach and others, confronted with *, suggested that it may be circular, and that it may not fulfil the role of plugging the gap between descriptive and normative factors because of the mention of scientists on its left hand (descriptive) side. But, for any methodology, * becomes circular only if one decides who is, and who is not, a scientist on the basis of the methodology concerned. But one can instead decide this on the basis of 'general informed opinion'. Now admittedly, (*a*) general opinion will be informed not by purely descriptive, empirical considerations, but by a mixture of empirical and normative considerations (this is argued in full below, pp. 169–70); and (*b*) the set of people whom 'general opinion' makes scientists will be 'fuzzy'. But admission (*a*) would affect my proposals only if the normative considerations which inform 'general opinion' were systematically taken from some explicit methodology (which they are not). As for (*b*), we can afford to be liberal about admission to the set of scientists provided that we included amongst the external factors, which may be invoked to explain divergences between their judgements and the judgements prescribed by the methodology, factors like a lack of the requisite intelligence or mathematical ability on the part of these scientists.

[225] Popper [1945], vol. 2, p. 260. See also Magee [1973], p. 28: 'Popper's view of science slides onto its history like a glove.'

title of crucial experiment have been Young's two slit experiment and his 1803 diffraction experiment. As we have seen, however, neither of these experiments did, as a matter of fact, convince scientists at the time at which they were (allegedly[226]) performed of the superiority of the wave theory. Thus, given either of these identifications of the crucial experiment, and given *, naive falsificationism can be saved only by an appeal to the inequality of things. This methodology is thus disconfirmed by my case study unless it can come up with a specific suggestion, backed by independent evidence, about what made things unequal.[227]

Could a falsificationist turn this disconfirmation into a confirmation? There are, as stated above, two ways in which he might do this. He could come up with a different contender for the role of crucial experiment between the two theories of light. Indeed several other contenders are already to be found in the literature. The most often mentioned of these is Foucault's experiment of 1850, which is taken as having established that the velocity of light in media denser than air is smaller than its velocity in air, as predicted by the wave theory, not larger than in air, as predicted by the corpuscular theory.

Given this re-allocation of the title of crucial experiment, falsificationism would say that the objective superiority of the wave theory was established only in 1850, and thus would no longer come into conflict (*via* *) with the attitudes of Young's contemporaries. Falsificationism would even be confirmed by the facts about the nineteenth-century revolution in optics if it turned out that scientists began to regard the wave theory as superior after the Foucault experiment had come out in its favour. For this is precisely what the methodology, given this identification of the crucial experiment, and given *, predicts. (Funnily enough, this descriptive prediction runs into the opposite difficulty. We have seen that if Young's experiments were crucial then, from the point of view of falsificationism, the revolution – in the sense of the time at which scientists started to regard the wave theory as superior – occurred twenty years too late. If the Foucault experiment (which was performed in 1850) was crucial then the revolution occurred twenty years *too early*. In the one case we should have to introduce external factors to explain scientists' irrational delay, in the other to explain their undue haste.[228])

Alternatively, the falsificationist may attempt to turn this historical

[226] As I pointed out earlier (§3, pp. 152–6), there is some doubt whether Young actually performed the two slit experiment.

[227] As I suggested earlier, we must not allow a methodology to get away with simply claiming that other things were not equal without making its claims historically testable by *specifying* the inequality. We must also require that the specific inequality-of-things hypothesis receive *independent* support – the falsificationist cannot be allowed to avoid disconfirmation by claiming, say, that Young's contemporaries *must* have been Newton-worshippers because they did not see that Young had crucially refuted Newtonian optics.

[228] Amongst the problems that someone who identified the Foucault experiment as the crucial one would face is that posed by the fact that the result Foucault is generally

disconfirmation into a confirmation by looking for a new 'external' factor, for whose obfuscating role there is independent historical evidence. Imagine, for example, that he decides that the really crucial experiment between the two theories of light was Young's two slit experiment; and that he conjectures that external factors intervened to prevent Young's contemporaries from knowing about this experiment. He finds that an account of this experiment was published *only* in Young's [1807] *Lectures*. Let us further imagine that historical research reveals that, because of some printing error, the copies of this book available up until the 1820s were produced with the very pages which ought to have contained the account of this experiment missing. This, it seems to me, would form a stunning confirmation of this falsificationist account.

These considerations apply quite generally to any methodology which ranks scientific theories. Any such methodology will, given $*$, be *disconfirmed* if it claims that theory A was better than theory B, yet theory B was accepted historically as better than A, and there is no independent support for the conjecture that external factors distorted scientists' judgement at that time. The methodology is *confirmed* either if its appraisals and scientists' intuitive appraisals go hand in hand, or if, in the case of divergence, independent evidence is produced for the existence of misjudgement-provoking external factors. Such confirmations may be particularly significant, if the same historical cases disconfirm other methodologies.[229]

On this account, the methodology of research programmes is confirmed by my historical case study. This methodology, conjoined with $*$ plus the fact that Young's wave theory was not preferred by the scientific community of the 1800s and 1810s, predicts that unless other things were

supposed to have established only in 1850 was already recognised as demonstrated beyond all serious doubt in the 1830s. Humphrey Lloyd, for example, writes in his [1833] of 'the *established fact* that the velocity of light is less in transparent bodies [than it is in air]...' (p. 392, emphasis supplied).

[229] Perhaps a more formal account will increase clarity. Writing 'CP' for the (*ceteris paribus*) assumption that other things are equal, '$A >_M B$' for the assertion that A is better than B according to the methodology M, and '$P(A, B)$' for the assertion that A was historically preferred to B, I claim that any methodology M should be regarded as asserting that

$$(*_M)\ CP \rightarrow (A >_M B \leftrightarrow P(A, B))$$

The first, straightforward kind of confirmation of M is where the 'initial condition' that $A >_M B$ is fed into the 'law' $*_M$, and the *ceteris paribus* assumption made, $*_M$ then yields $P(A, B)$, an assertion which may be confirmed by historical research.

$*_M$ implies:

$$(*'_M)\ (A >_M B \wedge \neg P(A, B)) \rightarrow \neg CP$$

This gives us the second kind of confirmation of M. Here the 'initial conditions' are '$A >_M B$' and '$\neg P(A, B)$' which, when fed into $*'_M$, yield $\neg CP$, an assertion which may be made specific and then independently historically tested.

If, however, all the historical evidence is that there were no particular perturbing factors at the time of the rivalry of A and B, then $\neg\neg CP$ should be regarded as historically confirmed and thus $*'_M$, and hence $*_M$ and with it the methodology M, are 'refuted'.

unequal, Young's theory was not better than its rival. And in fact Young's theory was not better than its rival according to the methodology of scientific research programmes. The successive versions of the wave theory developed by Young form a classic case of a degenerating problem shift: each new version did no more than deal *ad hoc* with (some of) its predecessor's refutations. Moreover Young was not really able, despite a few intelligent hints and gropings, to develop the heuristic machinery of the wave programme so as to provide clear guidelines for the further development of the programme and for how it should deal with the many problems (both theoretical and empirical) it was known to face. No wonder then that the majority of Young's fellow scientists preferred the (admittedly empirically degenerating but) heuristically powerful corpuscular programme.[230]

My account of how methodologies can be tested using history of science is, I hope, both a clarification of, and an improvement on, the account already given by Lakatos.[231] Many of Lakatos's readers have taken him as asserting that a methodology is the better, the more of the history of science it pronounces 'rational' or that it can explain 'internally'.[232] This would, however, make the Hegelian view that whatever is real is rational the supreme methodology. On my account, on the contrary, one methodology may be better confirmed than another even if it explains fewer historical developments 'internally'. For a methodology may be confirmed by historical cases of which it gives an *external* explanation, *provided* it can give independent historical evidence for the existence of the external factors it invokes.

I have assumed so far that there is such a thing as purely descriptive history against which methodologies can be tested. Several methodologists (notably Agassi[233] and Lakatos[234]) have claimed, however, that the writing of history is, whether or not historians realise it, informed by methodological considerations. This clearly raises the suspicion that my account

[230] Indeed, when the intellectual merits of the competing theories are assessed using the methodology of research programmes, this whole historical episode is explicable internally. Not only was the rejection of the wave programme, as Young had developed it, correct according to this methodology, so also, as I hope to show in a forthcoming paper, was the eventual acceptance of the programme as Fresnel had developed it. For not only did Fresnel make the wave programme progressive (it predicted new phenomena like conical refraction), he also revealed its heuristic power.

[231] See his [1971], reprinted in this volume.

[232] That is, in terms of the intellectual merits and demerits (as the methodology sees them) of the competing theories. (Perhaps I ought to stress again here that which factors are internal is, in the Lakatosian sense, methodology-dependent.) One of Lakatos's readers who interprets him as expounding the Hegelian view is R. J. Hall (see his [1971], p. 151). I should add that Lakatos explicitly denies that this is his position – see his [1971], p. 118 (reprinted above, p. 32).

[233] Agassi [1963].

[234] Lakatos [1971]; reprinted in this volume, pp. 1–39.

of the testing of methodologies against history involves a kind of circularity which may be vicious.[235] In order to remove this suspicion I shall now turn to the question of what history of science can learn from the philosophy of science, and, in particular, of how the writing of history of science may be affected by philosophical considerations. (This will also enable me to rebut the suggestion – made to me by intellectual friend and foe alike – that historical case studies undertaken for the purposes of investigating the merits of some methodology are likely to confirm that methodology; and to clear up some prevalent misunderstandings of Lakatos's account of 'rational reconstuctions'.)

A part of Lakatos's striking paraphrase of Kant's dictum holds that 'history of science without philosophy of science is blind'.[236] In arguing for this claim, Lakatos in fact argues for a pair of theses. The first thesis is that even history of science written without any explicit recourse to any methodology will in fact involve, or have implicit in it, some normative views. This thesis is susceptible of various interpretations (not always clearly distinguished by Lakatos). Given the thesis in some of these interpretations, my account of the testing of methodologies would indeed be vitiated by circularity. Fortunately under these interpretations the thesis is false. The second thesis (with which I unequivocally agree) is that by taking methodological considerations into *explicit* account, the historian will become aware of certain interesting problems (and, perhaps even more importantly, certain interesting shifts in problems) to which otherwise he would be blind. This second thesis does not imply any circularity problems for my account of the testing of methodologies.

What seems to me true about the first of these two theses is that a historian's selection procedures (and a historian may select only unconsciously although he *must* select) and the set of problems he finds interesting are circumscribed by normative considerations. Indeed, in simply deciding which people and what work to study (i.e. in deciding which people to regard as scientists, and which pieces of work to regard as scientific) the historian of science is implicitly relying on some normative methodological criteria. Of course, everyone will allow that the historian himself may be ignorant of any explicit philosophical or methodological views, and that he may simply follow 'general opinion' in deciding who was, and who was not, a scientist, what was, and what was not, scientific. But 'general opinion' on this point is not informed simply by empirical considerations – not everyone who was ever called, or whoever called himself, a scientist is held to be a scientist by general opinion,[237]

[235] Kuhn in his [1971] explicitly accuses Lakatos's account of the testing of methodologies of circularity.

[236] This paraphrase is, in full: 'Philosophy of science without history of science is empty; history of science without philosophy of science is blind.' (Above, p. 1.)

[237] For example, Lafayette Ron Hubbard (founder of 'Scientology') is not generally regarded as a scientist, although he and many others claim him to be one.

and similarly not every piece of work that was called scientific is so considered by general opinion.[238]

'General opinion' and with it the writing of history of science are informed by some (often rather uncritically accepted) normative views about science. But this, as I briefly indicated earlier,[239] does not introduce a circularity into the process of testing methodologies against history. Only if 'general opinion' were consistently and exclusively informed by a specific methodology so that, in doing history of science, one was, in fact, selecting as 'scientists' and as 'scientific advances' just those that conform with this methodology, would history be useless as a test of that methodology. But this is not the case and thus we have a set of scientists and scientific advances specified *independently* of any specific methodology.

Indeed, the situation here is analogous to that within science itself. The fact that our observations are made 'in the light of', or are directed by, theoretical considerations (a fact which Popper and others describe by saying that observation statements are 'theory-laden') does not mean that they cannot be used to test theories. Only if the observations were interpreted purely in the light of the theory would no real test of it by them be possible.

This is enough to deal with the circularity charge, but let us investigate the more general question of how history of science is informed by philosophical considerations by further pursuing this analogy with theory testing. Those philosophers who have argued that observation statements are 'theory-laden' were surely correct to stress that all observations are *directed by* theoretical expectations and considerations.[240] They were also surely correct in claiming that which observations and experiments a scientist makes will depend on which theories he seriously entertains. It does not, however, follow from any of this (as some of these methodologists may have thought) that our observations are 'theory-impregnated'.

This second claim *may* also be true, but its truth is not established by the argument that all our observations and experiments are made 'in the light' of theory; moreover, its analogue in the case of testing methodologies seems to me clearly false. The admission that normative considerations are, willy-nilly, involved in the historian's selection of material and of problems does *not* imply that all statements in history of science books are normative or have implicit in them normative views.

For example, the reasons historians of science have chosen to study Thomas Young and to read, say, his paper 'On the Theory of Light and Colours' are based on general considerations about what constitutes

[238] For example, as Paul Feyerabend likes to point out, witchcraft was once widely considered to be a paradigm of experimental science.

[239] Above, p. 165, footnote 224.

[240] The theory of perception indicates that this is true even of our non-systematic, everyday observations.

genuine science. But the statement: 'Thomas Young wrote a paper called "On the Theory of Light and Colours" which was published in the *Philosophical Transactions* for 1802' is obviously not normative. Coming to a decision on its truth value would involve only descriptive, and no methodological considerations.[241] More importantly, the (true) statement that Young's work did not evoke much immediate interest is factual and not normative, although it was methodological considerations which led historians to emphasise and comment on this fact – they assumed that Young was a great scientist and that his theory was great science and hence they found the neglect of his 'achievement' remarkable.[242]

As well as directing interest, methodological considerations may also penetrate descriptive history by providing certain terms of the historian's language. The historiography of science is littered with terms like 'crucial experiment', 'hard fact', 'verification', 'inductive method', 'simplicity' and the like. It is (partly) this which leads Lakatos to speak of 'normatively interpreted history'.[243] But again the fact that some of the historian's terms are imported from methodology does *not* mean that any history is bound to be 'normatively interpreted'. It may sometimes be difficult to know if a historian uses these terms whether he is making a descriptive or a normative claim, but that these are two quite separate kinds of claims is always clear. Say, for example, that a historian writes: 'Young performed a crucial experiment between the corpuscular and wave theories of light'. It may not be clear whether he means 'Young performed an experiment, the result of which was correctly predicted by the currect version of the wave theory, but incorrectly predicted by the current version of the corpuscular theory', or whether he means in addition that this experiment 'ought to have convinced scientific men of the truth of' the wave theory.[244] But it *is* clear that the first of these is a straightforwardly descriptive claim (though it involves claims about logical relations) whilst the additional claim is normative in character. So long as we can separate these two kinds of claims (and here, it seems to me, we can be completely 'operationalist', asking whether in arriving at a decision on the truth value of a sentence, we should need to see if some methodological criterion was satisfied), and so long as we use only

[241] This is not to deny that some statements which the historian makes will be normative. There is nothing to stop a historian making a statement like 'Thomas Young produced a theory of light and colours which was, in 1802, the best available theory'. It is uncontroversial that whether such a statement is true or false depends on what the correct criteria of scientific merit are. The issue is whether *all* history of science statements have implicit methodological commitments.

[242] This is beautifully illustrated by Crowther, who writes: '...it is essential to know why *so great a scientist as Young* had so little general impact' (Crowther [1973], p. 671).

[243] Lakatos [1971], p. 91, reprinted in this volume p. 1, ('normatively interpreted' occurs in brackets in the original).

[244] The last quote is from Whewell; see above, p. 109.

descriptive claims in testing a methodology, then again no question of circularity arises.[245]

I now come to the second of the pair of theses implicit in Lakatos's paraphrase of Kant. This is that history of science can be improved in various ways by an explicit use of a critically defensible methodology. That some of the historian's descriptive terms are supplied by methodology does not imply that all history of science is 'normatively interpreted' or 'soaked in methodology'; however, it does point to an important way in which history may be influenced and improved by an explicit appeal to methodology. For one methodology may supply a set of descriptive terms which enables the historian to express more concisely and more faithfully more of the facts about some historical episode. Indeed my case study suggests, I claim, that the methodology of scientific research programmes is an improvement on other methodologies in this respect. Other methodologies, for example, speak in terms of scientific theories, but there were no such things as *the* corpuscular theory of light, and *the* wave theory of light. There were instead a series of wave theories and a series of corpuscular theories. Moreover, the terms of these series were related in various clear ways. These are not methodology-induced facts, but straightforward historical facts. However, the notion of a research programme and those of positive and negative heuristics which provide the connecting links between the various theories issuing from the programme do direct the historian to these facts and enable him to describe them accurately and in detail.

A sort of corollary to this point is that problems may be generated for a historian simply by the lack of descriptive power in his language. For example, if the historian's language allows him to speak only in terms of theories, then he may well find it difficult not to consider Young as hard done by. After all, did Young not propound in the early 1800s the theory that light is a wave motion in an all-pervading medium? And was not this the theory for which Fresnel was much later given great credit? Why then did the scientists of the time have to wait for Fresnel to 'rediscover' the wave theory before accepting it? Once, however, the historian's language is enriched with the new descriptive notion of a programme, then there *need* be no mystery about this. It may have been (and, indeed, I claim

[245] In view of all this, Lakatos's claim that all histories ('histories$_2$' – sets of statements) are rational reconstructions (of 'history$_1$' – the set of historical events) is false, unless by rationally reconstructed history is meant simply history selected from some point of view. Lakatos uses the term 'rational reconstruction' in a variety of senses. His failure consistently to distinguish between the various senses has caused confusion (many have even taken him to be asking for licence to falsify history to fit his methodology!). In view of this, it is perhaps better to drop the term altogether, or to restrict it to its original Lakatosian use as a pedagogical device. (The discussion in his essay 'Proofs and Refutations' consists of 'rationally reconstructed history' from which the wrinkles have been ironed out and the less essential features omitted. These reappear in the footnotes where 'the *real history*... chime[s] in.' (Lakatos [1963–4], p. 7; my italics.))

that this is how it was) that the wave programme as it had developed up to Young, was not superior to the corpuscular programme as it then stood; whereas Fresnel developed the *same* wave programme into much the better of the two rivals.

This second Lakatosian thesis can be pursued further. Given that parts of the historian's terminology and many of his problems are supplied by (possibly unconscious) methodological considerations, it is clearly better if he can make these considerations explicit and so expose them to criticism. This will, at least, sharpen the problems the historian faces and will, very often, show him that the real problems are different from what he originally thought. For instance, I think I have shown in preceding sections that the problem of Young's neglect by his contemporaries is created by normative considerations – namely, the usual high appraisal of the merits and importance of Young's work. So long as the historian is not even conscious that this is an *assumption* that he is making, he is bound to continue to look for an explanation of Young's treatment in (what for us will be) 'external' terms. On the other hand, once this appraisal of Young's work and the methodological assumptions on which it rests are made explicit, then a new possibility appears. This is that this appraisal is mistaken; and if it is mistaken (as indeed I have argued it is) then the descriptive historical problem is (as we have also seen) radically shifted. The historian is bound to be blind even to this possibility if he insists that he can ply his trade without recourse to any methodology.[246]

There is a second way in which a historian's perceptiveness may be improved by making *explicit* use of methodology. A methodology may supply the historian with a heuristic – not only with a set of problems, but also with a stock of conjectured solutions of them. I shall illustrate this point with some examples.

The standard pattern of scientific development according to the methodology of scientific research programmes is of protracted rivalry between research programmes – a rivalry from which one eventually emerges victorious, not at a single blow ('crucial experiment' style), but rather after a sustained period of progress on its part and degeneration on its rival's part. The methodology does *not* predict that all (major and minor) scientific revolutions take this form: it is perfectly conceivable, for instance, that some programmes were never pursued into their degenerating phases. Nevetherless, the methodology does point to the possibility of such developments and was indeed developed to deal with

[246] Entirely similar considerations apply to sociologists. Any analysis of any of the social aspects of science is likely to be trivial unless account is taken of the intellectual merits and demerits of what scientists produce; but in order to assess these satisfactorily a critically defensible methodology is required. Moreover, very different sociological theories are likely to be produced when different methodologies are accepted as guides – what one says, e.g., about the 'authority structure' in early nineteenth-century physics is likely to depend very heavily on how one appraises Young's real intellectual achievement.

them. Thus, this methodology does not *predict* that where the secondary sources say there was a crucial experiment, the primary sources will reveal protracted rivalry and several alternative explanations of the experimental results based on the allegedly crucially refuted theory. It does, however, encourage the historian to look for these things in the primary sources. And (as we have seen in §3(*a*1)) they are, in fact, to be found in the case of Young's two slit and diffraction experiments.

Similarly, falsificationism does not logically imply that no experimental discoveries were made by chance. It does, however, suggest (as was pointed out by Agassi[247]) that most discoveries were arrived at as refutations of known and seriously entertained theories (especially where the discovery concerned was considered by scientists to be of major importance). Thus falsificationism provides a heuristic for the historian of science: take any famous 'chance discovery' and see if the discoverer did not in fact have a theory in mind, the testing (and refutation) of which led to the discovery.[248]

To take a third and final example, historical conjectures about scientists' decisions to work on one theory rather than another may also be suggested by systematic methodological appraisals. For although, as I argued above,[249] methodologies must not instruct the scientist to work on only the best available theory (and do not predict that all scientists will do so), there is no doubt that a scientist's intuitive appraisal of the present state of a theory and its competitors will often play a big role in his decision whether or not to try to develop that theory. Consider again the methodology of scientific research programmes (which is stronger than most other methodologies in this respect). It includes in its appraisal of the present state of a programme an appraisal of its present heuristic power. And it points out that in a progressive programme with a strong heuristic there will be plenty of new problems and well-articulated methods for attempting to solve them.[250] The creative leap required to invent a new progressive theory within such a programme is, in a clear, if intuitive sense, smaller than it is in less powerful programmes, where entirely new ideas may be needed. Hence this methodology *suggests* (but does not entail) testable historical conjectures of the following kind: if there was

[247] See his [1963].

[248] Agassi applies this heuristic to several interesting cases in his [1963]. Notice, by the way, that this heuristic could prove successful even if no new (i.e. hitherto unknown) historical facts were turned up in the process. For old material may be newly (and more plausibly) interpreted. For example, what was before considered a stray remark on which the scientist concerned, for some strange reason, laid a great deal of stress, may now become a central remark; what was before considered an entirely fortuitous change of conditions of some experiment may now be explained as made in a deliberate effort to refute a previously held theory. (This seems to me to deal with an argument for the inutility of philosophy in history of science presented by Pearce Williams in his stimulating [1975].)

[249] Pp. 162–4.

[250] For this, see especially Zahar's account of the relativity programme in his [1973]; reprinted below, pp. 211–75.

at some particular time a programme which was progressive and heuristically more powerful than other available programmes, most scientists will have worked on it.

(It may seem that in this third example of philosophy supplying a historical heuristic, I am taking back what I said earlier about appraisal and advice, and about a methodology having direct consequences only for scientists' appraisals of the merits of theories. I am not. To turn these suggestions about what scientists do into genuine *predictions*, into genuine deductive consequences of the methodology, extra historical assumptions have to be made. The methodology is by no means committed to truth of these extra assumptions in particular cases – as it *is*, I have argued, to the truth of *. Assume, for example, that there was, at some particular time, a programme which the methodology of research programmes appraises as more progressive and heuristically more powerful than any other available programme. If this appraisal is to yield the prediction that most scientists worked on this 'best' programme, we should have to add to it not only the * assumption that these scientists did intuitively make this appraisal, but also an extra premiss. This extra premiss would state roughly that the majority of scientists at that time were sufficiently enthusiastic about the progressive programme to want to devote their time to it, and that they were not self-confident enough – and not sufficiently motivated, say, by metaphysically-induced preferences – to try to instill new life into what they recognised to be a degenerating programme. But there seems no reason to regard the methodology as committed to these extra assumptions.[251] Thus, if some particular instance of this prediction turned out to be false, there would be no need for the methodology to invoke external factors which induced scientists to misjudge the intellectual situation. For, as I have pointed out, scientists may quite consistently try to develop a theory A, even while agreeing that there already exists a theory B, which *as things stand*, is better than A.[252] Whereas, as I argued above, if some particular instance of a prediction (based on a methodology and the relevant instance of *) about scientists' intuitive rankings of theories' merits turned out to be false, then the supporters of that methodology *would* be obliged to look for misjudgement-provoking external factors).

Finally, let me summarise this section. First, I tried to show that a methodology can be rendered historically testable by adding the relevant

[251] The fact that these extra assumptions do, however, seem to hold for most historical periods (i.e. that scientists do generally get excited enough about sufficiently powerful programmes, however 'absurd' their metaphysical presuppositions) provides the *explanation* (promised above, p. 163, footnote 221) of why most of the scientists work for most of the time in the most progressive programme available to them.

[252] For example, several scientists in the first few decades of the nineteenth century (Sir John Herschel and Sir David Brewster amongst them) seem to have been inclined to try to revitalise the corpuscular optics programme, although everyone agreed (Herschel and Brewster are on record as agreeing) that, as things stood, the wave optics programme had been put well ahead of its rival by Fresnel.

instantiation of *. Secondly, I tried to show that the writing of history of science is guided, for the most part implicitly, by normative considerations, but that this does not entail that all statements about the history of science are normative, nor that the attempt to test philosophies against history is circular. Lastly, I tried to indicate how explicit recourse to (normative) methodological considerations can improve history of science.

References

Agassi, J. [1963]: *Towards an Historiography of Science.*

Arago, F. [1819]: 'Rapport fait par M. Arago à l'Académie des Sciences au nom de la commission qui avait été chargée d'examiner les mémoires envoyées au concours pour le prix de la diffraction', *Annales de chimie et de physique*, **11**, May 1819.

Arago, F. [1859]: *Biographies of Distinguished Scientific Men*, translated by W. H. Smyth, Rev. Baden Powell and R. Grant.

Bechler, Z. [1974]: 'Newton's law of forces which are inversely as the mass: a suggested interpretation of his later efforts to normalise a mechanistic model of optical dispersion', *Centaurus*, **18**, pp. 184–222.

Biot, J. B. [1816]: *Traité de physique experimentale et mathématique*, four volumes.

Biot, J. B. and Arago, F. [1806]: 'Upon the Affinities of Bodies for Light, and particularly upon the Refractive Powers of different Gases', *Philosophical Magazine*, **26**.

Brewster, D. [1831]: *A Treatise on Optics.*

Brewster, D. [1832]: 'Report on the Recent Progress of Optics', *British Association Report* (*Second Meeting, 1832*).

Brougham, H. [1796]: 'Experiments and Observations on the Inflection, Reflection and Colours of Light', *Phil. Trans.*, **86**.

Brougham, H. [1803]: Review of Young [1802*a*] and [1802*b*], *Edinburgh Review*, January 1803, pp. 450–60.

Brougham, H. [1804]: Review of Young [1804*a*], *Edinburgh Review*, October 1804, pp. 97–103.

Cantor, G. N. [1970–1]: 'The Changing Role of Young's Ether', *British Journal for the History of Science*, **5**, pp. 44–62.

Cantor, G. N. [1971]: 'Henry Brougham and the Scottish Methodological Tradition', *Studies in History and Philosophy of Science*, **2**, pp. 69–89.

Clive, J. [1957]: *Scotch Reviewers. The "Edinburgh Review", 1802–1815.*

Cockburn, H. [1852]: *Life of Lord Jeffrey.*

Cohen, I. B. (ed.) [1958]: *Isaac Newton's Papers and Letters on Natural Philosophy.*

Crowther, J. G. [1968]: *Scientific Types.*

Crowther, J. G. [1973]: 'Light on Light Waves', *New Scientist*, 14 June 1973, pp. 671–3.

Duhem, P. [1906]: *La Théorie Physique, Son Objet et Sa Structure*; translated as: *The Aim and Structure of Physical Theory*, 1954.

Feyerabend, P. K. [1970]: 'Against Method', *Minnesota Studies for Philosophy of Science*, **4**.

Fresnel, A. [1822]: *De la lumière*; first published as a supplement to a French translation of Thomson's *Chemistry*, 1822.

Gouy, M. [1886]: 'Sur le mouvement lumineux', *Journal de physique théorique et appliquée*, 2e série, **5**.

Grimaldi, F. M. [1665]: *Physico-mathesis de lumine coloribus et iride.*

Grünbaum, A. [1975]: 'Falsifiability and Rationality', in J. Kockelmans, G. Fleming and S. S. Goldman (eds.): *Issues in Contemporary Physics and Philosophy of Science*.

Hall, R. J. [1971]: 'Can we use the history of science to decide between competing methodologies?', in R. C. Buck and R. S. Cohen (eds.): *Boston Studies in the Philosophy of Science* **8**, pp. 151–9.

Hardy, A. C. and Perrin, F. H. [1932]: *The Principles of Optics*.

Herschel, J. F. W. [1827]: 'Treatise on Light', *Encyclopaedia Metropolitana*, Article 788; all page references are to the version in the *Encyclopaedia of Mechanical Philosophy*.

Herschel, W. [1807]: 'Experiments for investigating the Cause of the Coloured concentric Rings, discovered by Sir Isaac Newton, between two Object-glasses laid upon one another', *Phil. Trans.*, pp. 180–233.

Herschel, W. [1809*a*]: 'Continuation of Experiments for investigating the Cause of Coloured concentric Rings, and other Appearances of a Similar Nature', *Phil. Trans.*

Herschel, W. [1809*b*]: 'Supplement to the First and Second Parts of the Paper of Experiments for investigating the Cause of Coloured Concentric Rings between Object Glasses, and other Appearances of a similar Nature', *Phil. Trans.*

Holton, G. and Roller, D. H. [1958]: *The Foundations of Modern Physical Science*.

Houston, P. [1938]: *A Treatise on Optics*, 7th edition.

Huygens, C. [1690]: *Treatise on Light*.

Jenkins, F. A. and White, H. E. [1957]: *Fundamentals of Optics*, 3rd edition.

Jordan, G. W. [1799*a*]: *An Account of the Irides or Coronae which appear around and contiguous to, the bodies of the sun, moon and other luminous objects*.

Jordan, G. W. [1799*b*]: *The Observations of Newton Concerning the Inflection of Light, etc.*

Jordan, G. W. [1800]: *New Observations concerning the colours of thin transparent bodies, showing these phenomena to be inflections of light, etc.*

Kuhn, T. S. [1971]: 'Notes on Lakatos', in R. C. Buck and R. S. Cohen (eds.): *Boston Studies in the Philosophy of Science*, **8**, pp. 137–46.

Lakatos, I. [1963–4]: 'Proofs and Refutations', *British Journal for the Philosophy of Science*, **14**, pp. 1–25, 120–39, 221–45, 296–342.

Lakatos, I. [1970]: 'Falsification and the Methodology of Scientific Research Programmes', in I. Lakatos and A. Musgrave (eds.): *Criticism and the Growth of Knowledge*, pp. 91–196.

Lakatos, I. [1971]: 'History of Science and its Rational Reconstructions', in R. C. Buck and R. S. Cohen (eds.): *Boston Studies in the Philosophy of Science*, **8**, pp. 91–136.

Laplace, P. S. de [1809]: 'Sur la loi de la réfraction extraordinaire dans les cristaux diaphanes', *Journal de Physique*, January 1809.

Laplace, P. S. de [1813]: *Exposition du Système du Monde*, 4th edition.

Larmor, J. [1934]: 'Thomas Young', *Nature*, **133**, February 1934.

Lloyd, H. [1833]: 'Report on the Progress and Present State of Physical Optics', *British Association for the Advancement of Science Reports*, **4**, pp. 295–413.

Mach, E. [1926]: *Physical Optics*.

Magee, B. [1973]: *Popper*.

Magie, W. F. (ed.) [1935]: *A Source Book in Physics*.

Moon, R. [1849]: *Fresnel and his Followers*.

Musgrave, A. E. [1976]: 'Method or Madness?' in R. S. Cohen, P. K. Feyerabend and M. W. Wartofsky (eds.): *Boston Studies in the Philosophy of Science. Imre Lakatos Memorial Volume*, forthcoming.

Newton, I. [1671–2]: 'New Theory about Light and Colours', *Phil. Trans.*, 1672, reprinted in Cohen [1958], pp. 47–59.

Newton, I. [1672]: 'Mr. Isaac Newton's Answer to some Considerations on his Doctrine of Light and Colours', *Phil. Trans.*; reprinted in Cohen [1958], pp. 116–35.

Newton, I. [1675]: 'Second Paper on Light and Colours', *Phil. Trans.*; reprinted in Cohen [1958], pp. 177–235.

Newton, I. [1729]: *Mathematical Principles of Natural Philosophy*; all page references to Motte–Cajori version, University of California Press.

Newton, I. [1730]: *Opticks*, 7th edition; page references to Dover paperback version.

Peacock, G. [1855]: *Life of Thomas Young.*

Pearce Williams, L. [1975]: 'Should Philosophers be allowed to write History?', *British Journal for the Philosophy of Science*, **27**, pp. 241–53.

Pettigrew, T. J. [1840]: *Medical Portrait Gallery*, **4**.

Popper, K. R. [1945]: *The Open Society and its Enemies*, two volumes.

Popper, K. R. [1963]: *Conjectures and Refutations.*

Popper, K. R. [1972]: *Objective Knowledge.*

Potter, R. [1833]: 'On the Modification of the Interference of two Pencils of Homogeneous Light produced by causing them to pass through a Prism of Glass, *etc*', *Philosophical Magazine*, 3rd series, **2**, no. 8, February 1833.

Prévost, P. [1798]: 'Quelques Remarques d'Optique, principalement relatives à la Réflexibilité des Rayons de la lumière', *Phil. Trans.*, **88**.

Priestley, J. [1772]: *The History and Present State of Discoveries relating to Vision, Light and Colours.*

Robison, J. [1788]: 'On the Motion of light, as affected by refracting and reflecting Substances which are also in Motion', *Transactions of the Royal Society of Edinburgh*, **2**.

Stuewer, R. H. [1970]: 'A Critical Analysis of Newton's Work on Diffraction', *Isis*, **61**, pp. 188–203.

Urbach, P. [1974]: 'Progress and Degeneration in the "I.Q. Debate" (I) and (II)', *British Journal for the Philosophy of Science*, **25**, pp. 99–135, 235–59.

Verdet, E. [1866]: 'Introduction aux Oeuvres d'Augustin Fresnel', in *Oeuvres Complètes d'Augustin Fresnel.*

Whewell, W. [1837]: *History of the Inductive Sciences.*

Whittaker, E. [1910]: *A History of Theories of the Aether and Electricity.*

Wollaston, W. H. [1802]: 'On the Oblique Refraction of Iceland Crystal', *Phil. Trans.* pp. 381–6.

Wood, A. [1954]: *Thomas Young, Natural Philosopher, 1773–1829*, completed by F. Oldham.

Wood, R. W. [1905]: *Physical Optics*, 1st edition.

Young, T. [1802*a*]: 'On the Theory of Light and Colours', *Phil. Trans.*; page references are to the reprint in Young [1855].

Young, T. [1802*b*]: 'An Account of some Cases of the Production of Colours not hitherto described', *Phil. Trans.*; page references are to the reprint in Young [1855].

Young, T. [1804*a*]: 'Experiments and Calculations relative to Physical Optics', *Phil. Trans.*; page references are to the reprint in Young [1855].

Young, T. [1804*b*]: 'Reply to the Animadversions of the Edinburgh Reviewers', reprinted in Young [1855].

Young, T. [1807]: *Lectures on Natural Philosophy*, volume 1.

Young, T. [1809]: Review of Laplace [1809] in *Quarterly Review*, **2**, November 1809, pp. 337–49.

Young, T. [1810]: 'Review of the "Mémoires de Physique et de Chimie de la Société d'Arcueil", Volumes I and II', *Quarterly Review*, **3**, May 1810; page references are to the reprint in Young [1855].
Young, T. [1817]: *Chromatics* (from the *Supplement to the Encyclopaedia Britannica*); page references are to the reprint in Young [1855].
Young, T. [1855]: *Miscellaneous Works of the late Thomas Young*, edited by George Peacock, volume 1.
Zahar, E. G. [1973]: 'Why did Einstein's Programme Supersede Lorentz's?' *British Journal for the Philosophy of Science*, **25**, pp. 95–123, 233–62.

Why did oxygen supplant phlogiston? Research programmes in the Chemical Revolution*

ALAN MUSGRAVE
UNIVERSITY OF OTAGO

Imre Lakatos showed how different philosophies of science provide different analytical tools with which to approach the history of science. And he showed how different philosophies of science could be evaluated by seeing how well they account for episodes in the history of science.[1] In this paper I shall reconsider a very famous episode in the history of science, the Chemical Revolution, and argue that Lakatos's methodology of scientific research programmes provides the best account of it. I will first outline the accounts of the Chemical Revolution given by other methodologies, and show that they are unsatisfactory: each of them must either deem the Chemical Revolution an irrational affair or falsify history so that it squares with its canons of rationality. Then I will argue that the actual story of the Chemical Revolution fits Lakatos's methodology like a glove.

We all know that it is easy to find 'confirmations' of a scientific theory if you look for them. Similarly, it is easy to find 'confirmations' of a methodology if you look for them in the history of science. So perhaps it is worth mentioning that I first become interested in the Chemical Revolution in order to try to refute Lakatos's methodology. The conclusions I have reached about the Chemical Revolution do not square with the preconceptions I had at the outset.[2] And while I have been, and still am, critical of some features of Lakatos's methodology, they are not features which need concern us here.[3]

The story of the Chemical Revolution is a famous one, which has been told and retold countless times. But just as new light can be shed upon well-known facts by a new scientific theory, so also new light can be shed upon well-known episodes in the history of science by a new methodological theory. My excuse for going over familiar ground once more is not that I have uncovered new facts (though there might be one or two of these), but that I want to put old facts in a new perspective. I think that the new perspective which Lakatos has given us is an important one: even

* I am indebted to Noretta Koertge for helpful comments on the earlier version of this paper which was read at the Nafplion Colloquium.

[1] This volume, pp. 1–39.

[2] My earlier preconceptions are still to be found, albeit in a rather battered state, in my [1972].

[3] See my [1974*a*] and my [1976].

the best accounts of the Chemical Revolution are tainted by erroneous methodological views often held almost unconsciously. Let us look briefly at some of these accounts, and at the difficulties they face.

1 *Inductivist accounts*

Inductivism claims that respectable scientific theories are derived from, and proved by, experiments. A naive inductivist account of the Chemical Revolution says that it consisted of replacing premature and erroneous speculation by proven truth. For example, Rodwell tells us, using classical Baconian terms, that phlogiston theory 'was built up of *idola theatri* collected from various sources; and these were cemented together by the particular *idola specus* of Becher and Stahl...their fault was a too hasty generalization'.[4] According to Herschel, in the phlogiston theory 'as if to prove the perversity of the human mind, of two possible roads the wrong was chosen'.[5] White regards phlogiston theory as an 'obstacle...which had to be swept away in the triumphal march towards truth'.[6] Partington regards it as a 'jungle' compared to 'the path of true discovery'.[7]

In our enlightened days one need not spend much time exposing the deficiencies of such accounts. The chief difficulty for inductivists is that Lavoisier's theory, allegedly proven truth, is actually false. Inductivist historians have to prune that theory of its mistaken branches (such as the caloric theory, or the thesis that all acids contain oxygen) in order to make their account plausible. Often Lavoisier's mistakes are not mentioned at all. Sometimes they are mentioned, only to be deemed irrelevant. We are told that 'Lavoisier's theory of combustion is as valid today as it was 150 years ago, even though we have abandoned the hypothesis of a caloric fluid which he advocated and used'.[8] Most historians would disavow a crude inductivist account of the Chemical Revolution: it survives explicitly only in illiterate prefaces to chemistry textbooks. And yet inductivist preconceptions still linger, and exert subtle influences, as we will see.

2 *Naive falsificationist accounts*

If experiments did not prove Lavoisier's oxygen theory, perhaps they *disproved* the rival phlogiston theory? A naive falsificationist account of the Chemical Revolution says that it was brought about by experiments

[4] Rodwell [1868], p. 28. [5] Herschel [1830], p. 300.

[6] White [1932], p. 11.

[7] Partington [1937], p. 84. Further examples of inductivist views are cited by Agassi in his [1963], footnote 36 (see also footnotes 31 and 32). Agassi's book is the best attack upon inductivist historiography, and upon the 1984-style rewriting of history which it entails.

[8] Cohen [1966], pp. 173–4.

which decisively refuted the phlogiston theory of combustion. The experiment usually selected for this decisive role is Lavoisier's experiment of 1772: phosphorus was burned in a volume of air confined over water, the volume of air was reduced, and the product of the combustion weighed more than the original. This decisively refuted phlogiston theory, which claimed that a burning substance emitted phlogiston and hence should decrease in weight. If overkill is required, the falsificationist can also cite Lavoisier's beautiful experiment of 1775: mercury was heated in air to form a calx (oxide), and again the volume of air was reduced and the calx weighed more than the original metal; the mercury calx was then heated strongly, expelling oxygen from it and restoring the original weight of the metal. This is a classic example of a crucial experiment, which directly refuted phlogiston theory (for which calcination is 'slow combustion') and, it is often added, equally directly established the oxygen theory. Thus, a falsificationist can write that phlogiston theory 'survived for almost a century, until Lavoisier exposed its fallacies by his study of the changes in weight during combustion'.[9]

This falsificationist account soon runs up against unpleasant historical facts. First of all, the result of Lavoisier's 1772 experiment (that combustion leads to a weight increase and to a reduction in the volume of air) was common knowledge long before the phlogiston theory was ever proposed. It was common knowledge that metallic calxes weigh more than the metals from which they have been prepared. Since calcination is slow combustion, if Lavoisier's 1772 experiment refutes phlogiston theory, then phlogiston theory was *born* refuted.

Falsificationists will have to say that phlogiston theorists ignored or twisted the facts.[10] And that is what they do say:

> In spite of these well-proved facts, the adherents of the theory of phlogiston ignored them, and it does not appear to have occurred to Becher or to Stahl that they were inconsistent with their theories.[11]

> From the very outset the theory was clearly inadequate, for it ignored two well established facts; first, that metals undergo an increase (instead of a loss) of weight when calcined; secondly, that a confined volume of air contracts (instead of expanding) when a body is burnt in it...A scientific theory ignoring such patent facts would not hold credence for a moment at the present day.[12]

[9] Hartley [1971], p. 7.

[10] They might, of course, emphasise the *new* feature of Lavoisier's 1772 experiment: that it showed a weight increase for a genuine combustic process, as opposed to a calcination or 'slow combustion'. But while this is important, its importance is *not* that Lavoisier's experiment refutes phlogiston theory while the earlier ones on calcinating metals do not. A falsificationist might also argue that Lavoisier's 1772 experiment was the first *unequivocal* demonstration that a combustic process leads to a weight increase, since reports about alcinating metals were confused and contradictory. But while there were early reports of weight *loss* in calcination, by the time the phlogiston theory was proposed these were discounted (see Gregory [1934], p. 221). In fact, even the weight increase of phosphorus had already been noted by phlogistonists.

[11] Ramsey [1905], p. 45.

[12] Read [1957], p. 121.

> [Phlogiston theory] was established in the face of facts which carried with them its refutation. When the first stage of its development was passed facts were adapted to the theory, and phenomena were tortured and garbled so as to fit in with it, by means of which the progress of science was somewhat retarded. Even when Lavoisier had conclusively proved the fallacy of the theory, this blind adherence shut the eyes of the phlogistians to the merits of the new system, and to the utter falsity of their own.[13]

In actual fact, however, phlogiston theorists neither ignored nor twisted these facts, they *explained* them: for example, Becher adopted Boyle's explanation, and Stahl invented a new one.[14]

A second difficulty for falsificationists is that Lavoisier's experiment of 1775, usually called the 'discovery of oxygen', was first performed by Scheele and Priestley, both of whom were phlogiston theorists and both of whom continued to adhere to the phlogiston theory. Falsificationists have to say that they were too stupid, dogmatic, or blinded by preconceived ideas, to appreciate the logical consequences of their own discovery: 'Priestley's blind adherence to the phlogiston theory in spite of his own effective discovery of oxygen and in spite of its obvious defects (such as the failure to account for the increase in weight on calcination) shows the hold that one conceptual scheme may have on the mind of an investigator.'[15] But *most* chemists continued to adhere to phlogiston theory long after 1775, when Lavoisier is supposed to have disproved it. Perhaps an *epidemic* of stupidity or dogmatism swept through European chemists, leaving only Lavoisier unscathed! Falsificationism is bankrupt if it must be supplemented with such dubious social psychology as this.[16]

These gross difficulties for the view that experiments disproved phlogiston theory and/or proved its rival are well known. And yet echoes of inductivism and of naive falsificationism are still be be found even among some of the best historians. It has been argued, for example, that *because* Lavoisier's oxygen theory is false, the Chemical Revolution cannot have consisted in the replacement of phlogiston theory by oxygen theory. The missing premiss here is the inductivist principle that a scientific revolution replaces falsehood by truth.[17]

[13] Rodwell [1868], p. 31.

[14] On Becher see Partington [1961], p. 650; on Stahl see Partington and McKie [1937–40], Part I, pp. 368–71. Partington and McKie conclude, from an examination of Stahl's rather obscure writings, that in his view phlogiston was an imponderable (or very light) substance which rendered lighter the substances it entered. Boyle's explanation of the weight increase of calxes is discussed below, pp. 189, 191–2.

[15] Conant [1950], p. 48.

[16] Some other bits of social psychology are often hinted at, but all are easily refuted: see below, pp. 206–7.

[17] For this argument see Siegfried and Dobbs [1968], p. 292. They claim that the Chemical Revolution really consisted in the triumph of Lavoisier's 'operational definition of chemical element'. But this definition was not new, and Lavoisier did not stick to it (as Siegfried and Dobbs point out on p. 291; see also below, p. 198, footnote 68). This is not, however, to deny their main thesis: that the victory of Lavoisier's programme focussed attention on the problem of the proliferation of chemical elements.

Many historians hint that the Chemical Revolution was brought about because Lavoisier had an accurate balance and used it carefully, insisting on the principle of conservation of weight and, more importantly, on the principle that only items which figured on the balance should figure in chemical explanations. His phlogistonist opponents, on the other hand, are supposed to have scorned accurate measurement, performed sloppy experiments, and structured their theory around a mysterious imponderable substance which could never be captured in a bottle.

Alas, this sort of tale is almost entirely mythical. The first accurate balance was made for Cavendish, a phlogistonist with a fetish about precision. Lavoisier did not get his most accurate balances until after 1785, when the battle against phlogiston was won.[18] Of course, more or less accurate weighings were made before then – but by *both* sides.[19] As for sloppy experiments, these too were not confined to phlogistonists. And as we will see, many phlogistonist experiments deemed 'sloppy' by dogmatist historians were not sloppy at all, but rather carefully conducted in the light of hypotheses which only *later* were found to be wrong. Phlogiston was *not* always claimed to be an imponderable substance which could not be isolated: some phlogistonists identified it with hydrogen which they isolated and weighed. Besides, an element called 'caloric' played an essential role in Lavoisier's own theory, and according to Lavoisier caloric was imponderable and could not be isolated.[20] Historians who like to emphasise Lavoisier's use of the balance engage in double-think when they tell us that he cleverly made caloric imponderable so that it never figured on his balance and could be ignored in the balance-sheets of his experiments. Lavoisier's imponderables were fine, phlogistonist imponderables sophistry and illusion![21]

3 *Conventionalist accounts*

Historians who realise that experiments neither proved the oxygen theory nor disproved its rival often fall back on *conventionalism* or *simplicism*. The two theories are regarded as alternative 'conceptual schemes' and it is tacitly admitted that both can accommodate the experimental results. It is then claimed that the oxygen theory superseded the phlogiston theory because it was a *simpler* 'conceptual scheme'. Simplicism is usually to be found mixed up with ingredients of inductivism and naive falsificationism, since most historians are eclectic about philosophy of science. But it is quite common to find statements like: 'The greater simplicity of Lavoisier's

[18] See Daumas [1963], p. 429.

[19] And even, on occasions, by Priestley: see Hartog [1941], p. 45.

[20] See Lavoisier [1790], pp. 6, 19, and Fourcroy's footnote on p. 202 of Kirwan's [1789].

[21] Students of this kind of double-think might care to consult Meyer [1906], pp. 177–8; Gregory [1934], pp. 190, 203, 209; Aykroyd [1935], p. 86; or Daumas [1963], p. 429.

system, and its greater coherence, undoubtedly prevailed over the complexities of phlogistic chemistry.'[22]

Conventionalism is a slippery philosophy. In our case the main problem in getting to grips with it lies in specifying the exact sense in which the oxygen theory is claimed to be *simpler* than the phlogiston theory. Conventionalist historians do not tell us. But there is a straightforward answer: the oxygen theory was simpler because the analyses which it gave of chemical reactions involved fewer elements than those provided by phlogistonists.[23] This can be supported as follows. According to phlogiston theory, when a metal calcinates it loses its phlogiston; but to explain the weight increase, some phlogistonists added that the calx also *gains* something. Lavoisier, on the other hand, simply dispenses with phlogiston and says that when a metal calcinates it combines with oxygen from the air. The phlogistonist analysis involves *three* elements, the anti-phlogistonist analysis only *two*. What better example of Occam's razor could there be?

Alas, this argument falsifies history. Lavoisier did not regard oxygen gas as a simple element, but as an 'igneous combination' of an 'oxygen base' and caloric. When a metal calcinates it seizes the oxygen base and sets free the caloric in the form of heat.[24] Here three elements are involved, as in the phlogistonist account. Indeed, the French chemists explicitly repudiated the two-element analysis of calcination:

> We do not therefore affirm, that vital air combines with metals to form metallic calxes, because this manner of speaking would not be sufficiently accurate: but we say, when a metal is heated to a certain temperature...it becomes capable of decomposing vital air, from which it seizes the base, namely *oxigene*, and sets the other principle, namely the caloric, at liberty.[25]

Thus, in this very simple sense of 'simple', the oxygen theory was no simpler than the phlogiston theory.[26] No doubt conventionalists could try to elaborate a more refined sense of the word 'simple' in which the oxygen theory was simpler than its rival. I shall not discuss this possibility, for it is high time that I turned to the account of the Chemical Revolution given by Lakatos's methodology. This methodology incorporates a sophisticated falsificationist position, and it also provides us with a new *historical* notion

[22] Gregory [1934], p. 211.

[23] Strangely enough, at one point Priestley used this crude notion of simplicity to claim that Lavoisier's theory was simpler than his own; but, he added, it was not so well supported by experiment (see Priestley [1779–86], vol. 3, p. 419).

[24] Thus, chapter 7 of Lavoisier's [1790] concerns the 'decomposition of oxygen gas' during calcination. Modern textbook writers, able to think of no higher praise, often say that Lavoisier's [1790] 'reads like a modern textbook'. But there is much in Lavoisier's exciting and revolutionary book which is not to be found in any modern textbook.

[25] Lavoisier *et al.* [1787], p. 288.

[26] Conventionalists might reply that in counting elements for simplicity comparisons we should ignore imponderables like caloric. But this conventionalist twist in conventionalist historiography will not help, for then we will have to ignore the imponderable phlogiston of some phlogistonists.

of simplicity (or rather, of the coherent development of a research programme). So we will, in the end, locate some grains of truth in falsificationism and conventionalism.

According to Lakatos, the study of the Chemical Revolution should reveal, upon inspection, competing research programmes. Each programme gives rise to a historical series of scientific theories, each of which has the same 'hard core' but with different 'protective belts' of auxiliary hypotheses. One happy result of this methodology should already be apparent: for it is implicit in what I have said so far that there was not one 'phlogiston theory' but many. Again according to Lakatos, we should expect to find that though experiments may refute particular theories within a research programme, they do not by themselves have the power to overthrow the entire programme. Adherents of a research programme will always try to accommodate such 'refutations' or 'anomalies' by developing new versions of the programme. According to Lakatos, a research programme can only be overthrown, and a scientific revolution occur, if there is a rival programme superior to the first. And one programme is superior to another, in the simplest case, if the first is progressing (each version makes novel predictions which are confirmed) while the second is degenerating (each version is *ad hoc* with respect to its predecessor). Crucial experiments between the two programmes can only be located with hindsight: they confirmed novel predictions of the progressive programme, and were only accommodated *ad hoc* in the degenerating one. At the time they were performed, such experiments could hardly have been regarded as crucial ones. These are the bare bones of Lakatos's methodology.[27] Let us now see how far the historical conjectures to which it leads are borne out in the story of the Chemical Revolution. I will begin with a brief look at the phlogiston programme before Lavoisier came on the scene.

4 *Phlogiston before Lavoisier*

The programme originated with Becher's claim that combustibles contained an 'inflammable principle' which they released upon burning. Since metals upon heating turned into powdery substances like ashes, it was also claimed that they too contained the 'inflammable principle', and that calcination was slow combustion. And since metallic ores, when heated with charcoal, turned into metals, it was said that in smelting ores we supply the 'inflammable principle' to them. The 'principle of inflammability' was thus transferred from a combustible (the charcoal) to a metal. But could it be transferred to another genuine combustible? Burning sulphur yielded vitriolic acid fumes. Stahl (who coined the term 'phlogiston') fixed these fumes in potash, and heated the resulting salt

[27] For further details see Lakatos [1970] and this volume, pp. 9–15.

(potassium sulphate) with charcoal to obtain 'liver of sulphur'. Since 'liver of sulphur' resulted from mixing sulphur and potash, Stahl concluded that the phlogiston from the charcoal had combined with the vitriolic acid fumes to produce sulphur.[28] Phlogiston, or the 'principle of inflammability', could be transferred from one combustible to another. Lavoisier later called this the 'great discovery of Stahl'.[29]

The phlogiston programme initially progressed: it gave a unified explanation of the apparently distinct phenomena of combustion and calcination, and Stahl managed to confirm a novel prediction. But there were many anomalies. Some well-known facts about combustion were not explained: why does combustion soon cease in an enclosed volume of air, and why is the volume of air reduced by it; why won't things burn at all in a vacuum?[30] Worse still, other well-known facts seemed to refute the theory: why, if calcination is the release of phlogiston, do calxes weigh *more* then the original metals?

The first two anomalies were dealt with by adding auxiliary hypotheses. Phlogiston must be carried away from a combustible by the air, and a given volume of air can only absorb a certain amount of it. Hence nothing will burn in a vacuum, and combustion soon ceases in a confined space. As for the reduction in volume of the air, we need only suppose that air saturated with phlogiston ('phlogisticated air') takes up less room than ordinary air (just as cotton-wool saturated with water takes up less room than ordinary cotton-wool).[31]

The third anomaly, the weight increase of calxes, was more troublesome, but it did not compel rejection of phlogistonism. We have here a nice example of the Duhem thesis. Phlogiston theory *alone* does not entail that calcination will lead to a weight loss. (By phlogiston theory I mean the emerging hard core of the phlogiston programme, the thesis that combustion and calcination involve the release of phlogiston.) To derive such a prediction we need the following additional premisses: phlogiston has weight, nothing weighty is added to the metal as it calcinates, and if something weighty is removed in a process, and nothing weighty added, then the result will weigh less than the original. The observed weight increase contradicts the conjunction of phlogiston theory with these additional premisses. One could resolve the inconsistency by rejecting

[28] See Watson [1781], vol. 1, pp. 168–9, or Partington [1937], pp. 87–8.

[29] See the 'Reflections on Phlogiston' (1783) in Lavoisier [1862], p. 625.

[30] The first anomaly is discussed, for example, by Bacon [1620], Book II, Aphorism I. Boyle used the second to refute the view that flames die in enclosed spaces because they are stifled by their own vapours.

[31] See the account given by Samuel Williams, cited by Conant [1950], pp. 15–16. The idea that phlogiston must be carried off by air perhaps explains Stahl's view that phlogiston 'cannot...be found by itself, outside of all compounds and unions with other bodies' (see Leicester and Klickstein [1952], p. 61). Scheele had the same view (see Scheele [1931], pp. 137–9), as did Watson (see Watson [1781], vol. 1, p. 167). But in the 1766 version of phlogistonism this view was denied.

phlogiston theory: Lavoisier did just that, and inaugurated a rival programme. But one could also stay within the phlogiston programme by rejecting one or more of the additional premisses.

Incidentally, nobody at the time bothered to spell out the additional premisses as I have done. They were, as Lakatos would put it, 'hidden lemmas' which first saw the light of day when they were made the target for the arrow of *modus tollens* and denied. In science, as in mathematics, theories get articulated under the impact of criticism. I have had to reconstruct this particular version of phlogiston theory from subsequent versions which were inconsistent with it.

Clearly, phlogistonists had several options for accommodating the weight increase. All of them, and a few more besides, were explored.[32] In 1772 de Morveau said that if phlogiston is lighter than air, then removing it from a body immersed in air will cause that body to weigh more (that is, he rejected the last additional premiss). Earlier, several phlogistonists harked back to Aristotelianism and ascribed negative weight or 'levity' to phlogiston (that is, they rejected the first additional premiss). This hypothesis of the levity of phlogiston has caused much levity among historians ever since; Scheele, Priestley, Cavendish and Kirwan thought it funny too, and would have nothing to do with it. The third broad option was to say that the weight of calxes was augmented by something added to them as their phlogiston was released (which involves rejecting the second additional premise). This third option leaves it open whether phlogiston is an imponderable substance or an extremely light one; it also leaves it open exactly what is the 'secondary augmenter' of the calx.

Before the phlogiston programme got underway, Boyle claimed that the weight of calxes was augmented by 'fire particles' which stuck to them: 'It is no wonder that, being wedged into the pores, . . . the accession of so many little bodies, that want not gravity, should, because of their multitudes, be considerable upon a balance.'[33] Several phlogistonists took over this ready-made solution to the anomaly. Earlier still, Rey claimed that the weight increase of calxes 'comes from the air, which in the vessel has been rendered denser, heavier, and in some measure adhesive, by the vehement and long continued heat of the furnace: which air mixes with the calx (frequent agitation aiding) and becomes attached to its most minute particles: not otherwise than water makes heavier sand which you throw into it and agitate, by moistening and adhering to the smallest of its grains'.[34] In the early 1770s Priestley incorporated Rey's explanation

[32] The various phlogistonist accounts of the weight increase are exhaustively surveyed by Partington and McKie [1937–40], especially in Parts I and II.

[33] Boyle [1673], 'Fire and Flame weighed in a Balance'.

[34] Rey [1630], Essay XVI, pp. 36–7. Bayen claimed as early as 1775 that Rey anticipated Lavoisier. And inductivist historians suffering from the who-got-the-truth-first syndrome echo him: 'It was nearly a century before Rey's conclusions were again arrived

into phlogistonism: the 'air' which is 'precipitated' into the calx is phlogisticated or 'fixed air' (together, perhaps, with some water) which has been formed by the calcination process.[35] The discovery that calxes yielded oxygen upon reduction to metals provided yet another candidate for the secondary augmenter of the calx.

The other open question, the imponderability or otherwise of phlogiston, seemed to be settled by the next phase of the programme. If metals contain phlogiston, and vitriolic acid is sulphur deprived of phlogiston and dissolved in water, then perhaps a metal will calcinate in vitriolic acid. But metals in acids effervesced and formed salts. In 1766 Cavendish immersed zinc, iron and tin in vitriolic and hydrochloric acids, and collected the gas given off. He found that it was eleven times lighter than common air, and highly inflammable. Cavendish concluded that when metals are immersed in acids 'their phlogiston flies off, without having its nature changed by the acid, and forms inflammable air'.[36] If this was correct, then metallic calxes immersed in acids should form the same salts, but no 'inflammable air' should be released. This was confirmed by experiment – a brilliant success.[37]

The identification of phlogiston with 'inflammable air' (hydrogen) did pose a few puzzles. 'Airs' rich in phlogiston were supposed to *inhibit* combustion, yet pure phlogiston burns! When things burn they are supposed to *release* phlogiston, so it would seem that when phlogiston burns it is released from itself! Finally, metals immersed in *concentrated* vitriolic acid yielded no 'inflammable air' at all. Cavendish explained this by saying that here the 'inflammable air' combines with some of the acid to produce 'volatile sulphureous acid', a half-way stage between vitriolic acid and sulphur. All this made Cavendish tentative about identifying 'inflammable air' with phlogiston; and in 1784 he proposed a new version of phlogistonism in which this identification was denied. Yet the appeal of his 1766 version of the theory was obvious. As Kirwan put it, phlogiston 'was no longer to be regarded as a mere hypothetical substance, since it

at' (Holmyard [1925], p. 61); Rey's explanation is 'closely in accord with Lavoisier's' (Conant [1951], p. 172). But this involves another falsification of history. Rey does not claim, as Lavoisier was to claim, that the addition of part of the air *forms* the calx: 'air' merely *mixes* with the calx, which is formed by heat. Rey makes it clear later that in combustion proper, and even in some calcinations, there is a weight loss because the 'air' gained is outweighed by what is lost (see Rey [1630], Essays XXVII and XXVIII pp. 53–4). What Agassi calls 'black-and-white history' is invariably falsified history. If Rey anticipated anybody, it was those later phlogistonists who said that calxes were formed by the release of phlogiston and that their weight was augmented by 'air' (or some part of it) which got mixed with the calxes.

[35] For this explanation, see Priestley [1775–7], vol. 1, pp. 192ff.

[36] Cavendish [1766], 'Part I: Containing Experiments on Inflammable Air', p. 147. Boyle had previously obtained an 'inflammable fume' by pouring acid on iron filings; see Boyle [1673], 'On the difficulty of preserving flame without air'.

[37] It is Holmyard, in his [1925], p. 60, who describes this as a 'brilliant success' for phlogiston theory. The experiment was not new, but the explanation was.

could be exhibited in an aerial form in as great a degree of purity as any other air'.[38] Priestley also came to accept the identification (along with Bergman, de Morveau and De la Métherie), and in the early 1880s provided spectacular confirmations of this version of the theory. We will come to these in due course, for we must now turn to Lavoisier and to the birth of the oxygen programme.

5 *The birth of the oxygen programme*

In 1772 Lavoisier burned sulphur and phosphorus in air confined over water, and noted the reduction in the volume of the air and the increase in weight of the sulphur and phosphorus. The experiment was not new, but Lavoisier's interpretation of it (contained in three sealed notes deposited with the secretary of the French Academy) was. In the first note (dated 10 September 1772) Lavoisier states that when phosphorus burns, air is absorbed. In the second note (dated 20 October 1772) he remembers the phlogiston theory, and says that as phosphorus releases phlogiston it absorbs air. Finally, in the third note (dated 1 November 1772) he takes the bold, though apparently obvious step of dispensing with phlogiston. Though he does not say so explicitly, the 'fixing' of a quantity of air is to explain both the burning and the weight increase:

> About eight days ago I discovered that sulphur in burning, far from losing weight, on the contrary gains it; it is the same with phosphorus; this increase of weight arises from a prodigious quantity of air that is fixed during combustion and combines with the vapours.
>
> This discovery, which I have established by experiments, that I regard as decisive, has led me to think that what is observed in the case of sulphur and phosphorus may well take place in the case of all substances that gain in weight by combustion and calcination; and I am persuaded that the increase in weight of metallic calxes is due to the same cause.[39]

Lavoisier next tests his theory of the weight increase of calxes against Boyle's theory, which had been adopted by some phlogistonists. On Boyle's theory, if a metal is calcinated in a closed container the weight increase comes from *outside* the container – on Lavoisier's, it comes from *inside* the container. Boyle had not weighed the entire container and its contents before and after the calcination, but only the metal and the calx. When Lavoisier did the former weighings, he found no overall weight increase: what augments the calx must come from inside the

[38] Kirwan [1789], p. 5. The story of Kirwan's *Essay on Phlogiston* is an amusing one. The first edition of 1777 was translated into French (probably by Mme Lavoisier) in 1788 along with critical footnotes by the French chemists. Whereupon Nicholson translated it back into English and Kirwan added new notes criticising the anti-phlogistic footnotes.

[39] Cited from Conant [1950], pp. 16–17. The analysis of the three sealed notes is due to Meldrum [1930].

container.[40] This was a success for Lavoisier, and a defeat for one version of phlogistonism.[41]

But how could Lavoisier explain the fact that combustion ceases before an enclosed volume of air is 'fixed'? He elaborates the theory further by postulating that ordinary air is composed of two very different substances, a 'pure part' which supports combustion and another part ('mephitic air') which does not.[42] His theory now predicted that if we reduce a calx to a metal, the 'pure part' of the air should be released. But the reduction of calxes always yielded 'fixed air', which will not support combustion. Undaunted, Lavoisier explains this apparent refutation away:

> The principle which combines with metals during their calcination, which increases their weight and constitutes them in the state of a calx, is nothing other than the...purest part of the air and such that, if the air, after having engaged in a metallic combination, becomes free again, it appears in an eminently respirable state...
>
> The majority of calxes are not to be reduced...without the immediate contact of a carbonaceous material...The charcoal which is used is completely destroyed in this operation if it is present in suitable proportion; whence it follows that the air which is evolved in metallic reductions with carbon is not a simple substance but in some manner is the result of the combination of the elastic fluid disengaged from the metal and that disengaged from the carbon. Therefore the fact that this fluid is obtained as fixed air gives us no right to conclude that it existed in this form in the metallic calx before its combination with the carbon.
>
> These considerations showed me that in order to clear up the mystery of the reduction of metallic calxes it would be necessary to experiment with those calxes which are reducible without the addition of anything.[43]

The reasoning is faultless, a new chemical composition for Black's 'fixed air' is suggested, and the discovery of the 'purest part of air' (oxygen) is predicted.

At this point it would be pleasant to relate that Lavoisier, having boldly predicted oxygen, went on to discover it. But in fact it was discovered before him by Priestley (and before Priestley by Scheele, who did not, however, publish his discovery until 1777). As Mark Twain said,

[40] See Lavoisier's 'On the Calcination of Tin in Closed Vessels and on the Cause of the Gain in Weight which this Metal acquires in this Operation', read to the French Academy in 1744; Lavoisier [1862], pp. 105–21. Actually, Lavoisier did detect a small overall weight increase, which he attributed to sooty deposits on the outside of the container. It was much less than the weight increase of the calx which, Lavoisier claimed, was equal to the weight of 'air' lost.

[41] Like most of Lavoisier's experiments, this one was not new: others, including Priestley, had already refuted Boyle in this way (see Gregory [1934], pp. 114–15 and 157–8).

[42] At the end of his 1774 memoir Lavoisier writes: 'the air of the atmosphere is not a simple substance at all but is composed of very different substances' (Lavoisier [1862], p. 120). In December 1773 he had written: 'there exists in the atmosphere a particular elastic fluid which occurs mixed with the air' (Lavoisier [1864], p. 620). Priestley's view was different: see below, p. 193.

[43] From the beginning of Lavoisier's memoir 'On the Nature of the Principle which combines with Metals during their Calcination and which increases their Weight', read to the French Academy on 12 April 1775; Lavoisier [1862], p. 123.

'in real life the right thing never happens at the right place at the right time – it is the business of the historian to remedy this mistake'. Remedying history's mistakes is called by Lakatos the 'rational reconstruction' of history. And to the complaint that rationally reconstructed history is often a caricature of actual history, Lakatos replied that actual history is often a caricature of its rational reconstruction.[44] However this may be, we can say that in rationally reconstructed history, having predicted oxygen, Lavoisier went on to discover it – while in actual history it was discovered by Priestley, who had not predicted it, but who taught Lavoisier how to discover it.

Not that Lavoisier did not try: he tells us that in 1774 he tried to obtain 'the purest portion of the air' by reducing iron calxes without charcoal, but without success.[45] He must have been distressed to learn of Bayen's report, in February 1774, that the red calx of mercury could be reduced by heat alone, and that *fixed air was obtained*.[46] If correct, Bayen's report refuted Lavoisier's prediction.

But Bayen's report confirmed Priestley's theory. To explain this, we must examine the version of phlogistonism with which Priestley was operating in the early 1770s. Priestley tried to use phlogistonism to explain the different 'kinds of air' which had been discovered, several of them by himself. His basic idea was that the different 'airs' were all composed of some 'earth', 'nitrous acid' and phlogiston in varying proportions. As he put it:

> Upon the whole, I think, it may safely be concluded, that the purest air is that which contains the least phlogiston: that air is impure...in proportion as it contains more of that principle; and that there is a regular gradation from *dephlogisticated air*, through common air, and *phlogisticated air*, down to nitrous air; the last species of air containing the most, and the first mentioned the least phlogiston possible, the common basis of them all being the nitrous acid; so that all these kinds of air differ chiefly in the quantity of phlogiston they contain...[47]

Priestley added that the weight of calxes was augmented by 'fixed air' formed by the phlogistication of the air during the calcination and precipitated into the calx. According to Priestley, then, calxes contain 'fixed

[44] This volume, p. **36**.

[45] Lavoisier [1862], pp. 123–4.

[46] Bayen's calxes, like some of Priestley's, must have been impure. Bayen also reported that he sometimes obtained an 'air' which, unlike 'fixed air', was not soluble in water – this was probably oxygen. Yet Bayen blamed *this* upon impurities in the calx, and discounted it (see Gregory [1934], p. 178).

[47] This is Priestley's summary of the theory, written *after* the discovery of oxygen in a letter to Sir John Pringle of 25 May 1775; see Lindsay [1970], p. 151. Priestley's 'dephlogisticated air' is oxygen, 'phlogisticated air' is carbon dioxide, and 'nitrous air' is nitrogen (the last is not to be confused with his 'nitrous acid' which is our nitric oxide). The same theory runs through Priestley's [1775–7]: there he claims that '*atmospherical air*, or the thing that we breathe, *consists of the nitrous acid and earth*, with so much phlogiston as is necessary to its elasticity' (vol. 2, section iii), and that 'fixed air' (carbon dioxide) is 'a modification of the nitrous acid' which contains more phlogiston than atmospherical air (vol. 1, pp. 34ff, vol. 2, pp. 55–8, 217, 263ff, 314ff).

air'.[48] And Bayen reported that red calx of mercury, reduced by a burning lens, yielded 'fixed air'.

Priestley obtained a lens to check Bayen's report. He soon found that Bayen was wrong: the air from the calx could not be 'fixed air', for it would not dissolve in water. After several false trails, each of which he reports in characteristic style, Priestley decides that the air from the mercury calx is a new kind of air, which supports combustion better than ordinary air. He calls it *dephlogisticated air* and says that he discovered it 'by chance'.[49]

Before his discovery was published, Priestley visited Paris and over the dinner table told Lavoisier how he had made it. Lavoisier repeated the experiment, misinterpreted it as Priestley at one stage had also done, and came to a rather curious conclusion which conflicted with his earlier prediction. He thought that the new gas was '*not only* common air but that it was more respirable, more combustible, and consequently that it was more pure than even the air in which we live', and that any calx reduced without charcoal would give 'common air' instead of 'fixed air'.[50] Priestley read this first version of Lavoisier's paper, in which he was not mentioned, just as his own account was being printed. He added a critique of Lavoisier, gently correcting him.[51] Lavoisier read the critique, promptly revised his paper before it appeared in the official proceedings of the Academy, and concluded what he had earlier predicted (again without mentioning poor Priestley). Combustion and calcination are combinations of substances with the 'pure part of the air'; 'fixed air' is a combination of the 'pure part of the air' with charcoal; any calx reduced without charcoal will yield the 'pure part of the air'.[52] Thus was oxygen discovered.

Here I would like to digress from the story to consider the concept o

[48] Priestley [1775–7], vol. 1, pp. 192ff.

[49] See Priestley [1775–7], vol. 2, section iii. Priestley's account is full of words like 'surprize' and 'astonishment'. At one point he apologises for 'the frequent repetition of the word *surprize*', only to continue 'the next day I was more surprized than ever I had been before'! He begins his story by saying that it illustrates how 'more is owing to... *chance*... than to any preconceived theory in this business', and objects to those who write '*synthetically*' instead of writing '*analytically* and ingenuously'. Lavoisier was, of course, a prime example of the 'synthetic' style. But the difference is not merely one of style: Lavoisier had predicted oxygen and Priestley had not, so that if Lavoisier had been prone to write 'analytically', words like 'delight' and 'satisfaction' would have replaced Priestley's 'surprize' and 'astonishment'.

[50] This is the conclusion of the first version of Lavoisier's famous 'Easter memoir', published in Rozier's *Journale de Physique*, May 1775. The revised version appeared in the 1775 volume of the *Mémoires de l'Académie des Sciences* which was not published until 1778 (see Lavoisier [1862], pp. 122–8). Translations of the two versions of the conclusion are printed side by side in Conant [1950], pp. 22–8.

[51] Priestley [1775–7], vol. 2, section xvi. Lavoisier was led to confuse oxygen with 'common air' by misinterpreting the result of applying Priestley's 'nitrous air test' to oxygen. Priestley had made exactly the same mistake before him, but by chance had managed to correct it. There is an excellent account of all this in Conant [1950].

[52] See Conant [1950], p. 27.

discovery. Historians of science devote much energy to questions like 'Who first discovered oxygen?' or 'Who first discovered the composition of water?' The questions are naive, I think, because implicit in them is the idea that discovery is an all-or-nothing affair for which the question 'Who did it first ?' makes sense. But discovering oxygen was not simply a matter of getting a sample of it into a bottle. Bayen seems to have done that in 1772, and Stephen Hales probably did it before him. And it might have been done, for all I know, by a chimpanzee let loose in a chemical laboratory (think of the same chimpanzee typing out a sonnet by random bashing of a typewriter). No, the discovery of oxygen also involved the *correct identification* of what had been isolated: it involved, in other words, a true hypothesis about what had been isolated.[53] So neither Hales nor Bayen discovered oxygen, though they might have isolated samples of it. Nor did Scheele, who thought his bottle contained 'fire air'. Nor did Priestley, who thought his bottle contained 'dephlogisticated air'. And nor, for that matter, did Lavoisier, who thought his bottle contained an igneous combination of an 'oxygen base' and caloric. But then who *did* first discover oxygen? I do not know and I do not care. All these men made important contributions towards its 'discovery'. We can say, if we like, that Lavoisier's discovery was the closest to what we think is the truth of the matter. (Notice that in speaking of Lavoisier's discovery, we must allow that 'discovery' ceases to be a success word, and that it makes sense to speak of *false* discoveries.) This apparently minor point has some major repercussions. For example, what seem to be cases of 'simultaneous discovery' may turn out, when their theoretical ingredients are taken into account, to be very different discoveries indeed (and so-called 'priority disputes' may turn out to be disputes about the truth or falsehood of what has been discovered).

Ignoring the digression, we can say that despite its comic aspects the discovery of oxygen was a dramatic success for the oxygen research programme. Was it an equally dramatic failure for phlogistonism? Priestley did not think so. He explained the experiment by saying that, since calxes contain 'fixed air' which contains phlogiston, when they are heated the phlogiston revives the metal and 'dephlogisticated air' is emitted.[54] But, said Priestley, this does not always happen: 'it will be

[53] Historians of astronomy have seen the point. It seems that both Uranus and Neptune had been seen and recorded before they were discovered, and historians refer to the 'pre-discovery observations' of them. The discoverer of a new planet is not the man who first gets it in his sights. Rather, he is the man who correctly identifies what he sees. (Adams and Leverrier are credited with the discovery of Neptune, yet I know of no evidence that either of them ever saw it!) It is a moot point, of course, whether Columbus discovered America.

[54] Priestley [1775–7], vol. 2, section iv. Priestley stuck to this explanation: 'I explain M. Lavoisier's experiments by supposing that *precipitate per se* contains all the phlogiston of the metal, but in a different state; but I can show other calxes which also contain more phlogiston than the metals' (Priestley to Franklin, 24 June 1782; see Bolton [1892], p. 37). Kirwan also accepted this explanation (see Hartog [1941], p. 47).

seen, in the course of my experiments, that several...[lead] calxes yield fixed air by *heat only*, without any addition of charcoal'.[55] Priestley could explain this *post hoc* by saying that not all of the phlogiston contained in the calx is needed to revive it, so that 'phlogisticated' or 'fixed' air is still driven out. Lavoisier could not explain it; if Priestley was right, then Lavoisier was wrong.

Lavoisier did not worry about Priestley's refutations. All he could say (though he did not, so far as I know, bother to say it) was that Priestley's calxes, and Bayen's before him, must have been impure in that they contained some source of carbon. And much later his followers showed that this was correct: lead calxes prepared by Priestley's methods often contain lead carbonates.[56] But in 1775 experiments on the reduction of calxes told as much against Lavoisier's theory as in favour of it.[57] Yet Lavoisier, emboldened by his successes and paying little heed to his failures, came out with his first direct attack on phlogistonism, his memoir 'On Combustion in General.[58]

Lavoisier's attack is an interesting one from a methodological point of view: it shows that he had a fine grasp of what Paul Feyerabend has called the *proliferation principle*, and a fine disregard for the great Newton's *Fourth Rule of Reasoning in Philosophy*. Lavoisier begins by admitting that the phenomena 'are explained in a very nice manner by the hypothesis of Stahl'. He then says that the 'system of Stahl will be shaken to its foundations' simply by the fact that the same phenomena 'may be explained in just as natural a manner by an opposing hypothesis'. And he proceeds to explain that hypothesis. The argument is quite correct. The proliferation of a rival hypothesis to explain the same phenomena does not, of course, show the existing hypothesis to be false. But it does shake the

[55] Priestley [1775–7], vol. 2, section xvi.

[56] See Conant [1950], p. 47. The first clue came in 1789 when Lavoisier showed that iron, and iron calxes, often contain carbon impurities (see Lavoisier [1862], p. 583).

[57] These were not the only experiments, of course, and Priestley's theory faced some special troubles of its own. According to Priestley, 'nitrous air' (nitric oxide, NO) contained more phlogiston than 'fixed air' (carbon dioxide, CO_2). Yet 'nitrous air' supported calcination while 'fixed air' did not, and 'nitrous air' is not the lightest air as it should be if the more phlogiston [an air] contains the lighter it is (Priestley [1775–7], vol. 2, section v). Priestley disposes of these anomalies with the *ad hoc* hypothesis that there is some difference in the 'mode of combination' of 'nitrous acid' and phlogiston in 'nitrous air'. But the trouble did not end there. If we calcinate a metal in 'nitrous air' we obtain what Priestley called 'phlogisticated nitrous air' (nitrous oxide or 'laughing gas', N_2O) which will *support combustion* (Priestley [1775–7], vol. 2, section iii). But how, if we further 'phlogisticate' the most phlogisticated of the airs, can we render it capable of supporting combustion? Wolff cannot believe that Priestley called a gas which supported combustion 'phlogisticated nitrous air', so he assumes a misprint and amends Priestley's text to read '*de*phlogisticated nitrous air' (see Wolff [1967], pp. 44, 55). Lindsay also uses this name on Priestley's behalf (see Lindsay [1970], p. 23). But how could calcination, which involves the *release* of phlogiston, dephlogisticate the 'air' in which it occurs? Priestley, dimly aware of the anomaly but too busy doing experiments to solve it, took to calling nitrous oxide '*modified* nitrous air'. But this hides the anomaly without solving it.

[58] Read to the French Academy on 5 September 1777; see Lavoisier [1862], pp. 225–33.

existing hypothesis to its foundations by showing that its alleged 'foundations in the phenomena' are spurious. Newton had seen the point, and proposed a rule banning the proliferation of hypotheses, adding 'This rule we must follow, that the argument of induction may not be evaded by hypotheses.'[59] Lavoisier inaugurated the Chemical Revolution by violating Newton's rule. (Notice that although Lavoisier had read Feyerabend by 1777, he had not yet read Lakatos; he talks about 'the system of Stahl' when the phlogiston programme had developed beyond Stahl's version of it. He did read Lakatos a little later, however.)

6 *The oxygen programme extended to acidity*

Having declared war on phlogiston, Lavoisier's next step was to try to conquer territory already occupied by the enemy.[60] The most important occupied territory was that of the *formation of acids*. Phlogistonists had realised that when certain substances are burned, or deprived of their phlogiston, they formed acids. It was fairly obvious what Lavoisier had to say: acids are formed when certain substances combine with oxygen.

Lavoisier's earliest experiments on the combustion of phosphorus and sulphur convinced him that 'acid of phosphorus' and 'vitriolic acid' contained 'air'. As we have seen, he soon realised that they contained only the 'pure part of the air'. Black's 'fixed air', also called 'aerial acid', contained it too (Lavoisier later called it 'carbonic acid'). After his 1775 experiments on calcination, he switched his attention to acidity. In 1766, by dissolving mercury in nitric acid solution, he decomposed nitric acid into 'nitrous air' (nitric oxide) and the 'pure part of the air'. He admitted that the experiments were borrowed from Priestley, but insisted that their 'consequences' were his: 'I hope that, if I am reproached for having borrowed my proofs from the works of this celebrated scientist, at least my property in the consequences will not be contested me.'[61] The 'consequence' in question was the general thesis that 'the purest part of the air enters into the composition of all acids without exception; that it is this substance which constitutes their acidity'.[62] He repeated his earlier experiments on the acids of phosphorus and vitriol, and claimed that they contained more than half their weight of 'pure or eminently respirable

[59] Newton's rule reads: 'In experimental philosophy we are to look upon propositions inferred by general induction from phenomena as accurately or very nearly true, *notwithstanding any contrary hypotheses that might be imagined,* till such time as other phenomena occur, by which they may either be made more accurate, or liable to exceptions' (Newton, *Principia*, Book III, my italics).

[60] In other words, Lavoisier wanted to show that his new theory involved no 'Kuhnian loss' in explanatory power. Kuhn's claim that replacing phlogiston by oxygen did involve a loss of explanatory power has been criticised, quite convincingly I think, by Noretta Koertge in her [1969].

[61] Lavoisier, 'On the Existence of Air in Nitric Acid, and on the Means of Decomposing and Reconstituting that Acid', read to the French Academy on 26 April 1776 (see Lavoisier [1862], p. 137).

[62] Lavoisier [1862], pp. 129–30.

air'.[63] Finally, in 1779 he coined the term 'oxygen' meaning 'acid-generator': 'Henceforth I shall designate dephlogisticated or eminently respirable air in a state of combination or fixity by the name of *acidifying principle*, or if a term of the same meaning is preferred in a Greek form, by the name of *oxygen principle*.'[64]

Lavoisier's oxygen theory of acidity predicts that the oxidation of organic compounds should produce acids. This prediction was so well confirmed that Lavoisier could later write that since 1766 'a new field of inquiry has been opened up to chemists; and, instead of five or six acids which were then known, near thirty new acids have been discovered'.[65] And yet, despite its success, the oxygen theory of acidity was false.

One of the best-known acids was 'acid of marine salt' or 'marine acid' or (as Lavoisier called it) 'muriatic acid'. Lavoisier tried and failed to extract oxygen from it.[66] His phlogistonist opponents also tried and failed, and they never tired of pointing to the anomaly.[67] Lavoisier was not impressed: in his *Elements of Chemistry* he insists that we 'cannot have the smallest doubt' that marine acid contains oxygen, though he later admits that he cannot extract it and reaches this indubitable conclusion 'from analogy with other acids'.[68] Lavoisier was, of course, wrong, as Davy showed in 1808: marine acid is hydrochloric acid, which contains no oxygen, while what Lavoisier called 'oxygenated marine acid' is actually an element, chlorine.[69] But in 1778 marine acid was a *minor* anomaly for Lavoisier, compared with some of the others.

[63] Lavoisier [1862], p. 143.

[64] Lavoisier, 'General Considerations on Acids', read to the French Academy on 23 November 1779 (see Lavoisier [1862], p. 249). In his *Elements of Chemistry* Lavoisier says that 'oxygen is an element common to... all [acids], which constitutes their acidity' (Lavoisier [1970], p. 65). There, of course, the theory had been developed into the theory of *degrees of oxygenation*: the lowest degree of oxygenation converts a body into an oxide; the second degree converts them into acids terminating in *-ous* (e.g. nitrous and sulphurous acids); the third degree converts them into acids terminating in *-ic* (e.g. nitric and sulphuric acids); the fourth degree converts them into oxygenated acids (e.g. oxygenated nitric acid). See Lavoisier [1790], p. 80.

[65] Lavoisier [1790], p. 191. The superiority of Lavoisier's oxygen theory of acidity over Stahl's more restricted phlogistonist one has been shown by Koertge [1969], p. 224. For a further defence of Lavoisier's theory, see Crosland [1973].

[66] In his notebooks Lavoisier recorded 'the possible exception of that [acid] from marine salt' (see Gillispie [1960], p. 227).

[67] Cavendish pointed out that marine acid was reluctant to part with its dephlogisticated air (Cavendish [1784–5], p. 38). Priestley claimed that 'the marine acid differs essentially from both the vitriolic and nitrous in this, that it cannot, by any combination whatever, be made to yield dephlogisticated air, at least with the degree of heat I was able to apply' (Priestley [1779–86], vol. 3, p. 424).

[68] Lavoisier [1790], pp. 71, 223. Lavoisier's theory of acidity led him to violate his own 'operational definition' of chemical element: he listed the 'base of muriatic acid' or 'murium', and the bases of fluoric and boracic acids, as elements although he never obtained them as the last points of chemical analysis (see Lavoisier [1790], p. 175).

[69] See Davy [1929]. Phlogistonists were right about the relationship between marine acid and chlorine: what Lavoisier regarded as 'oxygenated muriatic acid' they regarded as 'dephlogisticated marine acid'. For a while Davy flirted with the idea of trying to resurrect the phlogiston programme (on this see Siegfried [1964]).

In 1766 Cavendish had tentatively identified the 'inflammable air' released when metals dissolve in acids with phlogiston from the metal. What could Lavoisier say about this air? Clearly, it must be liberated from the acid as the metal seizes oxygen. It follows that if we restore oxygen to 'inflammable air' by burning it, an acid compound should be produced. Lavoisier did the experiment repeatedly, but could detect no compound: the 'inflammable air' just seemed to disappear into thin air![70]

While Lavoisier was failing, Priestley was having great success with the 1766 version of phlogistonism. This version predicted that combustibles and metals contain 'inflammable air' (now identified with phlogiston). Priestley managed to extract 'inflammable air' from charcoal by heating it with a lens and by passing steam over glowing charcoal. He wrote triumphantly:

> The experiment which, in my opinion, proves decisively that the principle which has hitherto been called phlogiston is a real *substance*, and even adds considerably to the *weight* of bodies, is that of the decomposition of charcoal by the heat of a burning lens, or by means of steam in a hot earthen tube.[71]

Priestley also claimed to have extracted 'inflammable air' by calcinating iron filings, and by heating 'finery cinder' (Fe_3O_4) with charcoal. But the most impressive experiment of all came in early 1783.

If 'inflammable air' is phlogiston, then we ought to be able to reduce calxes to metals by heating them in it. Priestley focussed his burning lens on lead calxes in 'inflammable air' confined over water. Much to his delight, the calxes 'ran in the form of perfect lead, at the same time that the air diminished at a great rate, the water ascending within the receiver'.[72] The conclusion was plain enough:

> I could not doubt but that the calx was actually imbibing something from the air; and from its effects in making the calx into metal, it could be no other than that to which chemists had unanimously given the name of *phlogiston*.[73]

This experiment was symmetrical with Lavoisier's 1772 experiment on the combustion of phosphorus, and so was Priestley's interpretation of it: the 'inflammable air' must be entering the calx and reviving it.[74]

Priestley had joined the Lunar Society of Birmingham in 1781, and found to his horror that its members were flirting with the French system of chemistry. They were even circulating amongst themselves a humorous

[70] Lavoisier's failures are recorded in his laboratory notebooks for 1777 and 1781–2; see Gregory [1934], p. 197 or Hartley [1971], pp. 32, 42–3.

[71] Priestley [1779–86], vol. 3, p. 422. The original experiments were done in 1782; see Priestley to Wedgwood, 16 September 1782, Bolton [1892], p. 39.

[72] Priestley [1783], p. 400.

[73] Priestley [1783], p. 402. Priestley can now afford to admit that 'The arguments in favour of [Lavoisier's] opinion, especially those drawn from the experiment...made on mercury, are so specious, that I own I was myself much inclined to adopt them' ([1783], p. 400).

[74] The symmetry of the experiments is pointed out by Toulmin in his [1957]. Bergman also dramatically confirmed 1766 phlogiston theory: see Stillman [1924], pp. 450–1.

piece entitled *The Birth, Life and Death of Phlogiston*.[75] Armed with his new experiments, Priestley soon whipped the faint-hearts back into line:

This simple experiment seems to prove, that what we have called *phlogiston* is the same thing with *inflammable air*.[76]

My experiments are certainly inconsistent with Mr Lavoisier's supposition of there being no such thing as phlogiston, and that it is the addition of air, and not the loss of anything that converts a metal into a calx...I reduce [calxes] to a perfect metallic state by nothing but inflammable air, which they imbibe *in toto*.[77]

Before my late experiments phlogiston was indeed almost given up by the Lunar Society, but now it seems to be re-established.[78]

Re-established it was: Mathew Boulton exclaimed:

We have long talked of phlogiston without knowing what we talked about, but now Dr Priestley hath brought ye matter to *light*. We can pour that Element out of one Vessell into another, can tell how much of it by accurate measure is necessary to reduce a Calx to a Metal, ...In short, this Goddess of levity can be measured and weighed like other Matter.[79]

Wedgwood was 'quite delighted with the resurrection of poor Phlogiston, as we had been old friends & I could not so well at my time of life supply his place with another'.[80] As late as 1788 Wedgwood was expressing to Priestley his 'particular satisfaction to find that my old favourite, Phlogiston, is likely to be restored to its former rank in the Chemical World'.[81]

Meanwhile, back in Paris, Lavoisier had enlisted Laplace's help to try once again to confirm his novel prediction that an acid compound should be produced when 'inflammable air' burns. Other help was on its way, however. And ironically enough , it came from Cavendish, the originator of the version of phlogiston theory which Priestley was busy confirming!

7 *The composition of water*

Several chemists, including Priestley, had noticed that when 'inflammable air' burned a 'dew' was deposited on the walls of the vessel. All had ignored it as an irrelevant side-effect.[82] But Cavendish, unhappy with the idea that when phlogiston burned it was released from itself, decided to investigate. Perhaps the 'dew' is what is left behind as phlogiston is

[75] See Schofield [1963], p. 290.

[76] Priestley to Wedgwood, 6 March 1782; see Bolton [1892], pp. 33–4.

[77] Priestley to Franklin, 24 June 1782; see Bolton [1892], pp. 36–8.

[78] Priestley to Wedgwood, 21 March 1782; see Bolton [1892], pp. 35–6.

[79] Boulton to Wedgwood, 30 March 1782; see Schofield [1963], p. 290.

[80] Wedgwood to Boulton, 8 April 1782; see Schofield [1963], p. 290.

[81] Wedgwood to Priestley, 17 October 1788; see Bolton [1892], p. 97, or Schofield [1963], p. 291.

[82] The 'dew' had been noticed by Scheele, Priestley, Macquer and de la Fond (see Gregory [1934], pp. 194–5). Priestley did the experiment to try to ascertain whether the heat of the explosion, radiated outside the vessel, had weight.

released from 'inflammable air'. The 'dew' turned out to be pure water. And Cavendish found that when 'inflammable air' was burned in 'dephlogisticated air', the 'airs' disappeared in the ratio two to one.[83] Cavendish developed yet another version of phlogistonism to account for this, as we will see.

In June 1783, before Cavendish's results were published, his assistant, Blagden, visited Paris and told Lavoisier about them. Lavoisier was delighted: he repeated the experiment the very next day, and announced its results (without mentioning Cavendish, but mentioning Cavendish's figure!). No wonder he was excited: the long-sought 'oxide of inflammable air' was actually *water*.[84] The outstanding anomalies facing his theory could now be solved.

Blagden also told Lavoisier about Priestley's experiments on the reduction of calxes in 'inflammable air'. Lavoisier knew immediately how to dispose of these apparent refutations. After giving a sympathetic account of Priestley's experiments, he continues:

> I observe that Mr Priestley has not paid attention to one capital circumstance which takes place in this experiment, that the lead, far from increasing in weight, on the contrary diminishes by almost a twelfth: it gives up, therefore, some substance; now this substance must be vital air, of which the minium contains about a twelfth. However there does not remain, after this experiment, any type of elastic fluid; not only does one not recover vital air in the bell-jar, but the inflammable air itself, with which it was filled, disappears: hence the products are no longer in an aeriform state; and since it is also obvious that water is a compound of inflammable air and dephlogisticated air, it is clear that Mr Priestley has formed water without suspecting it.[85]

Priestley had not noticed the water which was formed because he had done the experiment *over* water. A novel prediction was obvious: re-do the experiment over mercury and you will detect the water produced in it. Ironically, it was Priestley who confirmed this novel prediction in 1785. He was probably led to repeat his experiment over mercury not by Lavoisier but by Cavendish whose theory also demanded that water be produced in it.[86] Priestley found that his calxes 'began to revive, the inflammable air rapidly disappeared, and water was formed on the sides of the vessel in which the experiment was made'.[87] As a result, Priestley switched to Cavendish's 1784 version of phlogistonism, which he defended against his own and Kirwan's identification of phlogiston with hydrogen.[88]

[83] Cavendish [1784–5].

[84] Lavoisier's interpretation eventually appeared in his memoir 'On the Nature of Water and on Experiments that appear to Prove that this Substance is not Properly Speaking an Element, but can be decomposed and recombined', read to the French Academy on 12 November 1783 but published in its *Mémoires* for 1781 (see Lavoisier [1862], pp. 334–59).

[85] Lavoisier [1862], pp. 344–5.

[86] Cavendish's 1784 theory will be discussed below, pp. 203–5.

[87] Priestley [1785], p. 304. In 1787 Lavoisier was referring with glee to Priestley's 1785 experiments as 'overthrowing' phlogiston theory (see the passage cited in Kirwan's [1789], pp. 17–18).

[88] See Priestley [1785], pp. 299–300; also Gregory [1934], pp. 209–10.

Was Priestley's original experiment a 'sloppy' one? Was it careless of him to confine his 'inflammable air' over water, making it impossible to detect the water produced? It was not. Nobody dreamt in 1782 that water might be produced. So why not do the experiment over water, especially since neither 'inflammable air' nor the oxygen which Priestley's opponents expected to be produced was readily soluble in water? Priestley's experimental report of 1782 was false, but that does not mean that his experiment was a sloppy one. Lavoisier was quite right to point out that Priestley had 'not paid attention' to the decrease in weight of the calxes. But Lavoisier's own figures (that they decreased in weight 'by almost a twelfth') were not derived from weighings made in *this* experiment; they were guesses derived from Lavoisier's general theory and previous weighings. Priestley did not bother to weigh his calxes in this case because for him the *qualitative* result was enough to refute Lavoisier.

What of Priestley's other confirmations of 1766 phlogistonism, in which he extracted 'inflammable air' from combustibles? Lavoisier could explain these too: the combustibles must have contained some water which was decomposed into oxygen and 'inflammable air'. Priestley never accepted the general idea that 'inflammable air' always comes from decomposing water. He pointed out that he always heated his samples gently beforehand to expel any moisture they might contain. He added that if we pass steam *slowly* over glowing charcoal we get no 'fixed air' (as Lavoisier's theory demands) but *only* 'inflammable air'. He returned to these refutations again and again.[89]

Priestley was half right and half wrong. Charcoal *did* sometimes yield an 'inflammable air' without water being decomposed. But the 'inflammable air' was not hydrogen but a 'gaseous oxide of carbon' (carbon monoxide). Lavoisier's followers showed this only in 1801, and cleared up this anomaly.[90]

Again, it is sometimes said that Priestley's experiments were sloppy ones, and that it was careless of him not to have distinguished hydrogen and carbon monoxide. In fact, however, Priestley *did* distinguish between 'inflammable air from metals' (hydrogen) and 'inflammable air from charcoal' (carbon monoxide), including their different specific gravities.[91] For him, however, this was a distinction without very much of a difference: both were 'airs' with slightly different degrees of phlogistication, and so, having distinguished them, he could go on to treat them as one. The *point* is that Priestley had the wrong theory, not that he made sloppy experiments.

[89] They appear for the last time, for example, in his [1804].

[90] Carbon monoxide was distinguished from hydrogen independently by Désormes and Clement and by Cruikshank in 1801. Cruikshank's paper sparked off a controversy with Priestley (for details see Davis [1927] and Schofield [1966], p. 326). Interestingly, Lavoisier had listed 'oxyd of charcoal' as being 'unknown', 'acid of charcoal' being 'fixed air' (see Lavoisier [1790], p. 207).

[91] See Priestley [1785], and [1779–86], vol. 3, pp. 162–88.

The discovery of the composition of water also enabled Lavoisier to account for the production of inflammable air when metals are dissolved in acids. The metals oxidise, decomposing water and releasing inflammable air, and the oxide combines with the acid to form a salt. Calxes, being already oxidised, simply form the salt. Metals dissolved in *concentrated* nitric or sulphuric acid yield no inflammable air because here water is not decomposed (this was a real anomaly for phlogistonists, who said that the 'inflammable air' came from the metal).[92]

Finally, Lavoisier deduced a startling new prediction: that water, traditionally used to put out fires, should, since it contains oxygen, support slow combustion and yield hydrogen. Iron filings immersed in water did indeed rust and hydrogen was collected.[93]

Encouraged by all these successes, Lavoisier came out with his second attack on phlogistonism. Again, he displays his methodological sophistication: his chief complaint against phlogistonism is that it consists of a series of *ad hoc* devices, mutually inconsistent with each other. He concludes a beautiful summary of the history of the doctrine with the famous statement:

> Chemists have made a vague principle of phlogiston which is not strictly defined, and which in consequence accommodates itself to every explanation into which it is pressed. Sometimes this principle is heavy and sometimes it is not; sometimes it is free fire and sometimes it is combined with the earthy elements; sometimes it passes through the pores of vessels and sometimes they are impenetrable to it... It is a veritable Proteus which changes its form every minute.[94]

All these things, and more, had been said of phlogiston during its long history. In its struggle to keep up with the facts, especially those predicted and discovered by oxygen theorists, confused and contradictory properties had been built into the various versions of the theory. Lavoisier was groping towards the complaint that phlogistonism represented a *degenerating research programme*. (He had obviously been reading Lakatos between 1777 and 1783!)

But a degenerating programme can soldier on, and phlogistonism did just that. Proteus changed his form once more, as Cavendish developed a new version of phlogistonism in order to accommodate the discovery that 'inflammable air' exploded in 'dephlogisticated air' formed water. According to this 1784 version, 'inflammable air' is no longer to be identified with phlogiston. Instead, it is 'phlogisticated water', whereas

[92] See Lavoisier [1862], pp. 341ff., 509–27. It was Laplace who drew attention to this anomaly for phlogistonism (see Kirwan [1789], p. 169).

[93] Lavoisier [1862], p. 343. The reaction is the reverse of that in Priestley's experiments reducing calxes in hydrogen.

[94] Lavoisier, 'Reflections on Phlogiston, Serving to develop the Theories of Combustion and Calcination' (Lavoisier [1862], p. 640). Lavoisier was not here making fun of the idea of an imponderable and uncapturable element, as many historians think. After all, his own caloric was an element of this kind. He was, however, making fun of a theory which assigned contradictory properties to the same element in order to account for different phenomena.

'dephlogisticated air' is actually 'dephlogisticated water'. Exploding 'phlogisticated water' with 'dephlogisticated water' yields water:

We must allow that dephlogisticated air is in reality nothing but dephlogisticated water, or water deprived of its phlogiston; or, in other words, that water consists of dephlogisticated air united to phlogiston; and that inflammable air is...water united to phlogiston.[95]

Cavendish is well aware of Lavoisier's rival theory: he gives an excellent summary of it, and admits that it explains the phenomena 'as well, or nearly as well' as phlogiston theory. But, he insists, 'the commonly received principle of phlogiston explains all phenomena, at least as well as M. Lavoisier's'.[96] (Cavendish obviously had *not* been reading Lakatos; by the 'commonly received principle of phlogiston' he actually means his new version of the phlogiston programme.)

Now *provided* we allow that *post hoc* explanations are as good as predictions (which we should not), Cavendish's claim has substance. Consider the weight increase of calxes: according to 1784 phlogiston theory the weight of calxes is augmented by *water* precipitated out of the 'dephlogisticated air' ('dephlogisticated water') by phlogiston from the metal. Consider the reduction of calxes: according to the 1784 phlogiston theory the water in the calx is decomposed by heat into phlogiston (which revives the metal) and 'dephlogisticated water' ('dephlogisticated air') which is expelled. Consider the calcination of a metal in water: according to the 1784 phlogiston theory the phlogiston released from the metal phlogisticates some of the water to form 'inflammable air' ('phlogisticated water'), while more water enters the calx and augments its weight. Consider the reduction of calxes in inflammable air: according to the 1784 phlogiston theory the calx gives up its water, and takes phlogiston from the inflammable air ('phlogisticated water') to form the metal and leave more water behind. Cavendish can explain all the experiments qualitatively; he has only to assume that phlogiston is weightless, and all Lavoisier's quantitative results are accommodated too. As he himself put it:

As adding dephlogisticated air to a body comes to the same thing as depriving it of its phlogiston and adding water to it, and as there are, perhaps, no bodies entirely destitute of water, and as I know no way by which phlogiston can be transferred from one body to another, without leaving it uncertain whether water is not at the same time transferred, it will be very difficult to determine by experiment which of these opinions [his or Lavoisier's] is the truest.[97]

[95] Cavendish [1784–5], p. 22. Cavendish considers and rejects the view that inflammable air is 'pure phlogiston, as Dr Priestley and Mr Kirwan suppose'. His argument is that dephlogisticated air absorbs phlogiston readily from nitrous air (oxygen and nitric oxide readily form nitrogen peroxide), whereas heat is necessary to make dephlogisticated air absorb phlogiston from inflammable air; Cavendish thinks it odd to suppose that pure phlogiston will yield phlogiston less readily than a phlogiston mixture (Cavendish [1784–5], p. 22, footnote).

[96] Cavendish [1784–5], p. 37.

[97] Cavendish [1784–5], p. 37. It is perhaps worth worth adding that Cavendish's explanations do not involve more elements than Lavoisier's once we take into account the presence of caloric in the latter.

Cavendish gives two reasons for preferring his new version of phlogistonism. One is that neither 'marine acid' nor 'acid of tartar' appear to contain oxygen, as Lavoisier's theory demands. The other is that the 'infinite variety' of plants all decompose, on burning, into water, 'fixed air' and 'dephlogisticated air'; on the phlogiston theory the residue is *less* compound than the plants (since phlogiston is released also), whereas on the oxygen theory it is *more* compound; Cavendish finds the latter implausible.

Philosophers of science who think that the evidential support of a theory depends solely upon the *timeless logical relations* between theory and evidence will have to say that 1784 phlogiston theory had as much evidential support as 1784 oxygen theory. Both theories explained the main facts about combustion and calcination (and both faced some outstanding anomalies).[98] But the chemists of the late eighteenth century did not take this view.[99] They saw that 1784 phlogiston theory merely accommodated known facts, many of which had been discovered by testing predictions made within the oxygen programme. They saw that 1784 phlogiston theory was inconsistent with the previous version, and marked a return to the imponderable phlogiston of Stahl. They contrasted this incoherent development with the smooth development of the various versions of the oxygen programme. And one by one they changed their allegiance. The French chemists were the first to succumb, Berthollet in 1785, de Morveau and Fourcroy by 1787 (perhaps an 'external factor' such as their spatial proximity to Lavoisier explains this?). English chemists soon followed, Higgins in 1789, Darwin in 1790, Black in 1791, and Kirwan in 1792. By 1796 Priestley, who had emigrated to America, was complaining 'I have not heard of a single advocate of phlogiston'.[100] By this time the Chemical Revolution was over.

Moreover, it was a completely *rational* affair once we evaluate it using Lakatos's criteria of rationality. Between 1770 and 1785 the oxygen programme clearly demonstrated its superiority to phlogistonism: it developed coherently and each new version was theoretically and empirically progressive, whereas after 1770 the phlogiston programme did neither. Yet the oxygen theory was never established as true by experiment, as the inductivist requires. Nor was 1784 phlogistonism refuted by experiment,

[98] Lavoisier has still not accounted for the production of 'fixed air' when some calxes are reduced without charcoal (see above, p. 196), or for the production of 'inflammable air' from charcoal (see above, p. 199), or for the failure of 'marine acid' to yield oxygen (see above, p. 198).

[99] I have discussed various attempts to let the *time-order* of theory and evidence play a role in the theory of evidential support in my [1974*b*].

[100] See Forster [1969], Preface. Priestley was not entirely alone. Apart from Davy's flirtations with phlogistonism (see above, footnote 69), one Robert Harrington published in 1804 a book whose title is worth reading: *The Death Warrant of the French Theory of Chemistry, signed by Truth, Reason, Common Sense, Honour and Science, etc.* In 1819 Harrington gave us *An Elucidation and Extension of the Harringtonian System of Chemistry, explaining all the phenomena without one single anomaly*, together with a £100 prize to anybody who could refute him. (See White [1932], pp. 175–6.)

as the naive falsificationist requires (earlier versions were refuted, but the programme soldiered on). Nor was Lavoisier's 1784 oxygen theory in any obvious sense simpler than 1784 phlogiston theory, as the naive simplicist requires. Adherents of these methodologies will have to falsify the history of the Chemical Revolution if they want it to conform to their canons of rationality. Alternatively, they might deem the Chemical Revolution an irrational affair, to be explained by factors 'external' to their methodology.

No historian has explicitly taken the latter course. Yet hints that various external factors may have 'played a role' in the Chemical Revolution are to be found in many historians. Since 'external' explanations of scientific revolutions cannot be deemed false on *a priori* grounds, let us see whether these hints, taken singly or collectively, might add up to an adequate explanation.

Many historians mention the fact that the oxygen theory was commonly called the 'French System'.[101] Perhaps national sentiments played a role? Perhaps English and German chemists resisted the oxygen theory, despite the fact that that it had been proved and/or phlogiston theory disproved, simply because it was proposed in France? But the French chemists opposed it too until after 1785. And after 1785, when France was at war with the rest of Europe and nationalist feelings might be expected to have run deep, English and German chemists succumbed.

Perhaps the converts to the new system were the open-minded *young* chemists, who were able to see the plain truth when it was laid before them, whereas those who resisted were dogmatic old men like Priestley far too set in their ways?[102] This is refuted by Joseph Black, the grand old man of European chemistry, who was telling his students about the 'French system' in the late 1770s, who was responsible for Kerr's immediate translation of Lavoisier's *Elements of Chemistry*, and who wrote to Lavoisier in 1791 as follows:

> Having for thirty years been accustomed to believe and teach the phlogistic doctrine...I have felt for a long while at variance to the new ideas, which appeared at enmity with that which I had thought to be sound doctrine: however, this lack of sympathy, which only sprang from mere conservatism, has gradually diminished, conquered by the clarity of your demonstrations and the soundness of your scheme.[103]

It is also refuted by Kirwan who, after a lifetime spent in defence of phlogistonism, wrote to Berthollet in 1792 saying: 'I lay down my arms

[101] Lavoisier did not approve of the custom: 'This theory is not, as one hears said, the theory of French chemists: it is *mine*...' (Lavoisier [1862], p. 104).

[102] I do not mention this possibility as a joke. Max Planck *was* joking when he said that new ideas triumph because the adherents of the old ones die off (and Bohm-Bawerk *was* joking when he said that it was because old professors are forced to retire). But sociologists of science have no sense of humour. They erect these jokes into a biological theory of scientific change, whose major premises is 'All men are mortal'. I suppose we should be grateful that 'All men are mortal' has found a use after all these years.

[103] Cited by White [1932], pp. 132–3.

and abandon the cause of phlogiston.'[104] Priestley, it is true, did continue to resist, after flirting with the oxygen theory in the 1780s. But his resistance was always perfectly *rational*: he tried to argue against his opponents, and he did not hide the weak spots of phlogistonism. Moreover, he never lost his humour: he introduced his last defence of phlogiston to the French chemists by saying 'As as friend of the weak I have endeavoured to give [the doctrine of phlogiston] a little assistance', and signed off 'trusting that your political revolution will be more stable than this chemical one'.[105] These are not the words of a dogmatic old man, blinded by prejudice.

A third externalist explanation is less easy to dispose of. Perhaps the oxygen theory triumphed because of the new chemical *terminology* constructed in the light of it. Readers of the *Method of Chemical Nomenclature* of 1787 were insensibly inculcated into accepting the oxygen theory: they were, to put it bluntly, brainwashed rather than persuaded by rational argument. Now there is no doubt that the new nomenclature carried theoretical overtones (as did some of the old), and that it was difficult for some chemists to understand it:

> I wish M. Berthollet and his associates would relate their facts in plain prose, that all men might understand them, and reserve their poetry of the new nomenclature for their theoretical commentaries on the facts.[106]

But phlogistians were not *fooled* by the new nomenclature:

> I am much obliged to you for your advice to me to be converted to the true faith in chemistry; your principle argument in favour of which however is, not that it is *true*, but that is is becoming *fashionable*...What I dislike in the anti-phlogistians is their pedantry and presumption, in pretending that their system is proved...As to their language, it is formed on the supposition of their system being certain...I may use the old language, although I doubt of the theory on which it is founded, but I cannot use another theoretical language without taking for granted that the theory is true.[107]

Priestley had the right attitude to the new terminology when he said that whether chemists adopted the new system or not 'we are under the necessity of learning the new language, if we would understand some of the most valuable of the modern publications'.[108] Of course, the anti-phlogistians had no monopoly of the textbook market: there were phlogistic textbooks too. And more important, there were several books in which *both* systems were expounded and compared.

I conclude that, taken either individually or collectively, these factors do not add up to a satisfactory 'external' explanation of the Chemical Revolution. And they are unnecessary anyway, since the methodology of research programmes explains that revolution *internally*.

[104] Kirwan to Berthollet, 26 January 1792 (see Hartley [1971], p. 48). Priestley wryly remarked that his old ally had 'acquired more honour by this conduct than he could have done by the most brilliant discoveries that he could have made' (Priestley [1804], p. 104).

[105] Priestley [1804], pp. xiv, xv.

[106] Keir to Priestley, 1790; see Bolton [1892], p. 99.

[107] Keir to Darwin, 15 March 1790; see Schofield [1963], pp. 292–3.

[108] See Foster [1969], p. 21.

References

Agassi, J. [1963]: *Towards a Historiography of Science* (*History and Theory Beiheft* 2).

Aykroyd, W. R. [1935]: *Three Philosophers* (*Lavoisier, Priestley, and Cavendish*).

Bacon, F. [1620]: *Novum Organum*. (Library of Liberal Arts, New York, 1960.)

Bolton, H. C. (ed.) [1892]: *Scientific Correspondence of Joseph Priestley.*

Boyle, R. [1673]: *Essays on Effluviums*, in Thomas Birch (ed.): *The Works of the Honourable Robert Boyle* (London, 1772), volume 3.

Cavendish, H. [1766]: 'Three Papers, containing Experiments on Factitious Air', *Philosophical Transactions*, **56**, pp. 141–184. (Extracts reprinted in Leicester and Klickstein [1952], pp. 134–42.)

Cavendish, H. [1784–5]: 'Experiments on Air', *Philosophical Transactions*, **74**, pp. 119–53, and **75**, pp. 372–84. (Reprinted as: *Experiments on Air*, Alembic Club Reprints Number 3, 1926; page references are to this reprint.)

Cohen, I. B. [1966]: *Franklin and Newton.*

Conant, J. B. [1950]: *The Overthrow of the Phlogiston Theory: The Chemical Revolution of 1775–1789*. Harvard Case Histories in Experimental Science, Case 2.

Conant, J. B [1951]: *Science and Common Sense.*

Crosland, M. [1973]: 'Lavoisier's Theory of Acidity', *Isis*, **64**, No. 223, pp. 306–25.

Daumas, M. [1963]: 'Precision of Measurement and Physical and Chemical Research in the Eighteenth Century', in A. C. Crombie (ed.): *Scientific Change*, pp. 418–30.

Davis, T. L. [1927]: 'Priestley's Last Defence of Phlogiston', *Journal of Chemical Education*, **4**, No. 2, pp. 176–83.

Davy, H. [1929]: *The Elementary Nature of Chlorine. Papers by Humphry Davy* (*1809–1818*). Alembic Club Reprints Number 9.

Foster, W. (ed.) [1969]: *Considerations on the Doctrine of Phlogiston and the Decomposition of Water, by Joseph Priestley, and Two Lectures on Combustion and an examination of Dr. Priestley's considerations on the Doctrine of Phlogiston, by John MacLean*, 2nd edition.

Gillispie, C. C. [1960]: *The Edge of Objectivity.*

Gregory, J. C. [1934]: *Combustion from Heracleitos to Lavoisier.*

Guerlac, H. [1961]: *Lavoisier: The Crucial Year.*

Hartley, Sir H. [1971]: *Studies in the History of Chemistry.*

Hartog, Sir P. J. [1941]: 'The Newer Views of Priestley and Lavoisier', *Annals of Science*, **5**, pp. 1–56.

Herschel, J. F. W. [1930]: *Preliminary Discourse on the Study of Natural Philosophy.*

Holmyard, E. J. [1925]: *Chemistry to the Time of Dalton.*

Kirwan, R. [1789]: *An Essay on Phlogiston, and the Constitution of Acids. A New Edition. To Which are added, Notes Exhibiting and Defending the Antiphlogistic Theory; and annexed to the French Edition of this Work; by Messrs. de Morveau, Lavoisier, de la Place, Monge, Berthollet, and de Fourcroy: Translated into English. With additional Remarks and Replies, by the Author.*

Koertge, N. [1969]: *A Study of the Relations between Scientific Theories: A Test of the General Correspondence Principle*. University of London PhD Thesis (unpublished).

Lakatos, I. [1907]: 'Falsification and the Methodology of Scientific Research Programmes', in I. Lakatos and A. Musgrave (eds.): *Criticism and the Growth of Knowledge*, pp. 91–196.

Lavoisier, A. L. *et al.* [1787]: *Methode de Nomenclature chimique.*

Lavoisier, A. L. [1790]: *Elements of Chemistry in a new systematic order, containing all the modern discoveries*. Translated by Robert Kerr. (Dover edition, 1965.)

Lavoisier, A. L. [1862]: *Oeuvres de Lavoisier. Tome II. Mémoires de Chimie et de Physique.*
Lavoisier, A. L. [1864]: *Oeuvres de Lavoisier. Tome I.*
Leicester, H. M. and Klickstein, H. S. (eds.) [1952]: *A Source Book in Chemistry.*
Lindsay, J. (ed.) [1970]: *The Autobiography of Joseph Priestley.*
Meldrum, A. N. [1930]: *The Eighteenth-Century Revolution in Science – The First Phase.*
Meyer, E. [1906]: *History of Chemistry.* Translated by G. McGowan.
Musgrave, A. E. [1972]: 'The Role of Experiment in the Chemical Revolution', *Chemistry in New Zealand*, **36**, No. 5, pp. 151–62.
Musgrave, A. E. [1974*a*]: 'Falsification and Its Critics', in P. Suppes *et al.* (eds.): *Logic, Methodology and Philosophy of Science IV*, pp. 393–406.
Musgrave, A. E. [1974*b*]: 'Logical versus Historical Theories of Confirmation', *British Journal for the Philosophy of Science*, **25**, pp. 1–23.
Musgrave, A. E. [1976]: 'Method or Madness?' in R. S. Cohen, P. K. Feyerabend and M. W. Wartofsky (eds): *Boston Studies in the Philosophy of Science. Imre Lakatos Memorial Volume*, forthcoming.
Partington, J. R. [1937]: *A Short History of Chemistry.*
Partington, J. R. [1961]: *A History of Chemistry Volume II.*
Partington, J. R. and McKie, D. [1937–40]: 'Historical Studies on the Phlogiston Theory', Part I, *Annals of Science*, **2** (1937), pp. 361–404; Part II, *Annals of Science*, **3** (1938), pp. 1–58; Part III, *Annals of Science*, **3** (1938), pp. 337–71; Part IV, *Annals of Science*, **4** (1939–40), pp. 113–49.
Priestley, J. [1775–7]: *Experiments and Observations on Different Kinds of Air*, 3 volumes. Vol. 1, 1775; vol. 2, 1776; vol. 3, 1777. (Sections iii–v of vol. 2 are reprinted as: *The Discovery of Oxygen Part I*, Alembic Club Reprints No. 7, 1923.)
Priestley, J. [1779–86]: *Experiments and Observations relating to various branches of Natural Philosophy with a continuation of the Observations on Air*, 3 volumes. Vol. 1, 1779; vol. 2, 1781; vol. 3, 1786.
Priestley, J. [1783]: 'Experiments relating to Phlogiston', *Philosophical Transactions*, **73**, pp. 398–434.
Priestley, J. [1785]: 'Experiments and Observations relating to Air and Water', *Philosophical Transactions*, **75**, pp. 279–309.
Priestley, J. [1804]: *The Doctrine of Phlogiston Established, and that of the Composition of Water Refuted* (second edition).
Ramsey, Sir W. [1905]: *The Gases of the Atmosphere: The History of Their Discovery.*
Read, J. [1957]: *Through Alchemy to Chemistry.*
Rey, J. [1630]: *Essays on an Enquiry into the Cause wherefore Tin and Lead increase in Weight on Calcination.* Alembic Club Reprints Number 11, 1953.
Rodwell, G. F. [1868]: 'On the theory of phlogiston', *Philosophical Magazine*, **35**, pp. 1–32.
Scheele, C. W. [1931]: *The Collected Papers of Carl Wilhelm Scheele.* Translated by Leonard Dobbin.
Schofield, R. E. [1963]: *The Lunar Society of Birmingham.*
Schofield, R. E. [1966]: *A Scientific Autobiography of Joseph Priestley (1733–1804).*
Siegfried, R. [1964]: 'The phlogistic conjectures of Humphry Davy', *Chymia*, **9**, pp. 117–24.
Siegfried, R. and Dobbs, B. J. [1968]: 'Composition, a neglected aspect of the chemical revolution', *Annals of Science*, **24**, pp. 275–93.
Stillman, J. M. [1924]: *The Story of Early Chemistry.*
Toulmin, S. E. [1957]: 'Crucial experiments: Priestley and Lavoisier', *Journal of the History of Ideas*, **18**, pp. 205–20.
Watson, R. [1781]: *Chemical Essays.*
White, J. H. [1932]: *The History of the Phlogiston Theory.*
Wolff, P. (ed.) [1967]: *Breakthroughs in Chemistry.*

Why did Einstein's Programme supersede Lorentz's?*

ELIE ZAHAR

LONDON SCHOOL OF ECONOMICS AND POLITICAL SCIENCE

* This paper is an expanded version of a talk given before the British Society for the Philosophy of Science on 7 December 1970. I gratefully acknowledge the valuable criticisms and suggestions received from Clive Kilmister, Imre Lakatos, John Stachel, John Watkins and John Worrall – all of whom were burdened with previous versions of this paper.

Introduction

(*a*) *The Michelson–Morley Experimen*

Most of the answers to the question of why Einstein's programme superseded Lorentz's refer to the weaknesses of Lorentz's solution to the problem posed by Michelson's results. I shall argue (in §1) that these alleged weaknesses are illusory, but let me start by giving a schematic description in classical terms of the experiment which was first performed by Michelson in 1881 and then repeated with increased precision in 1887 and after.

Michelson used an interferometer consisting of two perpendicular arms: $BE = L$ and $BD = l$. At B a half-silvered mirror makes an angle of 45° with BE. A light source A emits a beam which is divided at B into two rays: a reflected ray R_1 which travels along BD, is reflected at D, then goes back to B; and a ray R_2 which is transmitted along BE, falls perpendicularly on a mirror at E, then returns to B where it is partially reflected before interfering with R_1. Suppose BE lies in the direction of the earth's motion through the ether and consider a frame of reference fixed with respect to the earth, then on the classical account $(c-v)$ and $(c+v)$ are the speeds of R_2 between B and E and between E and B respectively (where: v = velocity of the earth, and c = speed of light). Hence the time taken by R_2 to return to B is

$$t_2 = L/(c-v)+L/(c+v) = (2L/c)\,\beta^2$$

where $\beta = (1-v^2/c^2)^{-\frac{1}{2}}$. (Because of its central role in Relativity Theory, the coefficient $(1-v^2/c^2)^{-\frac{1}{2}}$ is denoted by a special symbol.)

Let u be the speed of R_1 along BD. The velocity of R_1 in the ether is $\boldsymbol{v}+\boldsymbol{u}$, from which it follows that $|\boldsymbol{v}+\boldsymbol{u}| = c$, i.e. $v^2+u^2 = c^2$; therefore $u = (c^2-v^2)^{\frac{1}{2}}$. Hence the time taken by R_1 to return to B is

$$t_1 = 2l/(c^2-v^2)^{\frac{1}{2}} = (2l/c)\,\beta$$

If the arms of the interferometer are equal, i.e. if $L = l$, then

$$t_2-t_1 = \frac{2L}{c}\beta^2-\frac{2L}{c}\beta = \frac{2L}{c}\beta(\beta-1).$$

If the apparatus is rotated through 90°, the time taken by R_1 to come back to B is increased by $(2L/c)\,\beta(\beta-1)$, while the time taken by R_2 is diminished by the same quantity. The total time difference is therefore:

$$\frac{4L}{c}\beta(\beta-1) = \frac{4L}{c}\left[\left(1-\frac{v^2}{c^2}\right)^{-1}-\left(1-\frac{v^2}{c^2}\right)^{-\frac{1}{2}}\right] \simeq \frac{2Lv^2}{c^3}.$$

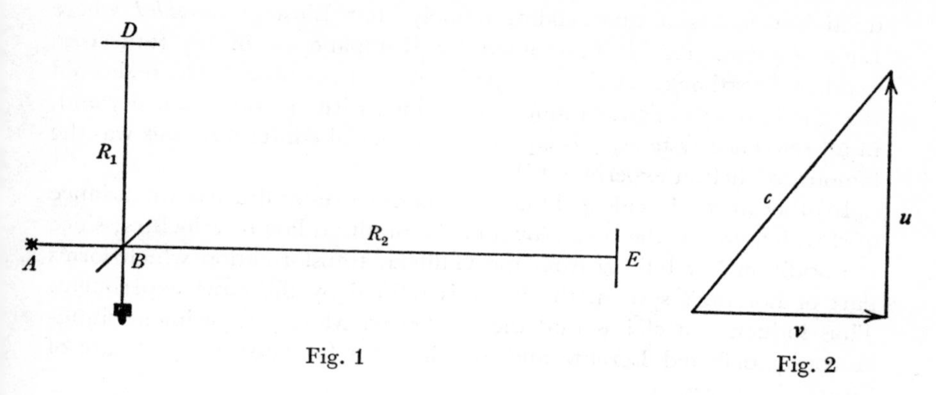

Fig. 1

Fig. 2

This time difference should cause a certain shift of the interference fringes. No such shift was observed and so Michelson claimed to have refuted Fresnel's original hypothesis of an ether at rest and confirmed Stokes's theory of total ether drag. However, in 1881 Michelson had taken the speed of R_1 to be c in both directions, thus obtaining a time difference of $4Lv^2/c^3$, which is twice as large as the correct one. Lorentz pointed out this mistake in Michelson's calculations and showed that the real shift fell within the limits of observational error.[1]

Michelson, together with Morley, repeated the experiment in 1887; they increased precision by making the light rays R_1 and R_2 travel several times between the mirrors. Still no shift of the fringes was observed. Lorentz was now convinced that the Michelson–Morley result was a serious difficulty, not just for Fresnel's but also for his own theory.

(*b*) *The standard accounts of the role of the Michelson–Morley experiment in Einstein's victory over Lorentz*

(*b1*) *The inductivist account*

Inductivists maintain that Einstein's second postulate concerning the invariance of c is a valid generalisation of Michelson's result.

According to Max Born '. . . the second statement, that of the constancy of the velocity of light, must be regarded as being experimentally established with certainty.'[2]

Reichenbach claims '. . .it would be mistaken to argue that Einstein's theory gives an explanation of Michelson's experiment, since it does not do so. Michelson's experiment is simply taken over as an axiom.'[3] In-

[1] Lorentz [1886]. There was a series of experiments performed by Michelson and not just one (I sometimes refer to the Michelson–Morley experiment, meaning the one performed in 1887); and, as Lakatos emphasized, in the long contest between the theoretician Lorentz and the experimenter Michelson, Lorentz always held the upper hand and eventually confused Michelson to such an extent that he gave up the hope of even interpreting his own experiments. (Cf. Lakatos [1970], **3** (*d1*).)

[2] Born [1962], p. 225.

[3] Reichenbach [1958], p. 201.

ductivism makes a dual claim, namely that Einstein *succeeded* where Lorentz *failed.* This is underlined by Kompaneyets in his (otherwise excellent) textbook: 'A direct experiment was performed which showed that the velocity of light cannot be combined with any other velocity and, in all reference systems, it is equal to a universal constant *c*. This was the famous Michelson experiment.'[4]

In other words, it is alleged that the experiment established the invariance of *c*, which entails the breakdown of the addition law of velocities; since the addition law follows from the Galilean transformation which forms part of Lorentz's system, the latter is refuted by the same experiment. Thus inductivism claims that the Michelson–Morley experiment simultaneously defeated Lorentz and established a fundamental postulate of Einstein's theory.

The inductivist account fails both for logical and for historical reasons. From the logical point of view, it has by now become a platitude that observation reports neither establish nor even probabilify high-level theories. None of Michelson's observational statements is equivalent to the proposition that in all inertial frames the speed of light is a universal constant independent of the velocity of the source. As far as history is concerned, as Michael Polanyi pointed out, 'Michelson's experiment had a negligible effect on the discovery of Relativity'.[5] Shankland's account supports Polanyi's claim: 'When I asked him [i.e. Einstein] how he had learned of the Michelson–Morley experiment, he told me that he had become aware of it through the writings of H. A. Lorentz, but only after 1905 had it come to his attention.'[6]

(*b2*) *The falsificationist account*

There is a falsificationist account of the role of the Michelson–Morley experiment which has something in common with the inductivist account, but which attributes Lorentz's failure and Einstein's success to their respectively *ad hoc* and non *ad hoc* responses to the experiment. On this account the 'crucial' experiment refuted the conjunction of Galilean kinematics, Newton's laws and Maxwell's equations for the ether. In order to explain away Michelson's result, Lorentz resorted to an auxiliary assumption, the Lorentz–Fitzgerald Contraction Hypothesis (hereafter referred to as the LFC), which was however *ad hoc*. Thus, for example, Popper writes: 'An example of an unsatisfactory auxiliary hypothesis would be the Contraction Hypothesis of Fitzgerald and Lorentz which had no falsifiable consequences but merely served to restore the agreement between theory and experiment – mainly the findings of Michelson and Morley.'[7]

Now Einstein, on his own admission, was familiar with part of Lorentz's work, in particular with the latter's [1892*a*] and his [1895]. Thus the

[4] Kompaneyets [1962], p. 191.
[5] Polanyi [1958], p. 10.
[6] Shankland [1963], pp. 47–8.
[7] Popper [1935], section 20.

falsificationist can easily explain what prompted Einstein to propose his own theory. Einstein, having realised that the LFC is *ad hoc*, proposed Special Relativity Theory (hereafter referred to as the SRT) as a better – non *ad hoc* – alternative.

(*b3*) *Holton's account*

In his [1969] Holton gives further support to the claim that the LFC was *ad hoc*: 'This saving *Hülfhypothese* [i.e. the LFC] is introduced completely *ad hoc*....No explicit comment is made which connects this assumed shrinkage with the Lorentz transformations in their still primitive form as published earlier in the book [Lorentz's [1895]].'[8]

Holton, however, attaches to *ad-hocness* a meaning different from that attached to it by Popper. For Holton the LFC is *ad hoc* because it is not integrated into the rest of Lorentz's system. It is, for example, not connected with Lorentz's transformation equations.

In the first section of this paper I propose to refute *all* the charges of *ad-hocness* which have been levelled at the LFC and show that Lorentz's programme progressed until after 1905.

1 *The progress of Lorentz's programme*

1.1 *Methodological preliminaries*

I shall appraise the progress of Lorentz's programme in the terms provided by the methodology of research programmes.[9] A scientific research programme is characterised by a hard core and by a heuristic. The hard core consists of assumptions which, by methodological decision, as it were, are kept unfalsified. Each theory in the programme is a conjunction of, on the one hand, the hard core and, on the other, of auxiliary hypotheses to which the modus tollens is directed whenever anomalies arise. A programme also has a heuristic which consists of a set of suggestions and hints which govern the construction or modification of the auxiliary hypotheses. The heuristic, which sets out a research policy, is less rigid than the hard core. A good example of a (progressive) research programme, as I shall argue, is Lorentz's ether programme. Its hard core consists of Maxwell's equations for the electromagnetic field, of Newton's laws of motion and of the Galilean transformation, to which Lorentz added his equation:

$$\boldsymbol{F} = e\left(\boldsymbol{D} + \frac{1}{c}\,\boldsymbol{v} \wedge \boldsymbol{H}\right)$$

for the so-called Lorentz force. The heuristic of the programme arises from the overall *metaphysical* principle[10] that all physical phenomena are governed by actions transmitted by the ether. Applications of the heuristic

[8] Holton [1969], p. 171.

[9] Cf. Lakatos [1968*b*], [1970], [1971*a*] and [1971*b*].

[10] For the connection between metaphysics and heuristic, cf. Watkins [1958].

to specific problems (which may or may not be set by 'refutations' or anomalies) generate a sequence of theories. We shall be mainly concerned with three consecutive theories belonging to Lorentz's ether programme. I shall refer to them as $\mathbf{T}_1$, $\mathbf{T}_2$ and $\mathbf{T}_3$.

$\mathbf{T}_1$ consists of the hard core as defined above together with the (tacit!) assumptions (*i*) that moving clocks are not retarded and (*ii*) that material rods are not shortened by their motion through the ether.

$\mathbf{T}_2$ is obtained from $\mathbf{T}_1$ by substituting the LFC for assumption (*ii*). According to the LFC a body moving through the ether with velocity $\boldsymbol{v}$ is shortened by the factor $(1-v^2/c^2)^{\frac{1}{2}}$.

$\mathbf{T}_3$ is the conjunction of the hard core, of the LFC and of the assumption, that, contrary to (*i*), clocks moving with velocity $\boldsymbol{v}$ are retarded by the factor $(1-v^2/c^2)^{\frac{1}{2}}$.[11]

I claim that both the shift from $\mathbf{T}_1$ to $\mathbf{T}_2$ and that from $\mathbf{T}_2$ to $\mathbf{T}_3$ were non *ad hoc*. This implies in particular that the introduction of the LFC which took Lorentz from $\mathbf{T}_1$ to $\mathbf{T}_2$ was not an *ad hoc* manoeuvre. Let me first clarify the various contrary claims. All the more or less vague charges of *ad-hocness*, including Holton's, which have been levelled at the LFC are captured by means of the three notions of *ad hocness* used to appraise research programmes.

Ad-hocness in research programmes is defined not as a property of an isolated hypothesis but as a relation between two consecutive theories. A theory is said to be *ad hoc*$_1$ if it has no novel consequences as compared with its predecessor. It is *ad hoc*$_2$ if none of its novel predictions have been actually 'verified'; for one reason or another the experiment in question may not have been carried out, or – much worse – an experiment devised to test a novel prediction may have yielded a negative result.[12] Finally the

[11] This is a slight simplication. We shall see that Lorentz deduced the LFC from the Molecular Forces Hypothesis. However, although he used an instrumental notion of 'local time', he did not realise that the MFH entails the Clock Retardation Hypothesis. Only after Einstein published his results in 1905, did it occur to Lorentz that the MFH implies that moving clocks are retarded by the factor $(1-v^2/c^2)^{\frac{1}{2}}$. In other words: before 1905 Lorentz did not give a realistic interpretation of 'local time'.

[12] Students of methodology will realise that my characterisation of the notion of '*ad hoc*$_2$' differs from that given by Lakatos. I characterise a theory as *ad hoc*$_2$ (at time t) if none of its excess content over its rivals has, at time t, been corroborated. Thus, on my characterisation, if a theory is non *ad hoc*$_2$, it has predicted a novel *fact*, i.e. a theory is empirically progressive if it is non *ad hoc*$_2$. Lakatos on the other hand characterises a theory as *ad hoc*$_2$ if all of its excess content has been *refuted*. (Cf. his [1968*b*].) Thus, on Lakatos's characterisation, a theory can be empirically non-progressive (none of its novel predictions have been corroborated) and at the same time non *ad hoc*$_2$ (not all of its novel predictions have been refuted). Apart from Lakatos's spoiling of symmetry (I take over from him the definitions: non *ad hoc*$_1$ = theoretically progressive, non *ad hoc*$_3$ = heuristically progressive) my characterisation is clearly more in the spirit of Lakatos's enterprise than his own. He (correctly) stresses dramatic confirmation as opposed to refutation; my '*ad hoc*$_1$' is simply the negation of his notion of 'empirically progressive'. (Lakatos's own notion of 'non *ad hoc*$_2$' seems to be simply a rechristening of Popper's 'third requirement' (Popper [1963], p. 242). But how can we ever establish (even tentatively) that *all* of a theory's excess content is false?)

theory is said to be *ad hoc*$_3$ if it is obtained from its predecessor through a modification of the auxiliary hypotheses which does not accord with the spirit of the heuristic of the programme.

Since 'ad-hocness' depends in an essential way on the notion of *novelty of facts*, this notion has to be examined in some detail.

Before embarking on a general discussion let us consider a few concrete examples of novel facts. Lakatos mentions the return of Halley's comet as a new fact anticipated by the Newtonian programme and, of course, I agree with him that the discovery of any new type of fact is the discovery of a novel fact. But, if we equate novelty simply with *temporal* novelty, we are driven into a paradoxical situation. We should, for example, have to give Einstein no credit for explaining the anomalous precession of Mercury's perihelion, because it had been recorded long before General Relativity was proposed. Similarly, we should have to say, contrary to informed opinion, that Michelson's experiment did not confirm Special Relativity and Galileo's experiments on free fall did not confirm Newton's theory of gravitation. Lakatos, who does not easily dismiss the judgements of physicists,[13] is aware of this difficulty and tries to avert it by shifting his original view and saying that, in the light of a new theory, some known facts may 'turn into' novel ones. For example, whereas Balmer merely 'observed' that the hydrogen lines obey a certain formula, Bohr connected these lines with the energy levels of the electron in the hydrogen atom.[14]

However, Lakatos's modified notion of 'novel fact' is open to the following fatal objection. Any theory is a set of propositions connecting different terms and relations. We can always define the properties of a physical entity like mass through the relations which 'mass' bears to other concepts and notions within a given theory. Consequently a new hypothesis will generally ascribe new meanings to old terms. For instance, any experimental consequence of relativity theory involving say mass, would trivially become the expression of a novel fact. Thus the fact that a steel ball rolling down a slope takes a certain time to reach the bottom could become a novel fact when the steel ball is considered as having relativistic mass. This is obviously absurd. Therefore, Lakatos's 1970 criterion for novelty is too liberal, while his 1968 criterion is too stringent.

Although Michelson's result contains a reference to 'length' which acquires a new meaning in Special Relativity, one must not claim that the result has thereby been changed into a novel fact. The Michelson result is indeed novel *vis-a-vis* Special Relativity, but its novelty does not rest on this reinterpretation of 'length'; since the 'crucial' experiment can be described in an 'observational' language which, though theory-

[13] Lakatos [1971*a*], pp. 120–2 and [1971*b*], pp. 179–80.

[14] Lakatos's shift from a bold purely temporal concept to a watered down concept of novelty becomes clear when one compares Lakatos [1968*a*], pp. 381–7, with Lakatos [1970], pp. 155–7. The question is whether this shift is *historiographically* progressive or *ad hoc* on his own criterion given in his [1971*a*] and [1971*b*].

laden, remains unaffected by theory change.[15] 'The arms of the interferometers have equal lengths' can for instance be replaced by 'The extremities of the two arms can be made to coincide by placing the two arms alongside each other'.

When then does a new prediction lend – if experimentally corroborated – *genuine* support to a theory, and when is such support only spurious? Consider the following situation. We are given a set of facts and a theory $T[\lambda_1, \ldots, \lambda_m]$ which contains an appropriate number of parameters. Very often the parameters can be adjusted so as to yield a theory T^* which 'explains' the given facts; it may even happen, given sufficiently many degrees of freedom, that new dramatic relations between old facts can be exhibited (or rather fabricated). For example, by expanding a harmonic function ϕ in terms of spherical harmonics and then adjusting the coefficients of the expansion, the precession of Mercury's perihelion can be accounted for within Newtonian physics.[16] In such a case we should certainly say that the facts provide little or no evidential support for the theory, since *the theory was specifically designed to deal with the facts*. In other words, the way in which a theory is constructed is relevant for an appraisal of its merits. If we are given only the end-product T^* which predicts facts a, b and c, we shall in general be unable to determine whether a, b and c lend genuine support to T^* or whether T^* was simply cleverly engineered to yield the known facts through an adjustment of parameters.

This suggests the following re-definition of the notion of 'novel fact'. *A fact will be considered novel with respect to a given hypothesis if it did not belong to the problem-situation which governed the construction of the hypothesis.* Consider two consecutive theories T_1 and T_2 in the same research programme; suppose that T_1 faces two anomalies e_1 and e_2 and that T_2 was specifically evolved in order to account for e_1; if it is then found that T_2 also explains e_2, e_2, in contradistinction to e_1, will be taken to provide evidential support for T_2. This proposal rests on the fact that ingenious and imaginative scientists can always construct theories which account for a finite number of known facts. Of course, under this definition, any temporally new type of experimental result e will be novel, since any theory which implies e could not have been proposed in the light of the evidence e. Temporal novelty in a research programme is then a sufficient but not a necessary condition for novelty.[17] A temporally new fact may have greater *psycho-*

[15] This was first pointed out by Reichenbach in the introduction of his posthumous [1965]. [16] See Adler, Bazin and Schiffer [1965], p. 202.

[17] John Worrall offered an amusing counter-example to the idea that temporal novelty is necessary: a socially isolated theoretician has an idea which, he realises, as well as explaining certain known facts, predicts a temporally novel fact. He asks an experimentalist friend to test the prediction without bothering to explain to him how he arrived at the prediction. The prediction is corroborated. The experimental result is submitted to journal J_1 and the theoretical idea to journal J_2. Journal J_2 takes two years longer to publish articles than does journal J_1, and so the fact is known to the scientific world for two years before it receives a theoretical explanation. Temporal novelty can be a mere accident.

logical impact that some known fact, but this, on its own, is irrelevant to the objective empirical support which it lends to a hypothesis.[18]

My re-definition of novelty amounts to the claim that *in order to assess the relation between theories and empirical data within a research programme, one has to take into account the way in which a theory is built and the problems it was designed to solve.*

This new criterion for novelty of facts also implies that the traditional methods of historical research are even more vital for evaluating experimental support than Lakatos had already suggested. The historian has to read the private correspondence of the scientist whose ideas he is studying; his purpose will not be to delve into the psyche of the scientist, but to disentangle the heuristic reasoning which the latter used in order to arrive at a new theory. Let us give an example. In Newton's time there was a well-known inverse square law for the intensity of light; Newton might have used some reasoning by analogy in order to propose that the gravitational 'intensity' is also distributed over the surface of a sphere and hence obeys an inverse square law; in this case Kepler's laws would support gravitational theory more strongly than if Newton had used them as his heuristic starting point.[19]

1.2 *Popper, Grünbaum and Holton on Lorentz and Einstein*

Having clarified the notion of '*ad-hocness*' and 'novel fact' let us now turn to Lorentz's theory $\mathbf{T}_2$[20] and examine whether it is *ad hoc* in any one of the three senses explained above.

We have seen that, in his [1935], Popper looked upon $\mathbf{T}_2$ as *ad hoc*, but in 1959 Grünbaum showed that the Lorentz–Fitzgerald Contraction Hypothesis does not constitute an *ad hoc* modification of the ether theory in the sense now under discussion, since its confirmation is possible in an experiment different from the Michelson–Morley type.[21] In the Kennedy–Thorndike-type experiment[22] described by Grünbaum the arms of the

[18] This amendment of the definition of novelty confirms Lakatos's views about the methodology of research programmes as applied to itself. (Cf. Lakatos [1971*a*], pp. 116–22.) Progress in the methodology of research programmes consists in proposing finer demarcation criteria between scientific progress and scientific degeneration. This, I hope, is what my definition of novelty does.

[19] The heuristic of certain programmes makes it very difficult (or even impossible) for them ever to achieve empirical progress. For instance Plato, through advocating the saving of phenomena by combinations of circular motions, condemned the Greek astronomical programme to degeneration; the epicycles acted as an infinite set of parameters which could be adjusted so as to account for *any* periodic motion, *after* the latter had been observed. There was no uniform method of epicycle-construction which was independent of the facts and hence capable of anticipating them. Also cf. Lakatos and Zahar [1976].

[20] Cf. above, p. 216.

[21] Cf. Grünbaum [1959] and [1963].

[22] Incidentally Kennedy and Thorndike thought that 'using this null result [i.e. the null result of the Kennedy–Thorndike experiment] and that of the Michelson–Morley experiment, [they could] derive the Lorentz–Einstein transformations, which are tantamount to the relativity principle'. (Cf. Kennedy and Thorndike [1932], p. 400.)

interferometer[23] have different lengths; $\mathbf{T}_2$ predicts that the difference $(t_2 - t_1)$ between the times it takes the two rays R_1 and R_2 to return to the half-silvered mirror is equal to:

$$\frac{2L}{c} \cdot \frac{1}{\beta} \cdot \beta^2 - \frac{2l}{c} \beta = \frac{2\beta}{c} (L - l)$$

This quantity is different both from the value predicted by Special Relativity and from the one predicted by the classical theory. These two theories respectively yield the following values for $(t_1 - t_2)$:

$$\frac{2}{c}(L-l) \quad \text{and} \quad \frac{2\beta}{c}(L\beta - l)$$

Popper accepted Grünbaum's criticism as valid.[24] Thus it is settled that Lorentz's $\mathbf{T}_2$ was not *ad hoc*$_1$. It will be more difficult to decide whether $\mathbf{T}_2$ was *ad hoc*$_2$ and/or *ad hoc*$_3$.

Was $\mathbf{T}_2$ *ad hoc*$_2$ in the sense that until 1905 nobody bothered to test its novel predictions? Later I shall show that it was not; but even had it been, this constitutes no damning criticism of the ether programme; it would mean only that its empirical progressiveness had not yet been shown. *Ad-hocness* in the second sense becomes a demerit of a research programme only if it is a lasting feature.

Is then $\mathbf{T}_2$ *ad hoc*$_3$? In actual scientific practice a hypothesis is intuitively judged to be *ad hoc* if it looks arbitrary or if it fits poorly into the research programme. Thus the introduction of a theory which is *ad hoc* in this sense destroys the organic unity of the whole nexus, since the various components of the resulting system are structured according to conflicting plans. For example, if, within the ether programme, a theory postulating some new instantaneous action-at-a-distance were proposed, the new theory would be found intuitively *ad hoc*.

Ad-hocness$_3$ is a good explication of this intuitive notion of *ad-hocness*. To repeat, a theory is said to be *ad hoc*$_3$ if it conflicts with the heuristic of the research programme. If it could be shown that $\mathbf{T}_2$ is *ad hoc*$_3$, then two results would be achieved at one stroke: from a *methodological* point of view Lorentz's programme would be shown to have had serious defects in 1905; from a *historical* point of view, it would become plausible that Einstein, who attributed so much importance to the criterion of 'internal perfection', was motivated to start his rival programme by the patched-up state of Lorentz's $\mathbf{T}_2$.

Holton seems to be claiming that Lorentz's $\mathbf{T}_2$ was both *ad hoc*$_2$ and *ad hoc*$_3$; and that Einstein's programme was triggered off by Einstein's recognition of these two defects. I shall show however that Holton is wrong on all counts and that his (rather Polanyiite) methodology misleads him into false history.

[23] Cf. above, p. 213. [24] Popper [1969], p. 51.

In his [1969] Holton writes:

This saving *Hülfshypothese* [the LFC] is introduced completely *ad hoc*... *No explicit comment is made which connects this assumed shrinkage with the Lorentz transformations in their still primitive form, as published earlier in the book*...

The contraction hypothesis when it was made was clearly and quite blatantly *ad hoc* – or, if one prefers to use the *patois* of the laboratory, *ingeniously cooked up for the narrow purpose which it was to serve*...

The important point to note is that '*ad hoc*' is not an absolute but a *relativistic term*. Postulates 1 and 2 [Einstein's two postulates in his [1905]] may be said to have been introduced *ad hoc* with respect to the Relativity Theory of 1905 as a whole... But these postulates were *not ad hoc with respect to the Michelson experiment*, for they were not specifically imagined in order to account for its results...[25]

Holton makes two distinct claims. His first claim is that the Contraction Hypothesis (LFC), which differentiates $\mathbf{T}_2$ from $\mathbf{T}_1$, was not connected with the rest of $\mathbf{T}_2$, in particular with the Lorentz transformation equations, and thus, on my terms, the LFC is *ad hoc*$_3$. His second claim is that the LFC was specifically engineered in order to account for Michelson's result; from this, together with the fact that for a long time the Michelson result was the only one which even seemed to support the LFC, I conclude that, in Holton's view, the LFC was not independently tested and therefore was *ad hoc*$_2$. But both of Holton's claims are false:

(i) Lorentz deduced the LFC from a deeper theory, namely from what I call the Molecular Forces Hypothesis (hereafter referred to as the MFH) and which can be loosely formulated as follows: 'Molecular forces behave and transform like electromagnetic forces.' Moreover, in his deduction of the LFC, Lorentz made use of his famous transformation, as is clearly indicated by the following passage from his [1895]:

For, if we now understand by S_1 and S_2 not, as formerly, two systems of changed particles but two systems of molecules – the second at rest and the first moving with velocity v in the direction of the axis of x – between the dimensions of which the relation subsists as previously stated; and if we assume that in both systems the x-components of the forces are the same, while the y- and z-components differ from one another by the factor $\sqrt{1-v^2/c^2}$, then it is clear that the forces in S_1 are in equilibrium whenever they are so in S_2... The displacement would naturally bring about this disposition of the molecules of its own accord and thus effect a shortening in the direction of motion in the proportion of 1 to $\sqrt{1-v^2/c^2}$ *in accordance with the formulae given in the above-mentioned paragraph*.[26]

I further maintain that for anybody prepared to accept the assumption of an ether at rest, the MFH is a plausible auxiliary hypothesis which

[25] Holton [1969], pp. 177–81; my italics. The term 'relativistic' was probably a slip of the pen and should read 'relative'. But I cannot make head or tail of Holton's sentence 'Postulates 1 and 2 may be said to have been introduced *ad hoc* with respect to the Relativity Theory of 1905 as a whole.' How can Einstein's two postulates, which constitute Special Relativity Theory or at any rate are part of Relativity Theory, be *ad hoc* with respect to Relativity Theory?

[26] Einstein and others [1923], p. 7; my italics.

introduces no alien elements into Lorentz's programme. Putting it more objectively, the theory $\mathbf{T}_2$ proposed by Lorentz is non *ad hoc*$_3$, because the MFH is structured in accordance with the heuristic of the ether programme, which requires that physical phenomena be explained in terms of actions propagated in the ether.

(ii) Moreover, *the MFH arose out of considerations which had nothing to do with Michelson's experiment.* The MFH arose out of mathematical considerations pertaining to the transformation properties of Maxwell's equations. *Hence Michelson's null result is a novel fact relative to the MFH*! The MFH is consequently non *ad hoc*$_2$; it constituted both theoretical and empirical progress.

Let me now briefly turn to the implications of my methodological theses for the historical accounts of the Einsteinian Revolution, and in particular for Holton's account. My claim is that the LFC is non *ad hoc*, and that Michelson's result, far from providing an obstacle for Lorentz's programme, in fact supported it. This clearly rules out all explanations of the genesis of SRT which depend on the assumption that Einstein *correctly* realised the LFC was *ad hoc* relative to Michelson's result. One such explanation is Holton's, even though he attributes only an indirect role to Michelson's result. He alleges that Einstein was dissatisfied with the LFC because it was blatantly *ad hoc*; it was 'cooked up for the narrow purpose which it was to serve'. But *ad hoc* relative to what?[27] Obviously Holton's claim is that the LFC was *ad hoc* relative to Michelson's result, since the 'narrow purpose which it was to serve' was precisely an explanation of the null result of the 'crucial' experiment. Holton adds that:

> the problem Einstein saw was not the logical status of the Contraction Hypothesis, not Michelson's experimental result (for it could be accommodated, even if not '*ohne Weiteres*') but the inability of Lorentz's theory to fulfil the criterion of 'inner perfection' of a theory.[28]

So Lorentz's theory lost its 'inner perfection' on the introduction of the LFC, which was contrived for the sole purpose of explaining Michelson's result. Thus, on this account, the 'crucial' experiment *did* play an important – if indirect – role in the genesis of SRT: the search for an explanation of Michelson's result compelled Lorentz to resort to an hypothesis whose *ad hoc* character provided Einstein with a good reason for starting his revolutionary new programme.

One might defend this Holtonian account against my arguments by assuming that Einstein appraised the LFC incorrectly.[29] Perhaps Einstein

[27] Holton correctly points out that 'a statement may be *ad hoc* relative to one context but not *ad hoc* relative to another'. (Holton [1969], p. 181.)

[28] Holton [1969], pp. 184–5.

[29] No doubt Holton would argue that the fact that my methodology admits of this possibility shows the folly of trying to appraise the actions of great scientists with the help of explicit general criteria. Fortunately, on my account, Einstein did not make such a mistake.

mistakenly regarded Lorentz's LFC as *ad hoc* in the sense of being engineered simply for the purpose of neutralising Michelson's result. But this assumption is highly implausible. Einstein read the *Versuch* in which Lorentz proposed the MFH and derived the LFC from it. Further Einstein could hardly have regarded the LFC as more *ad hoc* than his own light postulate. For, first, on Einstein's own criteria, the SRT, as presented in 1905, was far from being 'internally perfect'. It consisted of two heterogeneous parts which were on totally different levels: on the one hand a high-level, universal covariance principle and on the other a so-called light postulate which was both low-level and extremely counter-intuitive. And secondly, whereas Lorentz *explained* why a moving rod contracts, Einstein *bluntly asserted* that the speed of light is an invariant, an assumption from which Michelson's result trivially follows. Reichenbach was at least partially correct when he wrote that:

> it would be mistaken to argue that Einstein's theory gives an explanation of Michelson's experiment since it does not do so. Michelson's experiment is simply taken over as an axiom.

Reichenbach was right in the following sense: while *intuitively* the light postulate can be regarded as a low-level generalisation of Michelson's result, the result is *prima facie* unconnected with the MFH. (The fact that the light postulate is both low-level and counter-intuitive was recognised by many of Einstein's contemporaries and – understandably – gave rise to the myth that Einstein was a positivist who unquestioningly obeyed the dictates of experience.)

My view that Einstein could hardly have judged Lorentz's MFH as more *ad hoc* than his own light postulate is supported by the following fact. The proposition that in all inertial frames the measured speed of light must be equal to the same constant c is deducible from Lorentz's pre-1905 system, which includes the MFH.[30] In other words, *Lorentz's theory explains not only Michelson's null result but also the invariance of c.* Whatever meaning is attached to '*ad hoc* relative to a context', it cannot allow that the MFH should *both* imply the light postulate *and*, unlike the light postulate, be *ad hoc* relative to Michelson's experiment. Lorentz was justified in asserting that:

> ...the chief difference [is] that Einstein *simply postulates* what we have *deduced* with some difficulty and not altogether satisfactorily, from the fundamental equations of the electromagnetic field.[31]

Einstein's programme eventually proved superior to Lorentz's in a strictly objective sense,[32] but this superiority does not rest on the *ad hoc* character of Lorentz's system.

[30] Admittedly Lorentz performed the deduction only in 1909 (cf. below, p. 235). However, I am enough of a Polanyiite to find it implausible that Einstein failed to realise, prior to 1905, that, from an *intuitive* point of view, the MFH cannot be more *ad hoc* than his own light postulate.

[31] Cf. Lorentz [1909], p. 230; my italics.

[32] I shall argue this at length in §§2 and 3.

1.3 *The double heuristic role of mathematics in science*

I earlier claimed that the MFH had its origins in mathematical considerations. Before substantiating this claim, I shall examine in general terms the heuristic role which mathematics can play in the development of scientific research programmes.

It is well known that science has stimulated the development of mathematics. Physics sets problems for which an urgent mathematical solution is required; as a result, certain branches of pure mathematics receive a powerful impetus. For example, Newton invented the calculus specifically for the study of continuous and differentiable motion: the fluent variable was time and the fluxion, instantaneous velocity. Thus analysis, the theory which dominated pure mathematical thinking for over two centuries, owes its origin to physics. The study of differential equations and the development of what later came to be called the advanced calculus were also closely connected with the development of the Newtonian programme in the eighteenth century. A similar process took place in the nineteenth century when Faraday, using 'line of force' as a new physical concept, enunciated laws of which Maxwell later gave a mathematical formulation; this, together with hydrodynamics, contributed to the development of vector analysis.

These examples illustrate the heuristic function of physics with regard to mathematics. But what about the reverse process, namely the heuristic function of mathematics with regard to physics?

(*a*) *Increase of empirical content through translation into mathematical language and through the physical interpretation of mathematical entities*

There are two important ways in which mathematics furthers physical discovery.

The scientist may start from an intuitive physical principle. Through being 'translated' into one of the mathematical languages available at the time, the principle may be modified; in particular it may acquire additional structure and thus become a stronger physical assumption. For example, Fresnel set out to give a mathematical formulation of his conjecture that light is a wave process in the ether. He instinctively resorted to the periodic function with which he was most familiar, namely the sine function. His original assumption that light is a wave phenomenon was obviously weaker than the hypothesis he actually used, namely that the wave is representable by the function $\sin(2\pi t/T)$.[33]

Peierls gives another example which beautifully illustrates my thesis.

[33] In his [1913] Mach wrote: 'The sine form recommends itself on account of its simplicity, and the simplicity of the mechanical hypotheses which suffice for its explanation seemed to Fresnel to warrant such an assumption' (p. 216).

Peierls gives an account of the discovery of Maxwell's equations which runs as follows.[34]

Maxwell translated Faraday's intuitive physics into the theory of partial differential equations. The following relations summarise what was known about the electromagnetic field in Maxwell's time:

$$\nabla \cdot \boldsymbol{D} = 4\pi\rho \quad (1) \qquad \nabla \wedge \boldsymbol{E} = -\frac{1}{c}\frac{\partial \boldsymbol{B}}{\partial t} \quad (3)$$

$$\nabla \cdot \boldsymbol{B} = 0 \quad (2) \qquad \nabla \wedge \boldsymbol{H} = \frac{4\pi}{c}\boldsymbol{j} \quad (4)$$

Through taking the divergences of both sides of the fourth equation, we obtain $0 = \nabla \cdot \boldsymbol{j}$. This contradicts the law of conservation of charge which is expressed by the following equation:

$$\nabla \cdot \boldsymbol{j} + \frac{\partial \rho}{\partial t} = 0. \quad (5)$$

Instead of altogether rejecting either his own mathematical approach or Faraday's physical theory, Maxwell saw that he could restore the consistency of equations (1)–(5) by adding the extra term $(1/c)\,(\partial \boldsymbol{D}/\partial t)$ to the right hand side of (4). In this way he obtained a new theory consisting of (1)–(3) and:

$$\nabla \wedge \boldsymbol{H} = \frac{1}{c}\frac{\partial \boldsymbol{D}}{\partial t} + \frac{4\pi}{c}\boldsymbol{j}. \quad (4')$$

The new theory implies (5).

Thus, if this account is correct, Maxwell directly tampered with the mathematical form of the equations rather than trying first to modify Faraday's system and only then translating it into a new and hopefully consistent mathematical form. Of course, through altering the mathematical expression of the theory, Maxwell also modified its physical content. But mathematical considerations led the way.

There is a second way in which mathematics can play a fundamental role in physical discovery. The usual method in theoretical physics is to give mathematical expression to some physical hypothesis and then to use logico-mathematical techniques in order to draw consequences from the hypothesis. In doing so the physicist may have recourse to a number of mathematical operations; these operations are sometimes in the nature of 'tricks' or 'gimmicks' which may be needed to make the deduction possible. Duhem pointed out that it would be foolish to insist on giving a physical interpretation to all mathematical quantities and operations

[34] Peierls admits that he is giving a reconstruction for which he has no historical proof. Nevertheless he is fairly sure that the argument he describes 'was in fact, explicitly or implicitly, part of his [i.e. Maxwell's] reasoning' (Peierls [1963], p. 31). I am indebted to Lakatos and Worrall who drew my attention to these examples supporting my thesis. Also cf. Worrall, this volume pp. 107–79.

used in a scientific theory.[35] Duhem is evidently right: adding lengths corresponds to placing physical rods one after the other; multiplying lengths corresponds to the construction of rectangular areas; *but* multiplying the time t by $(-1)^{\frac{1}{2}}$, although useful in a pragmatic sense, does not seem susceptible of physical interpretation. However, through trying to find a *realistic* interpretation of certain mathematical entities which appear at first sight to be devoid of any physical meaning, the scientist may be led to a new physical conjecture. We shall see that Lorentz introduced his famous transformation as a mathematical tool for solving a certain differential equation.[36] Through interpreting the transformation as representing a physical dilatation of coordinates, Lorentz was led to the LFC or rather to a theory about molecular forces, the MFH, from which the LFC follows. Similarly Dirac proposed a relativistic equation which was found to possess negative energy solutions. *Prima facie* such solutions cannot be physically interpreted. Through insisting on interpreting the negative solutions, Dirac predicted the existence of the positron: the absence of an electron of charge $-e$ and energy $-E$ was interpreted as the presence of a positron, that is, of a particle of charge $+e$ and energy $+E$.

This dual heuristic role of mathematics will be apparent in the development of Einstein's programme;[37] Lorentz's programme, to which we now return, provides a fine example of the *second role* of mathematics.

(*b*) *An important illustration: the first version of Lorentz's Theory of Corresponding States arises out of the realistic interpretation of the Lorentz Transformation whose origins were purely mathematical*

In Lorentz's [1892*a*] there is no mention of the 'crucial' experiment first performed by Michelson in 1881, then repeated by Michelson and Morley in 1887. This should not surprise us once we have realised that the Lorentz transformation was originally used as a mathematical device of which Lorentz gave no physical interpretation, in much the same way as we might use the expression *ict* without attaching any physical meaning to the multiplication of the distance *ct* by $(-1)^{\frac{1}{2}}$.

Lorentz assumed the existence of the ether and of small particles, the electrons, which possess both material mass and electric charge. The electrons and their motion through the medium generate the field. Using Maxwell's equations for a frame fixed in the ether, one should be able to determine the electromagnetic field from the charge, position and state of motion of the electrons; that is from the electrical density and velocity distributions. Lorentz was thus led to write down the differential equation D: $[c^2\nabla^2-(\partial^2/\partial t^2)]f = G$ (f is the unknown and G is a known function of x, y, z, t), whose solution constituted a purely mathematical problem. This solution is valid only in a coordinate system which remains at rest in the

[35] Duhem [1906], part 2, chapter 1.

[36] Cf. below, p. 227.

[37] Cf. below, sections 2 and 3.

medium. Since the Earth is presumably moving through the ether and since we carry out our measurements relatively to the Earth, Lorentz was led to consider the field equations in a frame of reference attached to a moving body. He thus obtained relations which are more complicated than Maxwell's equations. Again the problem of computing the field quantities, given the charge and velocity distributions, forced itself on him. The differential equation D^*: $[c^2\nabla^2 - (\partial/\partial t - v(\partial/\partial x))^2]f = G$, which is the mathematical formulation of this problem, is more complicated than D. The Lorentz transformation was designed to reduce D^* to the form of D, so that any solution of D automatically yields a solution of D^*. The transformation equations are as follows:

$$x' = \beta x = \beta(x_1 - vt_1), \quad y' = y = y_1, \quad z' = z = z_1,$$
$$t' = t - v\beta^2 x/c^2 = \beta^2(t_1 - vx_1/c^2);$$

where $\beta = (1 - v^2/c^2)^{-\frac{1}{2}}$; x, y, z, t, are the Galilean coordinates in the moving frame. These equations, which carry the operator $[c^2\nabla_1^2 - \partial^2/\partial t_1^2]$ into $[c^2\nabla'^2 - \beta^2\,\partial^2/\partial t'^2]$,[38] constitute the classical Lorentz transformation to within the extra factor β in the expression of t'.[39] The origins of the Lorentz transformation were thus strictly mathematical, and had nothing to do with the Michelson–Morley experiment.

However, soon after writing his [1892*a*], Lorentz realised that the transformation equations lent themselves to an interpretation which he immediately set out in his [1892*b*] and which he then expounded in greater detail in his [1895].

Let us take a look at the equations:

$$x' = \beta x, \quad y' = y, \quad z' = z, \quad t' = t - v\beta^2 x/c^2$$

where $\beta = (1 - v^2/c^2)^{-\frac{1}{2}}$ and x, y, z, t are the Galilean coordinates in the moving frame S. x' is simply obtained by multiplying x by the factor β which is greater than 1. The variable t' is more difficult to interpret physically because it involves both the absolute time and the position. Fortunately, by considering a system of particles at rest in S, i.e. particles all moving with the same velocity $\mathbf{v}$ through the ether, Lorentz was able to simplify his problem. Now the field depends only on x, y, z, or alternatively on x', y', z'; so the 'local' time t' can be safely ignored. To the moving system S, Lorentz made correspond a system S' at rest in the medium; S' is obtained by expanding S by the factor β along the x-axis, while keeping the other two dimensions and the charge unaltered. Conversely, S is a contracted image of S', so the connection between this

[38] $\nabla_1 = (\partial/\partial x_1, \partial/\partial y_1, \partial/\partial z_1)$ and $\nabla' = (\partial/\partial x', \partial/\partial y', \partial/\partial z')$.

[39] One may wonder why Lorentz did not put $t' = \beta t''$ so as to obtain the full invariance of the operator $[c^2\nabla^2 - \partial^2/\partial t^2]$. It is clear that he was not at this stage interested in invariance as such but in a means of solving a particular mathematical problem. For an approach to Relativity Theory based on the invariance of $[c^2\nabla^2 - \partial^2/\partial t^2]$ cf. Stephenson and Kilmister [1958], chapter 1.

physical interpretation and Contraction Hypothesis, which Lorentz later inferred, becomes obvious.

The next step was to calculate the forces acting at corresponding points of S and S'. Lorentz found that, if

$$\boldsymbol{F} = (F_1, F_2, F_3) \quad \text{and} \quad \boldsymbol{F}' = (F_1', F_2', F_3')$$

denote the forces per unit charge in S and S' respectively, then:

$$\boldsymbol{F} = (1, 1/\beta, 1/\beta)\,\boldsymbol{F}', \quad \text{that is,} \quad F_1 = F_1', \quad F_2 = F_2'/\beta, \quad F_3 = F_3'/\beta.$$

These equations[40] play a fundamental role in the Theory of Corresponding States. Since each component of one force is proportional to the corresponding component of the other, the vanishing of one of the forces entails that of the other.

It is not surprising that the first step towards interpreting the Lorentz transformation should have been taken in electrostatics, where the time variable can be ignored. It is well-known that the time-coordinate and more generally the kinematical aspect of the whole problem caused Lorentz great difficulties and these were settled only by Poincaré and Einstein in 1905.

I have already mentioned that this interpretation of the transformation equations was first explained at length in the *Versuch* of 1895; but Lorentz had already used it in 1892 in order to derive the Contraction Hypothesis and, as a by-product, to account for the null result of Michelson's experiment.[41]

Both the transformation equations and the Contraction Hypothesis were proposed in 1892. In 1899, after developing a large part of his Theory of Corresponding States, Lorentz himself described the starting point of his investigations as follows: 'In the preceding investigations, I have assumed that all electrical and optical phenomena in ponderable bodies are produced by small charged particles (electrons).' He admitted that in the course of his investigations, '*Certain mathematical artifices have permitted me to arrive, by a concise argument, at conclusions to which, without these artifices, I should not have arrived except by considerably lengthier developments.*'[42]

[40] The relation $\boldsymbol{F} = (1, 1/\beta, 1/\beta)\,\boldsymbol{F}'$ follows from Planck's equation for the relativistic force $\boldsymbol{F} = \mathrm{d}/\mathrm{d}t(m_0\beta\boldsymbol{v})$, where the particle is instantaneously ar rest in the moving frame. $\boldsymbol{F} = (1, 1/\beta, 1/\beta)\,\boldsymbol{F}'$ contradicts Einstein's equation $\boldsymbol{F} = m_0(\beta^2 a_1, \beta^2 a_2, \beta^2 a_3)$, where $(a_1, a_2, a_3) = \boldsymbol{a} =$ acceleration. This agreement between Lorentz and Planck stems from the fact that both of them take the Lorentz-force as their paradigm of force. Today it is Planck's, and not Einstein's, equation which is generally accepted. In this sense Lorentz was ahead of Einstein. (Cf. Einstein [1905] and Planck [1906].) Professor Clive Kilmister pointed out to me that, in his [1905], Einstein makes a Lorentzian type of assumption about the properties common to all forces. Einstein writes: '. . . these results as to the mass are also valid for a ponderable material point, because a ponderable material point can be made into an electron (in our sense of the word) by the addition of an electric charge, no matter how small' (cf. Einstein and others [1923]: p. 63). Is this not Lorentz's assumption again? For, since one is allowed to have transverse and longitudinal masses, one might just as well expect different masses for electric and other forces.

[41] Lorentz [1892*b*].

[42] Lorentz [1899], (*Collected Papers*, **5**, p. 139); my translation and my italics.

1.4 *Lorentz derived the Lorentz–Fitzgerald Contraction Hypothesis from the Molecular Forces Hypothesis which is, in all senses, non* ad hoc*: the Michelson–Morley experiment lends dramatic support to the Molecular Forces Hypothesis, which conforms to the heuristic of the ether programme*

In 1892 Lorentz put forward the MFH, to which he was led, as we have just seen, by the coordinate transformation used in his [1892*a*]. The MFH asserts that, in passing from the stationary system S' to the moving system S, the molecular forces transform like the electrostatic ones; in other words the stationary and moving molecular forces are also connected by the equation: $\boldsymbol{F} = (1, 1/\beta, 1/\beta)\,\boldsymbol{F}'$. I shall now reconstruct Lorentz's deduction of the LFC from the MFH, using the extra assumption U that the equilibrium configuration of a system of particles is unique.[43]

Let $O'A' = L'$ be the length of the rod in the stationary system S'. Let $OA = L$ be the length of the same rod in the moving system S obtained by imparting to S' a uniform rectilinear motion with speed v along $O'X'$. We leave open the question of whether or not $L = L'$. Let $O'B' = \beta L$; i.e. $O'B'$ is obtained by expanding the rod OA by the factor β, while the charge and mass of corresponding elements remain the same. Let $\boldsymbol{G}$ be the sum of all the forces, molecular as well as electromagnetic, acting at some point P of OA and let $\boldsymbol{G}'$ be the force exerted at the corresponding P' of $O'B'$. Since P is in equilibrium $\boldsymbol{G} = 0$. The MFH implies $\boldsymbol{G} = (1, 1/\beta, 1/\beta)\,\boldsymbol{G}'$. Hence $\boldsymbol{G}'$ also vanishes and P' is in equilibrium. Since P' can be an arbitrary point of $O'B'$, the whole of the rod $O'B'$ must be in equilibrium. The same holds for $O'A'$, so by the uniqueness hypothesis U, $O'B' = O'A'$; i.e. $\beta L = L'$ or $L = (1/\beta)\,L'$; $O'A'$ is therefore contracted by the factor $1/\beta$.

Applying this result to Michelson's experiment[44] we find that, if $BD = L$, then BE equals not L but L/β. Hence

$$t_2 - t_1 = (2L/\beta c)\,\beta^2 - (2L/c)\,\beta = 0;$$

so no shift of the fringes will occur.

Thus we see that Michelson's experiment did not *refute* the conjunction of Newton's laws and Maxwell's equations. The central feature of this development from the point of view of my approach is that the MFH did not result from a consideration of the experimental result. Admittedly Lorentz knew of Michelson and Morley's result from 1887 onwards and confessed that it had been worrying him for some time;[45] but only in 1892, after finding his mathematical transformation equations, did he think of putting forward the MFH. In 1887 he might have simply postulated that

[43] Lorentz used U in 1892, but only in 1904 did he refer to it as an independent hypothesis.

[44] Cf. above, p. 212.

[45] 'The experiment has been puzzling me for some time' (Lorentz [1892*a*] in *Collected Papers*, **4**, p. 221).

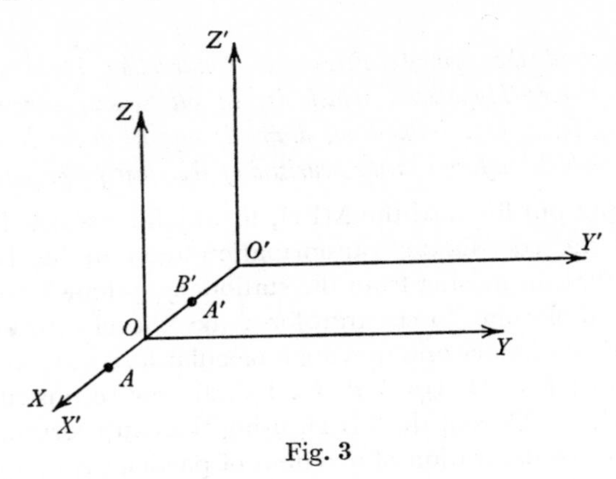

Fig. 3

one dimension of a body contracts in the direction of its motion through the ether, but he would have considered such a simplistic contraction hypothesis unacceptable.[46] Thus, since the discovery of the MFH was independent of the Michelson–Morley experiment, whose null-result the MFH implies, the experiment strongly supports the hypothesis. The Michelson result is, according to my amended definition of novelty,[47] a novel or unexpected prediction from the MFH. This hypothesis is consequently non *ad hoc*$_2$.

Did the MFH fit badly into Lorentz's theoretical system? Put more objectively, was the MFH *ad hoc*$_3$? Lorentz explained why it was not:

> Surprising as this hypothesis [i.e. the LFC] may appear at first sight, yet we shall have to admit that it is by no means far-fetched, as soon as we assume that molecular forces are also transmitted through the ether, like the electric and magnetic forces of which we are able at the present time to make the assumption definitely. If they are so transmitted, the translation will very probably affect the action between two molecules or atoms, in a manner resembling the attraction or repulsion between charged particles.[48]

The MFH is therefore non *ad hoc*$_3$ within the ether programme, whose heuristic requires that physical phenomena be explained in terms of contiguous actions through the medium. Molecular forces determining the shape of a given body are transmitted by the same medium as the electromagnetic field; since both types of force are states of the same substratum, why should they not behave and transform in the same way?

[46] In his [1969] Schaffner pointed out that the LFC is not as simple-minded a hypothesis as it is generally taken to be. Nonetheless he still regards the LFC as *ad hoc* (p. 500).
[47] Cf. above, p. 218.
[48] Einstein and others [1923], p. 5.

1.5 *The progress of Lorentz's programme after 1892*

(a) *The final version of the Theory of Corresponding States*

The last version of Lorentz's Theory of Corresponding States appeared in his [1904] which had the revealing title 'Electromagnetic Phenomena in a System Moving with any Velocity less than that of Light'.

The philosophical significance of the Theory of Corresponding States is that it could, as Poincaré showed, easily be turned into a theory observationally equivalent to Special Relativity.[49] Thus, in order to compare the merits of the two rival theories *anno* 1905, non-empirical criteria will have to be invoked. I shall suggest criteria which take into account the heuristic power of competing research programmes.[50]

Let met briefly outline the task[51] Lorentz set himself after 1895. This will clarify the philosophical problem mentioned above by showing in what way the Theory of Corresponding States can be regarded as equivalent to Special Relativity.

Maxwell's equations, we recall, hold in a frame of reference fixed in the ether. Consider a moving system $\bar{S}$ and an observer O carried along by the motion of $\bar{S}$ through the ether. If the instruments of O were unaffected by this motion, then the measured lengths, time intervals and field intensities would bear to one another relations more complicated than Maxwell's equations: in particular the measured velocity of light in $\bar{S}$ would not be the same in all directions and the observer would thus realise that he was moving relatively to the medium. The Theory of Corresponding States asserts that the instruments are distorted in such a way that the measured quantities (i.e. $t', x', y', z', \mathbf{E}', \mathbf{H}', \rho'$) *do* satisfy Maxwell's equations; hence the measured velocity of light is constant in $\bar{S}$. The observer will thus imagine himself to be within a system S' at rest in the ether, a system in which the *real* coordinates are the quantities t', x', y', z' measured by the distorted instruments. The 'fictitious' system S' is called the state corresponding to $\bar{S}$.[52]

Put in more technical language, Lorentz's project for a Theory of Corresponding States consisted of the following stages:

(1) Consider a frame S at rest and a frame $\bar{S}$ moving at constant velocity $\mathbf{v} = (v, 0, 0)$ through the ether. In $\bar{S}$: $\bar{x} = x - vt$, $\bar{y} = y$, $\bar{z} = z$, $\bar{t} = t$ (Galilean transformation). The components (D_1, D_2, D_3) and (H_1, H_2, H_3) of the electric and magnetic fields are the same in S and $\bar{S}$. However, in $\bar{S}$ the relations connecting $\mathbf{D}$, $\mathbf{H}$, $\bar{x}$, $\bar{y}$, $\bar{z}$, $\bar{t}$ are more complicated than Maxwell's equations, which only hold in S.

[49] Cf. Poincaré [1905] and [1906]; also cf. Ehrenfest [1923]. [50] Cf. §§2 and 3.

[51] I have used the word 'task' on purpose, because Lorentz did not fully achieve his aims.

[52] This informal description of Lorentz's theory is not to be taken as literally presupposing the presence of a conscious observer. The 'observer' is introduced for purposes of exposition and could well be replaced by a set of measuring instruments.

(2) The first problem is to determine in the frame $\bar{S}$ 'effective' variables t', x', y', z' and $\boldsymbol{D}'$, $\boldsymbol{H}'$, ρ' in such a way that with respect to t', x', y', z', the field quantities $\boldsymbol{D}'$, $\boldsymbol{H}'$, and ρ' satisfy Maxwell's equations.

(3) Construct a system S' fixed in the ether and in which the real variables are the 'effective' variables of $\bar{S}$. In other words, using Einstein's terminology, we consider the following correspondences between events:

$$\underset{\text{in } S}{(t, x, y, z)} \rightarrow \underset{\text{in } \bar{S}}{(\bar{t}, \bar{x}, \bar{y}, \bar{z})} \rightarrow \underset{\text{in } S'}{(t', x', y', z')}$$

To the moving system $\bar{S}$ corresponds the immobile system S'. Since ρ', $\boldsymbol{D}'$ and $\boldsymbol{H}'$ satisfy Maxwell's equations relatively to t', x', y', z', it follows that $\boldsymbol{D}'$ and $\boldsymbol{H}'$ are respectively the 'real' electric and the 'real' magnetic field in the 'fictitious' system S'.

(4) Examine hypotheses (such as the MFH and U) which imply that, if the system S' is set in motion with constant velocity v, it will rearrange itself so as to produce the system $\bar{S}$. One object of the exercise is now to show that the results of certain types of experiment, *or possibly of any experiment whatever*, are not affected by the motion of the earth through the ether. This is generally achieved by showing that, if certain quantities $\boldsymbol{P}$, $\boldsymbol{F}$, ... vanish in S', then the corresponding quantities will also vanish in $\bar{S}$.

(*b*) *Lorentz's failure to establish the full covariance of Maxwell's equations*

Neither in his (1904), nor in his (1909), did Lorentz completely achieve all these aims, except in the case of electrostatics. In all other cases he neglected second-order quantities.[53] Further, and relatedly, a great difficulty was posed by the problem of simultaneity: two events simultaneous in $\bar{S}$ may nonetheless occur in different 'local times' ('effective times'). Let me clarify these remarks. In his [1904] Lorentz writes Maxwell's equations as follows:

$$\text{(i)}\ \nabla\cdot\boldsymbol{D} = \rho \qquad \text{(iii)}\ \nabla\wedge\boldsymbol{H} = \frac{1}{c}\left(\frac{\partial\boldsymbol{D}}{\partial t}+\rho\boldsymbol{\nu}\right)$$

$$\text{(ii)}\ \nabla\cdot\boldsymbol{H} = 0 \qquad \text{(iv)}\ \nabla\wedge\boldsymbol{D} = -\frac{1}{c}\frac{\partial\boldsymbol{H}}{\partial t}$$

where ρ is the density and $\boldsymbol{\nu} = (\nu_1, \nu_2, \nu_3)$ is the velocity vector at time t at the point (x, y, z).

In a frame of reference moving with velocity $\boldsymbol{v} = (v, 0, 0)$ through the ether, the 'effective' coordinates are given by the Lorentz transformation equations as we know them today.[54]

$$x' = \beta(x-vt); \quad y' = y; \quad z' = z; \quad t' = \beta\left(t-\frac{v}{c^2}x\right) \tag{1}$$

[53] Cf. Einstein and others [1923], pp. 14–21. Also Lorentz [1909], p. 203.

[54] Lorentz interposes the Galilean coordinates: $\bar{x} = x-vt$, $\bar{y} = y$, $\bar{z} = z$, $\bar{t} = t$ between (t, x, y, z) and (t', x', y', z').

The vectors $\boldsymbol{H}' = (H_1', H_2', H_3')$ and $\boldsymbol{D}' = (D_1', D_2', D_3')$ are defined as follows:

$$D_1' = D_1;\quad D_2' = \beta\left(D_2 - \frac{v}{c}H_3\right);\quad D_3' = \beta\left(D_3 + \frac{v}{c}H_2\right) \tag{2}$$

$$H_1' = H_1;\quad H_2' = \beta\left(H_2 + \frac{v}{c}D_3\right);\quad H_3' = \beta\left(H_3 - \frac{v}{c}D_2\right)$$

Let us note in passing that equations (2) are identical with Einstein's transformation equations for the electric and magnetic fields.

Under the transformations (1) and (2), Maxwell's equations become:

$$\text{(i')}\quad \nabla'\cdot\boldsymbol{D}' = \left(1 - \frac{vu_1'}{c^2}\right)\rho' \qquad \text{(iii')}\quad \nabla'\wedge\boldsymbol{H}' = \frac{1}{c}\left(\frac{\partial\boldsymbol{D}'}{\partial t'} + \rho'\mathbf{u}'\right)$$

$$\text{(ii')}\quad \nabla'\cdot\boldsymbol{H}' = 0 \qquad \text{(iv')}\quad \nabla'\wedge\boldsymbol{D}' = -\frac{1}{c}\frac{\partial\boldsymbol{H}'}{\partial t'}$$

where $\boldsymbol{u}' = (\beta^2(v_1 - v), \beta v_2, \beta v_3)$ and $\rho' = \rho/\beta$. (3)

Except for the interpretation of ρ' and $\boldsymbol{u}'$, equations (i')–(iv') are exactly those obtained by Einstein in his [1905].

Although Lorentz's equation $\nabla'\cdot\boldsymbol{D}' = (1 - vu_1'/c^2)\,\rho'$ is correct, the transform of the charge density is not ρ' but $\sigma' = (1 - vu_1'/c^2)\,\rho'$. In fact:

$$\sigma' = (1/\beta)\,[1 - vu_1'/c^2]\,\rho = (1/\beta)\,[1 - (v\beta^2/c^2)\,(v_1 - v)]\,\rho = \beta(1 - vv_1/c^2)\,\rho$$ [55]

Here Lorentz seems to have made an easily corrigible mistake. Of course, in terms of the Galilean coordinates $\bar{x}$, $\bar{y}$, $\bar{z}$, we have: $x' = \beta\bar{x}$, $y' = \bar{y}$, $z' = \bar{z}$, from which it appears to follow that $\mathrm{d}x'\,\mathrm{d}y'\,\mathrm{d}z' = \beta\,\mathrm{d}\bar{x}\,\mathrm{d}\bar{y}\,\mathrm{d}\bar{z}$ and so ρ' = transformed density $= \rho/\beta$. But so to interpret ρ' is to forget that in general ρ depends on the time, so that $\rho' = \rho/\beta$ holds only for an electrostatic system. In order to determine the density σ' in S' at the space-time point (t', x', y', z') we consider an infinitesimal volume $\mathrm{d}V'$ enclosing the point (x', y', z') and count the charged particles lying within $\mathrm{d}V'$ all *at the same time* t'. We therefore have to take into account a number of events which, while being simultaneous in S', are not necessarily simultaneous in S. *Lorentz's mistake in the transformation equation of* ρ *is therefore deeply significant. It stems from the difficulties presented by the physical interpretation of local time.* In 1904 Lorentz had not yet realised that the local or effective time t' was in fact the time measured by a moving clock synchronised in accordance with Einstein's convention. Yet the equations obtained by Lorentz in 1904 are so similar to Maxwell's that one wonders why Lorentz did not simply postulate:

$$\sigma' = (1 - v\cdot u_1'/c^2)\,\rho' = \text{transformed density}$$

and thus obtain the full invariance of Maxwell's equations. However, had he taken this step, he would still have had to face the problem posed by equation (iii'), which now becomes:

$$\nabla'\wedge\boldsymbol{H}' = \frac{1}{c}\left[\frac{\partial\boldsymbol{D}'}{\partial t'} + \sigma'\boldsymbol{u}'\Big/\left(1 - \frac{vu_1'}{c^2}\right)\right]$$

[55] This is precisely the value given by Einstein to the transformed density.

He would have had to interpret $\boldsymbol{u}'/(1-vu_1'/c^2)$ as an effective *velocity*, and for this he needed to have developed a general kinematical framework. *This is where Einstein's approach, which starts with general kinematical considerations, proved far superior to Lorentz's. Whereas Einstein sorts out his kinematics* before *imposing the condition of Lorentz-covariance on all physical laws and in particular on electromagnetic theory, Lorentz painfully struggles to arrive at a new kinematics* via *electromagnetism.*[56]

(*c*) *Poincaré's contribution and the 'observational equivalence' of Special Relativity and the Theory of Corresponding States*

Using Lorentzian methods, Poincaré established the full covariance of Maxwell's equations by determining the correct transformation rule for the density as follows.[57] Consider an electron whose charge is e, moving with velocity (ξ, η, ζ) in the ether. The equation of a sphere of radius · centred on the electron is:

$$(x-\xi t)^2+(y-\eta t)^2+(z-\zeta t)^2 = r^2$$

Using the effective coordinates, we obtain:

$$x = \beta(x'+vt'), \quad y = y', \quad z = z', \quad t = \beta(t'+(v/c^2)\,x')$$

Hence:

$$[\beta(x'+vt')-\xi\beta(t'+vx'/c^2)]^2+[y'-\eta\beta(t'+vx'/c^2)]^2 \\ +[z'-\zeta\beta(t'+vx'/c^2)]^2 = r^2.$$

Consider this equation for some given time t_0', say $t_0' = 0$,

$$[\beta^2(1-\xi v/c^2)^2]\,x'^2+[y'-\eta\beta vx'/c^2]^2+[z'-\zeta\beta vx'/c^2]^2 = r^2$$

This represents an ellipsoid whose volume is:

$$\tfrac{4}{3}\pi r^3/\beta(1-\xi v/c^2)$$

By the law of conservation of charge:

$$e = \rho\cdot\tfrac{4}{3}\pi r^3 = \rho'\cdot\tfrac{4}{3}\pi r^3/\beta(1-\xi v/c^2)$$

Hence

$$\rho' = \rho\beta(1-\xi v/c^2)$$

This is exactly Einstein's transformation equation for ρ.

Poincaré considers the electron as a compressible body actually flattened by pressure exerted on its outer surface. The contraction along the direction of motion is not brought about, as Einstein was later to maintain, by the peculiar structure of space-time, but by a real physical force. This is why I said that Poincaré used Lorentizian methods.[58]

[56] Cf below, §§2 and 3.

[57] Poincaré [1905] and [1906]. Also cf. Kilmister [1970], p. 145.

[58] But despite this, one hesitates to count Poincaré among classical physicists such as Maxwell and Lorentz. Poincaré was the first scientist to recognise the group character of the transformation equations and probably also the first clearly to enunciate a physical

The connection between Special Relativity and the Theory of Corresponding States, as amended by Poincaré, can be described as follows.[59] For Einstein the 'effective' variables t', x', y', z' are the 'real' coordinates in $\bar{S}$. The equations $x' = \beta(x-vt)$, $y' = y$, $z' = z$, $t' = \beta(t-vx/c^2)$, $\beta = (1-v^2/c^2)^{-\frac{1}{2}}$, which were discovered by Lorentz, were used in Einstein's [1905] to relate the coordinates in the two inertial frames S and $\bar{S}$. Einstein completely abolished the system S'.[60] Lorentz realised after 1905 that his 'effective' variables are in fact the *measured* lengths, time intervals and field intensities in the moving frame $\bar{S}$. Because his rods are shortened, an experimenter in S obtains as a measure of the x-coordinate not $\bar{x} = x-vt$ but $x' = \beta\bar{x} = \beta(x-vt)$; moreover, his clocks are retarded by the factor $(1/\beta)$; if he adopts Einstein's convention for clock-synchronisation, he obtains as the measure of time in $\bar{S}$ not t but $t' = \beta(t-vx/c^2)$.

It can be easily verified that $[(dx/dt)^2+(dy/dt)^2+(dz/dt)^2 = c^2]$ holds if and only if $[(dx'/dt')^2+(dy'/dt')^2+(dz'/dt')^2 = c^2]$ holds. Thus, in Lorentz's system, the *measured* velocity of light is the same in all inertial frames.

Einstein differs from Lorentz in that he regards the 'effective' variables in $\bar{S}$ as the real ones and totally abolishes the Galilean transformation, i.e. the mapping $(t, x, y, z) \rightarrow (\bar{t}, \bar{x}, \bar{y}, \bar{z})$. The Theory of Corresponding States is 'observationally equivalent' to Special Relativity because experimental results involve only measured, that is, 'effective', quantities. Since the latter satisfy Maxwell's equations, we are unable, whether we adopt Lorentz's or Einstein's theory, to decide on empirical grounds whether our frame of reference is in motion or at rest in the 'ether'.

1.6 *The rationality of Lorentz's pursuing his own programme after 1905*

The subsequent success of Relativity Theory can easily give one the impression that Lorentz was wrong-headed, not to say crankish, in not immediately accepting Einstein's ideas, that he was too slow to see the light. But Lorentz was the champion of the classical electromagnetic programme started by Faraday and articulated by Maxwell. Needless to say, the mechanics of this programme was borrowed from Newton. Its hard core consisted of Newton's three laws of motion and Maxwell's equations. Newton had a classical Principle of Relativity, according to which the laws of mechanics are the same in any two frames moving with uniform velocity relatively to each other. However, Newton assumed the existence of Absolute Space, that is of a preferred frame, although the

principle of relativity. (Einstein is supposed to have carefully read 'Science and Hypothesis' before 1905.) Whatever the case may be, Poincaré showed, using Lorentz's own approach, how the Theory of Corresponding States could be made observationally equivalent to Special Relativity.

[59] Cf. above, pp. 231–4.

[60] More precisely, Einstein identified S' with $\bar{S}$.

latter could not be determined on the basis of mechanics alone. The Absolute Space Hypothesis was an idle component of Newtonian theory in the sense that any inertial frame could be considered as immobile in Absolute Space. With the wave theory of light, which seemed to presuppose a medium of transmission, arose the possibility of turning the Absolute Space Hypothesis into an empirically testable theory.[61] Lorentz's ontology consisted of an infinite immobile ether in which charge was continuously distributed. The electrons were spherical regions of the ether where the charge and possibly the mass densities differ from zero. The total amount of charge remains constant, but the movement of the electrons creates a field which travels in free space (the ether) at a constant finite speed c. Lorentz tentatively assumed that the electron possesses no material mass; the electron engenders a field which acts back on the source and decelerates its motion; this capacity for resisting change of motion is the electromagnetic mass which varies with the speed and accounts for the total inertia of the particle. Thus we see how fundamental is the role played by charge, Absolute Time and Absolute Space (i.e. the ether) in Lorentz's approach. They are the ultimate constituents of a physical world closely resembling Newton's and which we may call the classical world.

Thus, for Lorentz to have switched to the Relativity programme would have involved a major change in his metaphysical outlook. But why should he have made the change? If Einstein's theory had immediately thrown up new facts which Lorentz's system either could not account for, or could only account for in an *ad hoc* way, then Lorentz's adherence to the classical ontology could not be characterised as rational. But, on the contrary, Lorentz's own approach, based on his classical ontology, enabled him to make theoretical and empirical progress – often in advance of Einstein. For example, Lorentz explained Michelson's result in a non *ad hoc* way; he was first to discover the transformation laws for the electromagnetic field; he described the way in which the inertia of the electron depends both on its energy and on its velocity; and he explained the invariance of c. Thus, Lorentz's continued adherence to his own programme after 1905 was completely rational.

The ontology presupposed by Einstein's theory is radically different from the classical one. Some positivists, among them Bridgman,[62] claim that Einstein had no 'metaphysical' commitments, his theory being a mere description of actual physical operations. But this is an illusion. In his [1905], Einstein implicitly posits a domain of events, each of which can be referred to by coordinates (t, x, y, z) in any one of infinitely many

[61] I do not mean that the Absolute Space Hypothesis was to become testable in isolation but that its addition to existing theories would increase the number of testable consequences.

[62] Bridgman [1936], chapter 2.

equivalent inertial frames. *Events* are therefore the constituents of the Einsteinian universe. At the beginning it was very difficult for Lorentz to acquiesce in this radical change of world view, especially since his theory and Einstein's explain the same facts. Because of the 'observational equivalence' of the two theories, one might be tempted to think that 'metaphysics' is irrelevant to this whole issue. We shall, however, see that ontologies, which might be unimportant in the case of individual theories, can play an essential heuristic role in the development of research programmes by providing different regulative principles.[63] This can be properly appreciated only after examining Einstein's programme in greater detail.

What I have established so far is that one cannot explain the success of Einstein's Special Relativity Theory in terms of the demerits of Lorentz's rival theory. Lorentz's programme was non *ad hoc* in all senses of the term. The adjustments to the theory in the 1890s were not made in the light of Michelson's result and thus were not *ad hoc* relative to it. The adjustments were both theoretically and empirically progressive and they were made in conformity to the heuristic of the classical programme. Thus, if the eventual acceptance by the scientific community of Einstein's theory in preference to Lorentz's was rational (i.e. if there are acceptable general criteria according to which Einstein's theory was objectively better than Lorentz's), that rationality must lie in the *extra merits* of Einstein's theory. I now turn to the Einsteinian programme and a consideration of its merits. Let me say that I shall argue that the acceptance of Einstein's programme *was* rational, although, given that Lorentz's and Einstein's theories were *anno* 1905 'observationally equivalent', my claim may well appear doubtful at this stage.

2 *Einstein's heuristics*

In §1 I showed that Lorentz's classical programme was progressive until after 1905 – the year in which Einstein published his Theory of Special Relativity (hereafter referred to as SRT). In the next two sections I shall try to deal with the following three questions. First what were Einstein's reasons for objecting to the classical programme and hence for starting his own? (I have already shown in §1 that these reasons could not have been of an empirical kind.) My second question is this. Once Einstein's programme was launched, why did other scientists like Planck, Lewis and Tolman work on Einstein's programme rather than on Lorentz's?[64]

[63] It should be noted that a 'metaphysical principle', as I use the term, is now an integral part of a scientific research programme and relates either to its hard core or to its heuristic (or both). Ontological propositions are, in turn, part of this scientific metaphysics. All these components, in the last analysis, are appraised in terms of the overall progress or degeneration of the whole programme.

[64] My answer will also show that Kuhn's theory of paradigm-change is not applicable to the Einsteinian Revolution. (Cf. below, pp. 250–2.)

Thirdly I shall try to answer the question, at what stage, if any, did the relativity programme empirically supersede Lorentz's.

2.1 *Einstein's appraisal of classical physics*

Why did Einstein object to classical physics? Let me immediately say that the answer to this question will not be a psychologistic answer; I shall not for example be indulging in speculations about Einstein's childhood. What I shall try to show is that certain (unfalsifiable) metaphysical beliefs – at first sight rather vague and empty – which Einstein held, correspond to heuristic prescriptions which, when skilfully applied to particular cases, become specific and powerful tools for the invention of scientific theories. Thus metaphysics can play an important role in starting a new programme, especially when the existing one is empirically successful. Of course, the *triumph* of a programme can be achieved only by empirical means. However interesting its metaphysics, the programme will ultimately be judged by its ability to anticipate facts. I should like to formulate, as clearly as I can, two devices which formed part of Einstein's heuristics. To these devices correspond metaphysical beliefs which Einstein articulated in his later years.

(I) Theories have to fulfil the so-called internal requirement of coherence.[65] *Science should present us with a coherent, unified, harmonious, simple, organically compact picture of the world.* The mathematics used in the theory should reflect the degree of internal perfection of the world. 'The aim of science is, on the one hand, a comprehension as complete as possible of the connection between the sense experiences in their totality, and on the other hand, the accomplishment of this aim by the use of a minimum of primary concepts and relations. (Seeking as far as possible, logical unity in the world picture, i.e. paucity in logical elements.)'[66]

Einstein went as far as asserting that reality, although independent of the mind, was nonetheless knowable *a priori*. His so-called aestheticism was not meant in any subjective sense but was linked to a definite metaphysical position. Because Nature *is* simple, scientific hypotheses ought to be organically compact. Simplicity or coherence are not aimed at because they please our minds or because they effect economy of thought, but because they are an index of verisimilitude.

If it is true that the axiomatic foundations of theoretical physics cannot be derived from experience but have to be freely invented, can we at all hope to find

[65] There is also an external requirement on theories, namely that they be consistent with empirical results. Thus Einstein writes: 'The first point of view is obvious: the theory must not contradict empirical facts' (cf. Einstein [1949], p. 21). Also: 'The great attraction of the theory [General Relativity] is its logical consistency. If any deduction from it should prove untenable, it must be given up. A modification of it seems impossible without destruction of the whole.' (Einstein [1950], p. 110. For Einstein, 'logical consistency' meant 'coherence' or 'organic compactness'.) Fortunately Einstein did not follow this rule.

[66] Cf. Einstein [1950], p. 62.

the right way? Or worse still: does this 'right way' exist only as an illusion...To this I answer with complete confidence that this right way exists and that we are capable of finding it. In view of our experience so far we are justified in feeling that Nature is the realisation of what is mathematically simplest...It is my conviction that we are able, through pure mathematical construction, to find those concepts and the law-like connections between them, which yield the key to the understanding of natural phenomena...The really creative principle is in mathematics. In a certain sense I consider it therefore to be true – as was the dream of the Ancients – that pure thought is capable of grasping reality.[67]

I shall illustrate the importance of prescription (I) in §2.2.

(II) The second heuristic device is more difficult to formulate. Its metaphysical underpinning is the claim that since God is no deceiver, there can be no accidents in Nature. All observationally revealed symmetries signify fundamental symmetries at the ontological level. Hence the heuristic rule: *replace any theory which does not explain symmetrical observational situations as the manifestations of deeper symmetries – whether nor not descriptions of all known facts can be deduced from the theory.* This will become much clearer with two examples.

(*a*) *The induction experiment*

If we move a magnet with respect to the ether while keeping a conductor fixed, then, due to the variation of the magnetic field with time, an electric field arises in the whole of space. Let P be any point of the conductor at which an electron may be situated. In view of the Lorentz formula:

$$\boldsymbol{F} = e\left(\boldsymbol{D} + \frac{\boldsymbol{v}}{c} \wedge \boldsymbol{H}\right), \quad \text{where} \quad \boldsymbol{D} \neq 0 \quad \text{and} \quad \boldsymbol{v} = 0$$

the electron will experience a force which generates a current in the conductor.

We now keep the magnet fixed and move the conductor with velocity $\boldsymbol{v}$ with respect to the medium. No electric field is created because H is static, i.e. independent of the time. The situation is very different from the previous one, so we might expect the current in the loop either not to arise at all or at any rate to be different from what it was in the first case. However, in view of $\boldsymbol{F} = e\left(\boldsymbol{D} + \frac{\boldsymbol{v}}{c} \wedge \boldsymbol{H}\right)$, where now $\boldsymbol{D} = 0$ but $\boldsymbol{v} \neq 0$, a current does arise, and, if the relative motion between the conductor and the magnet is the same as in the previous case, the current also turns out to be the same. This result is wholly explained by Maxwell's theory; in other words, if we assume the existence of a preferred frame and accept Maxwell's equations, we can infer that the outcome of the experiment depends solely on the relative motion of the magnet and the conductor, and not on their absolute motion with respect to the ether. Hence, this time without the aid of any auxiliary hypothesis, an ether theory yields the undetectability of the ether.

[67] Cf. Einstein [1934], p. 116; my translation.

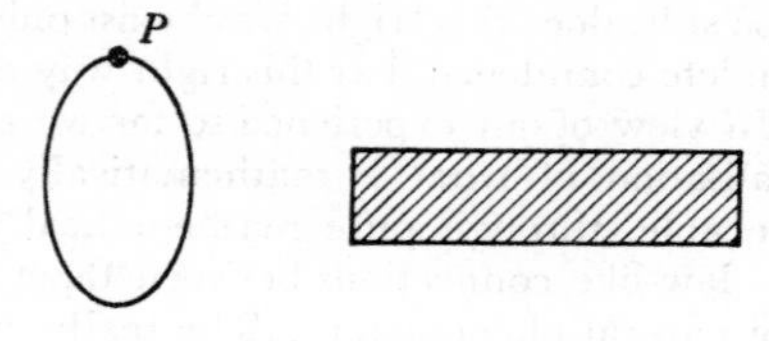

Fig. 4

Thus, in classical electromagnetism there is a basic *ontological* difference between a situation in which a magnet moves in the ether (presence both of a magnetic and of an electric field) and one in which the same magnet is stationary (presence of a magnetic field alone). However, when we *apply* Maxwell's equations to compute the current due to the motion of a conductor in the field created by the magnet, the result depends only on the relative motion between the magnet and the conductor. Thus, there exists at the 'observational' level, a symmetry between the following two situations: (*a*) magnet moving towards the conductor, and (*b*) conductor moving towards the magnet. This conflicts with the asymmetry obtaining at a higher level. Special Relativity eliminates the asymmetry: equations of exactly the same form apply, whether we choose the magnet or the conductor as our frame of reference. There are no separate electric and magnetic fields but one anti-symmetric tensor which transforms globally.[68]

(*b*) *Equality of gravitational and inertial masses*

In Newtonian theory the inertial mass m_i of a body represents its laziness, i.e. its capacity for resisting acceleration. Inertia is a primary irreducible property of matter which appears in the fundamental laws of motion. The gravitational mass m_g is a measure of the body's receptiveness to the gravitational field. According to Newton, gravity is not a primary quality to be treated on a par with inertia or impenetrability. Hence inertia and gravity ought to be independent properties. One should, for instance, be able to

[68] Exactly similar considerations as apply to the classical explanation of the induction experiment apply to Lorentz's explanation of the Michelson result (cf. p. 230). Once we accept the existence of an ether as the carrier of the electromagnetic field, we are led to look upon the latter as a state of the substratum. Molecular forces are transmitted by the same medium, so they also form part of its state; we thus have a good reason for supposing that molecular and electromagnetic forces are similar, i.e. for accepting the MFH (Molecular Forces Hypothesis). From this assumption follow the LFC (Lorentz–Fitzgerald Contraction Hypothesis) and Michelson's null result. There is something paradoxical in that, through postulating the ether as a universal medium, we are driven to the conclusion that it must be undetectable. Was it not dissatisfaction with this paradox, so closely connected with the crucial experiment, which caused Einstein to look for another explanation? The answer is that Einstein had become aware of the paradox independently of Michelson, as is indicated in the first paragraph of his [1905] where the induction experiment is mentioned. What from Einstein's point of view, was an unsatisfactory feature of classical physics is already evinced by Maxwell's account of the induction experiment and is, in this sense, completely independent of Michelson.

alter the gravitational 'charge' m_g without affecting the inertia of the body, in the same way that one can alter the electric charge e while keeping the inertial mass m_i constant. However, this is not the case: a doubling of m_i is instantaneously matched by a doubling of m_g. Newton postulated that the two masses, m_i and m_g, are equal, but did not explain why. In other words, there is a symmetry between doubling m_i and doubling m_g, which is at odds with the disparity between the two properties of inertia and gravity. Let us note that the observational 'cash-value' of '$m_i = m_g$' is the proposition that all bodies fall with the same acceleration in a given gravitational field.

The problem can be put a little differently as follows. If a moving train suddenly decelerates, the passengers, being thrown forward, imagine that they are subject to a field of force to which they respond proportionately to their inertial masses. The Newtonian physicist will tell them that this field is a fictitious inertial field due to an inappropriate choice of coordinate system (the train); in fact, by virtue of their inertia, the passengers are still moving uniformly in Absolute Space (that is, if we neglect the attraction of the earth).

According to the Newtonians this fictitious inertial field differs fundamentally from the 'real' gravitational field created by the earth; it is only by accident, namely because $m_i = m_g$, that all objects respond to the two fields in exactly the same way.

Einstein eliminates this asymmetry between gravity and inertia by proposing that all gravitational fields are inertial; i.e. that all gravitational fields are created by a (local) acceleration of the frame of reference. To put it crudely: being thrown forward in a moving train and being attracted by the earth are basically one and the same phenomenon.[69] It is no wonder that all bodies fall with the same acceleration, since it is the common frame which is accelerating under their feet.

These prescriptions may be susceptible of a more precise formulation, but I leave this question open.[70] Whatever the case may be, the lack of a more accurate rendering in no way entails that the propositions in question must be given a subjective (or psychological) interpretation. Einstein's metaphysical statements are admittedly vague, yet they may still correspond to real properties of an external world independent of the scientist's mind, of his private feelings about harmony, perfection and the like. My main object will now consist in examining the role the above prescriptions played in the genesis of SRT. In this specific context it turns out that these otherwise vague rules and propositions assume a very precise form, leaving no doubt as to their intended objective meaning.

[69] This is not strictly speaking true. In the case of the train the field is globally eliminable, whereas in the case of the earth the field is irreducible.

[70] Einstein himself thought '*that a sharper formulation would be possible*. In any case it turns out that among the augurs there usually is agreement in judging the inner perfection of the theories and even more so the degree of external confirmation' (Einstein [1949], p. 23).

2.2 *The discovery of Special Relativity Theory: removal of the asymmetry between classical mechanics and electrodynamics*

Let us look at the more general features of Einstein's objections to classical physics. According to Einstein, one of Maxwell's and Faraday's greatest contributions to science was the introduction of the field as a constituent of physical reality to be treated on a par with other constituents such as corpuscles and electric charge.[71] Lorentz's electromagnetic theory confronts us with a dualism to whose removal Einstein was to devote much of his life: on the one hand there are discrete charged particles whose motions are governed by Newton's laws, and on the other hand a continuous field obeying Maxwell's equations. It is true that the charged corpuscles and their motions generate the field; but, once started, an electromagnetic disturbance propagates itself with velocity c independently of its source; the field may act back on the particles, thereby modifying their motion. Fields and particles are therefore ontologically on a par. One way of resolving this dualism is to explain the behaviour of the field in terms of the mechanical properties of an all-pervading medium. Lorentz clearly recognised that all such attempts throughout the nineteenth century had failed; he was about to try a solution in the opposite direction, and in particular to explain inertial mass in electromagnetic terms.

Einstein was clearly dissatisfied with this dualism, as is apparent from the following passage:

> If one views this phase of the development of the theory critically, one is struck by the dualism which lies in the fact that the material point in Newton's sense and the field as continuum are used as elementary concepts side by side. Kinetic energy and field-energy appear as essentially different things. This appears all the more unsatisfactory inasmuch as, in accordance with Maxwell's theory, the magnetic field of a moving electric charge represents inertia. Why not then *total* inertia? Then only field-energy would be left and the particle would be merely an area of special density of field-energy. In that case one could hope to deduce the concept of the mass-point together with the equations of the motions of the particles from the field-equations, – the disturbing dualism would have been removed.[72]

This lack of unity in the physical foundations, which violates prescription (II), was reflected in the mathematical formulation of the theory. Einstein explains:

> The weakness of the theory lies in the fact that it tried to determine the phenomena by a combination of partial differential equations (Maxwell's field equations for empty space) and total differential equations (equations of motion of point masses), which procedure was obviously unnatural.[73]

The dualism was made far worse by Newton's classical Principle of Relativity which applies to mechanics but apparently not to electrodynamics.[74] In view of the Galilean transformation which physicists took

[71] Einstein [1934], p. 160.
[72] Einstein [1949], p. 36.
[73] Einstein [1950], p. 75.
[74] Newton [1686], p. 20.

for granted, Maxwell's equations seem to presuppose the existence of an ether, or at any rate of a unique frame of reference in which they would hold good. Assessing Lorentz's work, Einstein wrote:

For him [i.e. for Lorentz], Maxwell's equations concerning empty space applied only to a given system of co-ordinates, which, on account of its state of rest, appeared excellent in comparison to all other existing systems of co-ordinates. This was a truly paradoxical situation since the theory appeared to restrict the inertial systems more than classical mechanics.[75]

The Absolute Space Hypothesis, i.e. the assumption that among all inertial frames there exists a privileged one, is an idle metaphysical component of classical mechanics. That its elimination does not reduce the empirical content of classical dynamics was clearly recognised by Newton who wrote: 'The motion of bodies included in a given space are the same among themselves, whether that space is at rest or moves uniformly forward in a right line without any circular motion.'[76] One could further maintain that the Absolute Space Hypothesis was scientifically useless in that one could not even in principle define the Absolute Frame as that in which Newton's laws of motion hold good; for if these laws are true in one of the inertial frames, they are automatically true in all.[77]

With the advent of the wave theory of light, of Fresnel's and Lorentz's postulation of a stationary ether,[78] the situation changed dramatically. One could now define the Absolute or Ether Frame as that in which Maxwell's equations are true. Given the old kinematics and in particular the Galilean transformation, this definition singles out a unique frame in which, because Maxwell's equations hold in it, light propagates itself in all directions with the same speed c. The ether frame was taken to be inertial, so that in all other frames, whether inertial or accelerated, light would not have a constant velocity. This implied the possibility of devising experiments which might detect the 'absolute' motion of ponderable bodies. The experiment would be such that its outcome tells us whether the body in question was in motion or at rest in the ether. In this connection Michelson's experiment is typical: a null outcome would tell us that the earth is at rest in the ether, and from a shift of the fringes it would be concluded that the earth moves. In this particular case, however, we know that the earth changes its velocity with respect to the inertial frame determined by the stars, so the earth must at one point of its trajectory be moving in the ether. Hence we can predict that the experimental result must be positive.

Why did Einstein find such developments in the evolution of physics 'paradoxical'? We have seen that Einstein disliked the dualism of particles and fields. The fact that the laws of mechanics (which govern the motions

[75] De Haas-Lorentz [1957], p. 7.
[76] Newton [1686], p. 20.
[77] In the 'Science of Mechanics', which Einstein carefully read, Mach attacked the concept of absolute space and went as far as proposing that even the distinction between inertial and non-inertial frames ought to be abolished (Mach [1883], chapter 2, vi).
[78] For more exact details, cf. Lakatos [1970], pp. 159–65.

of particles) obey the Principle of Relativity, while Maxwell's equations (which govern the behaviour of the field) do not, makes the dualism much worse. Given the problem-situation, there were to my mind two courses of action open to a unificationist like Einstein: he could maintain either that the Relativity Principle applies *neither* to mechanics *nor* to electrodynamics or else that it applies to *both* at the same time. In the first case he could have modified mechanics in such a way that it only holds in the ether frame; in the second case he would have to extend the Relativity Principle to electrodynamics. In its Galilean form, the Relativity Principle is inapplicable to electromagnetic theory. At this point, however, Einstein's critique of the induction experiment proved crucial in that it tipped the balance in favour of extending Relativity to electrodynamics and *thereby* modifying classical kinematics.[79]

To repeat, in the induction experiment there is complete symmetry between the two experimental results, which is at odds with the asymmetry introduced by the 'theoretical' explanation. To put it more pedantically, the observational statements describing the behaviour of the currents in the conductor are identical in the two cases, but the high-level explanations in terms of the accepted theory differ widely. There would be nothing intrinsically wrong in this state of affairs, had the asymmetry not been introduced through considerations of absolute motion which the Relativity Principle forbids. Seen from that angle, however, the experiment suggests that an extension of the Relativity Principle to include electrodynamic phenomena might abolish the 'theoretical' asymmetry; it promises to make the symmetry between the two experimental outcomes appear, not as a fortuitous result, but as a direct manifestation of a general principle, the principle of Lorentz-covariance. In this he was following prescription II.

In his [1905] Einstein concluded that:

> examples [like the induction experiment] together with the unsuccessful attempts to discover any motion of the earth relatively to the light medium, suggest that the phenomena of electrodynamics as well as of mechanics possess no properties corresponding to the idea of absolute rest. They suggest rather, as has already been shown to the first order of quantities, that the same laws of electrodynamics and optics will be valid for all frames of references for which the equations of mechanics hold good. We will raise this conjecture (the purport of which will hereafter be called the Principle of Relativity) to the status of a postulate and also introduce another postulate, which is only apparently irreconcilable with the former, namely that light is always propagated in empty space with a definite velocity c which is independent of the state of motion of the emitting body.[80]

Note the two references made to mechanics, underlining the important part which classical relativity played in Einstein's thinking prior to 1905.[81]

[79] I do not, of course, mean that the experiment was 'crucial' in the traditional sense of refuting one theory while confirming another.

[80] Einstein and others [1923], p. 38.

[81] *This was later confirmed in his more philosophical writings.* (See de Haas-Lorentz [1957], quoted above, p. 243. Also cf. Einstein [1950], p. 55.)

In this passage Einstein alludes to the absence of any first-order effects of absolute motion, which Lorentz had explained in the *Versuch* through an early version of the theory of corresponding states. This first-order equivalence between observers, which runs counter to the preference given to a unique frame, must have increased Einstein's suspicion that the Relativity Principle applies to electrodynamics as well as to mechanics; under the new theory the absence of first-order effects, instead of being a stray fact, would directly reveal the presence of a universal principle.

What commended the Relativity Principle was therefore its universality, its unifying role in subsuming mechanics and electrodynamics under the same law and in providing a unified explanation for various features of phenomena such as the symmetry in the induction experiment and the absence of first-order effects due to the earth's motion.

The phrase: '... together with the unsuccessful attempts to discover any motion of the earth relatively to the light medium', has given historians and philosophers of science some problems.[82] It also seems inconsistent with the thesis of §1 that the Michelson experiment played a negligible role in the genesis of SRT.

Einstein might be referring in the quoted phrase to Michelson's experiment, which must have been in the back of his mind, if only through Lorentz's [1895].[83] This is perfectly compatible with his assertion that the experiment came to his *attention* only after 1905. To my mind the above phrase is no more than a casual allusion to a number of results which he had registered without any surprise, for they anyway followed from his own conjectures. On this point I agree with Holton;[84] for otherwise Einstein would certainly have cited Michelson's result in support of his second postulate, the Light Principle *P*2. This postulate presents us with a new difficulty.

Unlike his first postulate (the Relativity Principle) *P*1, the Light Principle *P*2 is thrown out with no justification whatever. Moreover, on the face of it, *P*2 runs counter to Einstein's prescription (I):[85] there seems to be no connection at all between the fundamental properties of space–time and those of light. Why should purely kinematical considerations involve c? The light principle is quite a low-level statement which is not as yet integrated into a more general system. It is precisely for this reason

[82] Grünbaum, for example, says: 'Unless they provide some other consistent explanation for the presence of the latter statement in Einstein's text of 1905, it is surely incumbent upon all those historians of Relativity Theory who deny the inspirational role of the Michelson–Morley experiment to tell us specifically what other "unsuccessful attempts to discover any motion of the earth relatively to the light medium" Einstein had in mind here.' (Cf. Pearce Williams [1968], p. 114.)

[83] He admitted to Shankland that 'he had also been conscious of Michelson's result before 1905, partly through his readings of the papers of Lorentz and more because he had simply assumed this result of Michelson to be true' (Holton [1969], p. 154).

[84] Cf Holton [1969], pp. 164–5.

[85] Cf. above, p. 238.

that philosophers and scientists supposed that Einstein was obeying the dictate of experience, basing his second postulate on Michelson's result.[86] Later on in his [1905], Einstein does say that the light principle is in agreement with experience:[87] he had after all heard of various experiments trying to detect the earth's absolute motion. Nowhere, however, does he assert that experience had suggested the second postulate, or even made it look plausible to him.

I think the problem can be solved simply by examining more carefully Einstein's later writings – in particular his autobiography – and then comparing them with his [1905]. In his [1934] Einstein writes:

> Then came the Special Theory of Relativity with its recognition of the physical equivalence of all inertial systems. In conjunction with Electrodynamics or the law of propagation of light, it implied the inseparability of space and time.[88]

Perhaps the most illuminating passage occurs in Einstein's [1950]:

> The second principle on which the Special Relativity theory rests is that of the constancy of the velocity of light in the vacuum. Light in a vacuum has a definite and constant velocity, independent of the velocity of its source. *Scientists owe their confidence in this proposition to the Maxwell–Lorentz theory of electrodynamics.*[89]

Also, in his [1949], Einstein tells us about a thought-experiment in which, at about the age of sixteen, he imagined himself to be following a ray of light at speed c:

> If I pursue a beam of light with a velocity c (velocity of light in a vacuum), I should observe[5] such a beam of light as a spatially oscillatory electromagnetic field at rest. However, there seems to be no such thing, whether on the basis of experience or according to Maxwell's equations. From the very beginning *it appeared to me intuitively clear that, judged from the standpoint of such an observer, everything would have to happen according to the same laws as for an observer who, relatively to the earth, was at rest.* For how, otherwise, should the first observer know, i.e. be able to determine, that he is in a state of fast uniform motion?[90]

What is most striking about this passage is the conclusion which Einstein draws from his thought-experiment. He does not restrict himself to what seems warranted by the experiment, namely that c is an unattainable speed or that the addition law of velocities must break down. He immediately jumps to a general conclusion, or rather puts forward the sweeping conjecture: the laws of physics – more specifically those of electromagnetism – would have to be the same for the moving and for the stationary observers. Both historically and epistemologically speaking,

[86] Cf. below.

[87] Einstein [1905], p. 40.

[88] Einstein [1934], p. 143 (my translation). In his [1949] Einstein again says 'The Special Theory of Relativity owes its origin to Maxwell's equations of the electromagnetic field. Inversely the latter can be grasped formally in satisfactory fashion only by way of the Special Theory of Relativity. Maxwell's equations are the simplest Lorentz invariant field equations which can be postulated for an anti-symmetric tensor derived from a vector field' (p. 62).

[89] Einstein [1950], p. 56; my italics.

[90] Einstein [1949], p. 53; my italics.

Einstein's second starting point – the first one being the Relativity Postulate – is not the Light Principle but the proposition:

($P3$) *Maxwell's equations express a law of nature*;

in virtue of $P1$, they must therefore assume the same form in all inertial frames. Maxwell's equations imply that, within each coordinate system in which they hold, electrodynamic disturbances propagate themselves with velocity c, which velocity must therefore be an invariant. Thus $P1$ and $P3$ imply $P2$.

In the electrodynamical part of the 1905 paper Einstein does in fact suppose that Maxwell's equations are Lorentz-covariant and then deduces the transformation laws for $\boldsymbol{E}$ and $\boldsymbol{H}$. He does not try to infer $P3$ from $P1$ and $P2$; so the electrodynamic part, by exhibiting a transformation which makes Maxwell's equations covariant, simply established that the latter are compatible with the Relativity Postulate and the Light Principle. It is a consistency proof. Although $P3$ is a stronger statement than $P2$, it is more plausible and incidentally less counter-intuitive. In accordance with (I), $P3$ derives its plausibility from being a unified, well-knit theory in which the primitive concepts (electric field, magnetic field, charge density) are all closely connected; also it had been tested for a whole generation prior to 1905. Thus the logical order is reversed through *a priori* heuristic considerations: $P3$ is more plausible, though stronger, than $P2$.

Another piece of evidence which confirms the view that Einstein approached the problem of relativity through Maxwell's equations and their covariance is to be found in Lorentz's [1895]. In a part of this work which is completely independent of Michelson's experiment and of the Contraction Hypothesis Lorentz had proved that no first-order effects of the earth's motion can be detected. Neglecting all terms in $(v/c)^2$, he used a limiting case of the Lorentz transformation; he then found transformation laws for the field $\boldsymbol{E}$, $\boldsymbol{H}$, under which the equations take on a form very similar to, and in some cases identical with, the form assumed in the ether frame. It is not far-fetched to suppose that Lorentz's techniques made a strong impression on Einstein; they might well have led him to wonder whether a more general transformation would yield complete covariance together with the result that no effects whatever arise from uniform rectilinear motion. Lorentz himself was to attempt this solution in his [1904], which Einstein did not read before publishing the 'Electrodynamics of Moving Bodies'.

However, Lorentz's programme for a Theory of Corresponding States was outlined in the *Versuch* and one is struck by the similarity between the methods used by Lorentz and by Einstein. In his [1905] the latter first constructed a transformation law for the coordinates x, y, z, t; then, assuming the covariance of Maxwell's equations, he deduced the transformation laws for $\boldsymbol{E}$, $\boldsymbol{H}$ and ρ. Lorentz's influence on Einstein cannot be overrated; it was not Michelson, the experimentalist, but Lorentz, the theoretician, who played a considerable inspirational role in the genesis of Special Relativity. This is indicated in the all too brief second paragraph of Einstein's [1905] by the clause: 'as has already been shown to the first order of small quantities'.

We have seen that Einstein rejected Lorentz's *classical* approach, but he made use of Lorentz's tremendous technical achievement, albeit under very different kinematical assumptions. I have mentioned that Lorentz's [1892*a*] already contains the full Lorentz transformation up to a constant factor in the expression of t'.[91] Einstein's greatest contribution was to extend Lorentz's methods and give the transformed quantities a *realistic* interpretation in the 'moving' system.[92]

One might still wonder why Einstein did not start by postulating $P1$ and $P3$ instead of $P1$ and $P2$. He had, I think, at least two good reasons for presenting the new theory in the way he did. On the one hand it is preferable, from the logical point of view, to use the weaker assumption ($P2$) which, in conjunction with $P1$, suffices for developing a new kinematics and deriving the Lorentz transformation. On the other hand Einstein had come to the conclusion that Maxwell's equations, although true of macroscopic phenomena, did not provide the ultimate foundations for the whole of physics.[93] He might therefore have wanted to make his space–time system independent of electrodynamics.[94]

[91] Cf. above, p. 227.

[92] It is now finally clear why Einstein could rightly claim that Michelson's experiment had been quite irrelevant to his work and that he could easily have anticipated its null outcome. That almost nothing is cited in support of the Light Principle may be due to the fact that it follows from a well-corroborated hypothesis (Maxwell–Lorentz equations) together with the Relativity Postulate $P1$, for whose acceptance Einstein had already argued. (Cf. above, pp. 222–3.)

[93] Einstein did not accept Lorentz's (tentative) assumption that all physical phenomena could be explained in terms of charges and fields governed by Maxwell's equations.

[94] He wrote: 'I knew only of Lorentz's works in 1895 – "*La Théorie Electromagnétique de Maxwell*" [this is in fact Lorentz's [1892*a*]] and "*Versuch einer Theorie der electrischen und optischen Erscheinungen in bewegten Koerpern*" – but not Lorentz's later works, nor the consecutive investigations by Poincaré. In this sense my work of 1905 was independent. The new feature of it was the realisation of the fact that the bearing of the Lorentz transformation transcended its connection with Maxwell's equations and was concerned with the nature of space and time in general. A further new result was that Lorentz-invariance is a general condition for any physical theory. This was for me of particular importance because I had already previously found that Maxwell's theory did not account for the micro-structure of radiation and could therefore have no general validity.' Einstein indicates that a connection between the Lorentz transformation and Maxwell's equations clearly existed but was then transcended. The logical picture seems to be as follows:

$P1$ and $P3 \Rightarrow P1$ and $P2$,

$P1$ and $P2 \Rightarrow$ new kinematics and Lorentz transformation equations,

$P1$ and Lorentz transformation equations $\Rightarrow$ requirement of Lorentz-invariance for all physics.

The connection between Maxwell's theory and the Lorentz transformation is given by:

$P1$ and $P3 \Rightarrow$ Lorentz equations.

This connection is transcended by the result that, from the Relativity Principle $P1$ and

One question remains unanswered. Einstein faced a problem forced on him by the incompatibility, if taken together, of the following three hypotheses:

(*P1*) The Relativity Postulate,
(*N*) Newton's second law of motion,
(*P3*) The Maxwell–Lorentz equations.[95]

His solution consisted in modifying *N*, or rather in replacing *N* by a new theory *N′* such that *P1* and *N′* and *P3* are consistent. Considering that the (Galilean) Relativity Principle was first shown by Newton to hold for mechanics, it is puzzling that Einstein seems never to have envisaged keeping *N* and substituting for *P3* a new set of equations *P3′* covariant under the Galilean transformation. *P3′* would of course have had to yield *P3* as a limiting case.[96]

This is all the more intriguing because he realised that Maxwell's equations were not as fundamental or as ultimate as Lorentz had taken them to be.[97] In other words: why did Einstein throw in his lot with Maxwell rather than with Newton? In the end Einstein guessed; it was not, however, as uninformed a guess as it might at first appear.

Einstein was dissatisfied with the further and fundamental dualism between fields and particles which beset Lorentz's theory. In virtue of his prescription I, it seemed obvious to Einstein that one component of the dual structure ought to be reduced to the other. But which one? He found,

the Lorentz transformation equations, there follows a new structure of space–time and a condition of Lorentz-invariance which applies not only to Maxwell's equations but to the whole of physics. See Born [1956].

[95] Of course, there is the underlying and common assumption that the law of inertia should hold in all allowable frames. This is precisely why these frames are called 'inertial'.

[96] W. Ritz adopted this approach. Rather than adjusting the whole of physics to electrodynamics, he tried to alter electrodynamics so as to make it Galileo-covariant. He looked upon the field quantities $\boldsymbol{E}$ and $\boldsymbol{H}$ as intermediate quantities which enable one to compute the Lorentz force $\boldsymbol{F} = e(\boldsymbol{E}+\boldsymbol{v} \wedge \boldsymbol{H}/c)$. In the last analysis only the particles, their masses, charges and relative velocities are real. Everything else is scaffolding. Ritz then considers the equation found by Schwarzschild for the Lorentz force which one accelerating charged particle exerts on another. (Cf. Ritz [1911], p. 378.) This equation is not Galileo-covariant because, among other things, it involves the 'absolute' velocities of the particles, i.e. their velocities relative to the underlying frame of reference. Since such absolute velocities are altered by a Galilean transformation, Ritz sets out the following programme: alter Schwarzschild's equation in such a way that the new equation involves only the accelerations and the relative velocities of the particles. In his general equations Ritz left three functions: ϕ, ψ and χ totally undetermined; he then proposed to adjust these functions so as to account for known experimental results. Ritz's programme did not attract many disciples (cf. O'Rahilly [1965], chapter 11); a Kuhnian might be tempted to attribute this to his early death, but then there are in fact good *objective* reasons for the lack of interest among other scientists. First, in view of the proposed adjustment of parameters (which in this case happen to be functions), this programme held out no promise for empirical progress. Secondly, even when he did adjust the parameters, he did so only in order to match results which Lorentz had already obtained. (Cf. Ritz [1911], p. 416.)

[97] It is also intriguing due to the fact that, according to McCormmach, Einstein was initially inclined to regard mechanics as the most fundamental branch of physics (cf. McCormmach [1970]).

as he explained in his autobiography, that all attempts at a mechanical explanation of the behaviour of the field had failed.[98]

> If mechanics was to be maintained as the foundation of physics, Maxwell's equations had to be interpreted mechanically. This was zealously but fruitlessly attempted, while the equations were proving themselves increasingly fruitful.[99]

We have seen *how* Einstein arrived at his programme and hence at SRT. We shall see that this programme finally superseded Lorentz's in the strictly empirical sense in 1915.[100] But was Einstein's programme objectively superior to Lorentz's in 1905? Did Lorentz's programme, as is generally claimed, really collapse in the face of SRT?[101]

2.3 *The heuristic superiority in 1905 of the Relativity Programme: Einstein's covariance versus Lorentz's ether. The power of Einstein's heuristics: derivation of a new relativistic law of motion and of $E = mc^2$*

As I have already shown,[102] Lorentz's theory $\mathbf{T}_3$ is observationally equivalent to the SRT; Einstein's transformed coordinates can be interpreted as the measured coordinates in Lorentz's moving frame. In the latter the 'real' coordinates are still the Galilean ones: $x_r = x - vt$, $y_r = y$, $z_r = z$, $t_r = t$; but, due to the contraction of measuring rods, to time-dilation and to the synchronisation of clocks through light signals, the measured coordinates are

$$x' = \beta(x - vt), \quad y' = y, \quad z' = z, \quad t' = \beta(t - vx/c^2),$$

where $\beta = 1/(1 - v^2/c^2)^{\frac{1}{2}}$.

[98] Einstein adduces from Planck's quantum hypothesis a second reason for abandoning mechanics in favour of electrodynamics. Although this reason seems to be a *post hoc* rationalisation, I shall quote Einstein in full: '[Planck's] form of reasoning does not make obvious the fact that it contradicts the mechanical and electrodynamic basis, upon which the derivation otherwise depends. Actually, however, it presupposes implicitly that energy can be absorbed and emitted by the individual resonator only in quanta of magnitude h, i.e. that the energy of a mechanical structure capable of oscillations as well as the energy of radiation can be transferred only in such quanta – in contradiction of the laws of mechanics and electrodynamics. *The contradition with dynamics was here fundamental: whereas the contradiction with electrodynamics could be less fundamental. For the expression for the density of radiation energy, although it is compatible with Maxwell's equations, is not a necessary consequence of these equations.*' (Einstein [1949], p. 45; my italics.)

This passage indicates that Einstein gave precedence to Maxwell over Newton and took his starting point with electrodynamics rather than with mechanics. Nevertheless, having accepted Planck's quantum hypothesis, Einstein could not regard Maxwell's equations as fundamental. Thus Einstein had enough reservations about electrodynamics to avoid making it into a cornerstone of his kinematics. Although *P*3 implies and lends plausibility to the Light Principle, the latter is still more fundamental in the sense of applying both to micro- and macroscopic phenomena; Einstein's lucky and unexplained guess was that the invariance of c was a universal principle which transcends its obvious dependence on Maxwell's equations.

[99] Einstein [1949], p. 25.

[100] Cf. below, §3.

[101] Cf. above, p. 231.

[102] Cf. above, pp. 234–5.

Thus, as he indicated at the end of his *Theory of Electrons*, Lorentz was in a position so to reformulate his theory that no 'crucial' experiment between his system and Einstein's could have been devised in 1905.[103]

In view of this situation, why did brilliant mathematicians and physicists like Minkowski and Planck abandon the classical programme in order to work on Special Relativity? Given the lack of any crucial experiment, a Kuhnian account of the 'conversion' of Planck and others may seem plausible. But the idea of a new bandwagon is highly implausible. Firstly, Einstein was a relatively unknown figure while Lorentz was a recognised authority. Secondly, Lorentz's theory was eminently intelligible whereas Einstein's involved a major revision of our most basic notions of space and time. Thirdly, there was no build-up of unsolved anomalies which Einstein's theory dissolved better than Lorentz's.[104] Moreover, at the time when Planck was converted, that is in 1906, no bandwagon had started.[105] Nor did the leading protagonists of the old paradigm die out unconverted as Kuhn claims is generally the case. Lorentz himself was eventually converted to the new outlook. In the *Theory of Electrons*, first published in 1909, he gives essentially the same account of the theory of corresponding states as in his *Electromagnetic Phenomena* of 1904; however, the footnotes indicate that by 1915 he had already accepted the Relativity Principle.

[103] For the argument that follows, I do not even need the assumption that $\mathbf{T}_3$ and SRT are observationally equivalent. It is enough that: (i) between 1905 and 1908 no 'crucial' experiment between the two rival theories was carried out; and (ii) neither hypothesis logically implies the other. (i) is a historical fact; as for (ii), Lorentz proposed a 'model' of the electron as a spherical distribution of charge in the ether, while Einstein remained agnostic as to the shape, charge density and mass of the electron; on the other hand Einstein asserted that *all* physical laws are Lorentz-covariant whereas Lorentz restricted his attention largely to electrodynamics (and did not fully establish the covariance of Maxwell's equations).

[104] Did Lorentz face insuperable difficulties which were known to his contemporaries? We have seen that Lorentz used the 'Molecular Forces Hypothesis' in order to obtain the appropriate laws about rod-contractions and clock-retardations. He had thereby assumed a transformational similarity between electromagnetic and molecular forces. In view of his programme, his next most natural step would have been to give a precise classical law of force for molecular and atomic interactions. We have also seen that, in order to explain the variation of the inertia with the velocity, Lorentz accounted for the mass of an electron in electromagnetic terms (electromagnetic longitudinal and transversal masses). In other words, in producing the revolutionary results which either matched or even anticipated Einstein's, Lorentz had to give a classical account of elementary particles. Lorentz was therefore unlucky in his choice of problems: he was straight away involved in difficulties which were to defeat Einstein himself, but at a much later stage. With hindsight we can see why Lorentz would probably have failed anyway; he was overtaken by the quantum theorists who realised that classical laws, and in particular Maxwell's equations, did not explain atomic stability. However, in 1905 there was hardly any indication that Lorentz could go no further in developing his programme and that no satisfactory classical account would be given of the behaviour of elementary particles. Yet, already in 1906, a physicist of Planck's stature and conservatism abandoned the classical approach, knowing very well that the SRT might well have been refuted by Kaufmann's experiment. Planck's choice, if rational, must have been guided by considerations different from the ones just given.

[105] A Kuhnian might fall back on individual Gestaltswitches: but if so, the Gestaltswitches would be *different* for Planck, for Minkowski, for Sommerfeld, for Lorentz!

Kuhnian explanations of the victory of SRT do not work. Another explanation is Whittaker's.[106] He tackles the difficulty by considering Lorentz and Poincaré as the real authors of Special Relativity, leaving to Einstein the merit of proposing a new theory of gravitation (i.e. General Relativity). Thus Lorentz's ether programme was not defeated by, but developed into, the relativity programme. However interesting and plausible this explanation may seem in the light of the foregoing discussion, it is unacceptable. As will be shown, the two programmes possess *very different heuristics*.

The most commonly held explanation is the third one. According to this, Einstein's theory represented the success of positivism in ridding classical physics of redundant metaphysics.[107]

I shall both develop this positivist claim and present my answer to it by comparing the Einsteinian revolution with the Copernican (or rather the Keplerian) one. This comparison will point to a feature which has often been a symptom if not a cause of decline in the heuristic of research programmes. This is a divorce between the empirical content and the mathematical formulation of certain scientific hypotheses: these hypotheses contain a large number of (physically) uninterpreted mathematical entities.[108] According to positivists like Mach, the mere elimination of such entities increases simplicity and thereby constitutes progress. My claim is that such eliminations are *by-products* of new research programmes whose heuristic eliminates certain entities. This may be accompanied by a – contingent – increase in simplicity. Let us take the example of the Copernican revolution.

The Platonic programme of saving the phenomena by the use of circular and spherical motions was initially successful: to each mathematical entity corresponded a physical one. Each planet was fixed on a physically real crystalline sphere which performed a number of axial rotations. It was, however, discovered that the distance between the earth and a given planet varied, so the astronomers resorted to eccentrics, epicycles and equants in order to account for the new phenomena. The physical problem was to determine the motion of the heavenly bodies relative to the earth. Since the paths of the planets are non-circular and since their motion is non-uniform, a widening gap appeared between the physical problem and the mathematical methods, which allowed only for circular motions. Although the earth allegedly occupied the centre of the universe, the paths of the planets about the earth were not dealt with directly; epicycles, deferents and equants, *all of which had no 'physical reality'*, were introduced in order to

[106] Cf. Whittaker [1953], chapters 2 and 5.

[107] Cf. Bridgman [1936], pp. 7–9, von Laue [1952], p. 6, Eddington [1939], pp. 70–5. Also Eddington [1920], pp. 1–16.

[108] This is so to speak the obverse of the point made earlier about the second heuristic role of mathematics in physics. There we saw how new physical theories can be constructed by interpreting hitherto uninterpreted mathematical entities (cf. above, pp. 224–7).

predict astronomical data; both the centre of an epicycle and the *punctum equans* are empty points in space.

Copernicus did not heal this rift between the physical picture and the mathematical description. True, he got rid of the equant; but, although his problem was to determine the motion of the planets with respect to a fixed sun, he interposed between the sun and the planets roughly as many epicycles with as many empty centres as were involved in the Ptolemaic system. It was left to Kepler to investigate the direct relation between the sun and the planets, to abolish epicycles and to find that the planets describe ellipses with one focus at the centre of the sun.

Let us now return to Lorentz. We have seen that the Lorentz transformation is always carried out in two steps.[109]

The first step yields the Galilean coordinates:

$$x_r = x - vt, \quad y_r = y, \quad z_r = z, \quad t_r = t$$

The second one gives us the effective coordinates:

$$x' = \beta(x - vt), \quad y' = y, \quad z' = z, \quad t' = \beta(t - vx/c^2)$$

The Galilean coordinates are interposed between the absolute coordinates and the effective ones in the same way as various epicycles were placed between the earth or the sun on the one hand and the planets on the other.[110]

In a moving frame only the Galilean coordinates are taken by Lorentz to be ontologically 'real' in the same way that before Kepler only circular motions were considered permissible from a metaphysical point of view. These metaphysical assumptions were naturally reflected in the mathematics: in the Galilean transformation used by Lorentz and in the epicycles used by Ptolemy and Copernicus. The Galilean transformation is a vestige of the original aim of the Classical programme, namely the aim of giving to the ether frame a privileged status. (Because of the Galilean transformation, Maxwell's equations hold good only in the ether frame.) The assumption of an ether frame no longer has any observational cash-value. Similarly the Ptolemaic epicycles were reminders of a hope which had long vanished, the hope of finding that the motions of the planets are both uniform and circular.

[109] Cf. above, p. 231.

[110] By drawing a parallel between the Galilean transformation on the one hand and a system of epicycles on the other, I do not want to suggest that, in Lorentz's theory $\mathbf{T}_3$, the Galilean transformation is physically uninterpreted. In fact, even the epicycles can be interpreted in the following trivial way: God, in contemplating his creation, sees it as a huge system of interlocking circles. Similarly, in Lorentz's case, God would perceive an infinite extended substance, the ether, in which any two events are separated by an absolute time interval. Such interpretations, which do not increase the empirical content of existing theories, could conceivably be made useful by indicating how they are to be heuristically exploited in order to construct new physical theories. Lorentz did not give such an indication in connection with the Galilean transformation. (Cf. below, p. 256.)

Copernicus was aware that the motions of the planets are neither circular nor uniform and Lorentz later realised that the effective coordinates, and not the Galilean ones, are the *measured* quantities in the moving frame.

I have drawn a parallel between Copernicus and Lorentz. Kepler and Einstein can be similarly compared. Kepler's greatest contribution to astronomy allegedly consisted in eliminating epicycles and in showing that the 'real' paths of the planets are ellipses with one focus at the sun. Similarly, according for instance to Bridgman and to von Laue, Einstein's chief merit lay in abolishing the Galilean transformation and in identifying the effective or measured coordinates as the only real ones. In equating 'to be' with 'to be perceived or measured' Einstein is supposed to have carried out a positivistic revolution in physics. However, if the merit both of Kepler and of Einstein *only* consisted in ridding physics of unnecessary 'epicycles', then the importance of these two physicists in the history of science is very much overrated: Copernicus and Lorentz did all the creative work, and Kepler and Einstein only applied Occam's razor in order to demolish the expendable metaphysical scaffolding used by their predecessors. Moreover, Copernicus knew that the paths of the planets were not circular, hence that his epicycles were part of the scaffolding; Lorentz realised that he did not need the Galilean coordinates in order to deduce the null results which he set out to explain. *If so, Kepler and Einstein contributed to the economy of thought and not to the growth of knowledge.*

This is an unacceptable conclusion. Let us start with Kepler. Copernicus's account of the motion of heavenly bodies had been largely Aristotelian in character: because the planets are perfect spheres, their natural motion is both uniform and circular. Through trying to give a dynamical explanation of the motion of heavenly bodies, Kepler provided classical astronomy with its heuristic. He proposed to determine the forces which emanate from the sun and directly act on the planets. He abolished epicycles *because* he wanted nothing but forces to mediate between the sun and the planets. Circles centred on empty points did nothing but conceal the 'true' relation which linked one heavenly body to the other. Kepler proposed a dynamical theory which is now largely forgotten because it was contradicted and supplanted by Newtonian astronomy. But in forgetting Kepler's dynamical theory we should not forget that Kepler created the programme which culminated in the Newtonian system; Kepler's method consisted in trying to discover the law of force responsible for the periodic motion of the planets round the sun. *Getting rid of Copernican epicycles was not an end in itself: it was subordinate to the needs of the new heuristic.*

Einstein, like Kepler, created a programme, not only an isolated theory. We shall see that Einstein's heuristic is based on a general requirement of Lorentz-covariance for all physical laws; we recall that the Lorentz transformation sends (x, y, z, t) *directly* into (x', y', z', t') without passing by the Galilean coordinates x_r, y_r, z_r, t_r. *The new heuristic therefore requires the abolition of the Galilean transformation which plays the role of a cumbersome epicycle. The parallel with Kepler is complete.*

After these criticisms of the Kuhnian, Whittakerian and positivist 'explanations' of the Einsteinian revolution, let me venture my own. In my view the main difference between Lorentz and Einstein lies in the difference between the heuristics of their respective programmes. *The ether programme did not collapse but was superseded by a programme of greater heuristic power.* This greater heuristic power explains why Planck and others joined Einstein's programme before it became empirically progressive. The difference between the two theories cannot be appreciated by taking an instantaneous look at Lorentz's and Einstein's systems. One has first to embed them in their respective programmes. In this way one realises that the two theories are similar because they stand at the intersection of two research programmes which later diverged. It will further be shown that the difference between the two approaches did not emerge with hindsight but guided the deliberate choice of scientists at the beginning of this century.[111]

Lorentz, unlike Einstein, did not create the heuristic of his own programme. The heuristic of Lorentz's programme consisted in endowing the ether with such properties as would explain the behaviour both of the electromagnetic field and of as many other physical phenomena as possible. In view of the overwhelming success of Newtonian dynamics it is hardly surprising that the ether was supposed to possess primarily mechanical properties. *The ether programme developed rapidly in certain respects, yet towards the end of the nineteenth century its positive heuristic was running out of steam.* A succession of mechanical models for the ether were proposed and abandoned. One serious difficulty was the presence in these models of longitudinal as well as transversal waves.[112] Lorentz faced a daunting problem of a different sort: in order to explain certain electromagnetic phenomena he postulated an ether at rest. He considered a portion of the ether, calculated the resultant $\boldsymbol{R}$ of the Maxwellian stresses acting on its

[111] I have reached the seemingly paradoxical conclusion that both Einstein (and Planck) on the one hand and Lorentz on the other were perfectly rational in doing what they did, i.e. in doing opposite things. Let me immediately add that they were rational, *given* their metaphysical positions. The conflict between Lorentz and Einstein is, among other things, the age old conflict between two metaphysical doctrines which, Polanyi notwithstanding, do not belong to the tacit component but can be articulated. Lorentz held that the universe obeys intelligible laws (e.g. wave processes presuppose a medium, there exists an absolute 'now' etc.) and Einstein held that the universe is governed by principles which can be given a mathematically coherent form (e.g. *all* laws are covariant). All major scientific revolutions were accompanied by an increase of mathematical coherence together with a (temporary) loss of intelligibility. (This applies to the Copernican, to the Newtonian, to the Einsteinian and to the quantum-mechanical revolutions.) It can moreover be argued that intelligibility is a time-dependent property, while mathematical coherence is not. We still consider Newtonian astronomy more coherent than Ptolemaic astronomy; but action at a distance was unintelligible before Newton, became perfectly intelligible at the end of the eighteenth century, and again unacceptable after Maxwell.

[112] Cf. Whittaker [1951], chapter 5.

surface and found that ***R*** is generally non-zero. Hence, if he was to assume that the ether was anything like an ordinary substance, he would have also to suppose that it was in constant motion. But this contradicted his original assumption of an ether at rest. He concluded 'that the ether is undoubtedly widely different from all ordinary matter' and that 'we may make the assumption that this medium, which is the receptacle of electromagnetic energy and the vehicle for many and perhaps for all the forces acting on ponderable matter, is, by its very nature, never put in motion, that it has neither velocity nor acceleration, so that we have no reason to speak of its mass or of forces that are applied to it'.[113] In other words Lorentz had reached a point where the behaviour of the electromagnetic field dictated what properties the ether ought to have, no matter how implausible these properties might be: for example the ether was to be both motionless *and* acted upon by non-zero net forces. The ether was nothing but the carrier of the field. *This involved a reversal of the heuristic of Lorentz's programme: instead of learning something about the field from a general theory of the ether, he could only get at the ether post hoc by way of the field.* In the case of the MFH for example, Lorentz *first* studied the transformational properties of the electromagnetic field; only *then* did he extend these properties to other molecular forces. Instead of positing *one medium* endowed with certain properties from which all forces inherit some *common* characteristic, we have an electromagnetic field acting as the archetype which determines the respects in which all forces are similar.

I do not claim that the ether programme was beyond redemption. Of course there was no obvious reason why the postulation of some non-mechanical properties of the ether should not account both for electromagnetic phenomena and for molecular interactions. All I claim is that the heuristic, *as it stood*, had petered out and that the ether programme was in need of a 'creative shift'[114] – a shift which, as a matter of fact, Lorentz did not provide.

Einstein based his heuristic on the requirement that all physical laws should be Lorentz-covariant; i.e. all theories should assume the same form, whether they are expressed in terms of x, y, z, t or in terms of x', y', z', t'. But it would be practically impossible to discover new laws simply by looking out for all the equations which are covariant under the Lorentz transformation. A good method is to start from well-tested laws whose past success would anyway have to be explained by any new theory. *Thus the heuristic of Einstein's programme is based on two distinct requirements:* (i) *a new law should be Lorentz-covariant and* (ii) *it should yield some classical law as a limiting case.*

We have just seen that Lorentz used the ether in order to extend certain

[113] Lorentz [1909], p. 30.

[114] This is a technical term in the methodology of scientific research programmes: cf. Lakatos [1970], p. 137.

properties of the electromagnetic field to molecular forces. His methods were effective in explaining Michelson's and other null results. By requiring that all forces and not only the electromagnetic and the molecular forces obey the same transformation laws; by taking Maxwell's equations and imposing their transformation properties on the whole of physics, Einstein both strengthened those Lorentzian methods which had proved effective in particular cases and turned them into a heuristic of general applicability. In this sense Einstein's programme displayed greater heuristic power than Lorentz's.

Let us give a more formal rendering of these two requirements. Let $R(a_1, a_2, \ldots, a_n) = 0$ be an equation which constitutes a physical law in some inertial frame I. If I' is any other inertial frame in which the quantities $a_1, a_2, \ldots, a_n$ assume the values $a'_1, a'_2, \ldots, a'_n$, respectively, then by the Relativity Principle:

$$[R(a'_1, a'_2, \ldots, a'_n) = 0] \Leftrightarrow [R(a_1, a_2, \ldots, a_n) = 0] \qquad (1)$$

But as Kretschmann pointed out to Einstein, every empirical law can be given not only a Lorentz-covariant but also a generally covariant expression (of course, general covariance implies Lorentz-covariance).[115] Thus, on the face of it, the most distinctive requirement of Einstein's heuristic is empty. However, the requirement is only trivialised if one is allowed complete freedom in reformulating the law. If one is restricted to a given number of entities $a_1, a_2, \ldots, a_n$, then the covariance requirement, far from being empty, becomes a stringent condition. As we shall see, in each particular case in which the heuristic is applied, the entities involved in the covariant law are precisely those involved in the corresponding classical law.[116]

Now we consider the requirement that a new relativistic law should yield the corresponding classical theory as a limiting case. In the most general case laws will involve the speed of light, the velocities $\boldsymbol{v}_1, \ldots, \boldsymbol{v}_n$ of a finite number of particles or processes and some other quantities $a, b, \ldots$. If $R = 0$ and $K = 0$ are the relativistic and classical laws respectively, we require that:

$$R \to K \quad \text{as} \quad (v_1/c, v_2/c, \ldots, v_n/c) \to (0, 0, \ldots, 0).$$

There are at least two ways of letting v_m/c tend to zero for $m = 1, 2, \ldots, n$. First we take c to be a constant and let $(v_1, \ldots, v_n)$ approach zero. In this case we put $\boldsymbol{w}_m = \boldsymbol{v}_m/c$, for all $m = 1, 2, \ldots, n$ and consider both R and K as functions of $c, \boldsymbol{w}_1, \ldots, \boldsymbol{w}_n, a, b, \ldots$. In other words, we write:

$$R = R(c, \boldsymbol{w}_1, \ldots, \boldsymbol{w}_n, a, b, \ldots) \quad \text{and} \quad K = K(c, \boldsymbol{w}_1, \ldots, \boldsymbol{w}_n, a, b, \ldots)$$

[115] Cf. Kretschmann [1917] and Einstein [1918].

[116] This problem arises also in the case of General Relativity where a different set of restrictions again render the covariance principle non-empty. (Apart from the energy tensor $T_{\mu\nu}$ only the $g_{\mu\nu}$'s and their first- and second-order derivatives can occur in the field equations.)

We then make

$$R(c, \boldsymbol{w}_1, \ldots, \boldsymbol{w}_n, a, b, \ldots) - K(c, \boldsymbol{w}_1, \ldots, \boldsymbol{w}_n, a, b, \ldots)$$

approach zero as $w_1, \ldots. w_n$ simultaneously tend to zero. It is of course tacitly assumed that R and K are continuous functions. Hence

$$R(c, \boldsymbol{w}_1, \ldots, \boldsymbol{w}_n, a, b, \ldots) - K(c, \boldsymbol{w}_1, \ldots, \boldsymbol{w}_n, a, b, \ldots)$$

approaches

$$R(c, 0, \ldots, 0, a, b, \ldots) - K(c, 0, \ldots, 0, a, b, \ldots)$$

as $(w_1, \ldots, w_n)$ approaches $(0, \ldots, 0)$. Thus the second requirement reduces to the equation:

$$R(c, 0, \ldots, 0, a, b, \ldots) = K(c, 0, \ldots, 0, a, b, \ldots).$$

In this first case the function R, which is to be determined, will therefore be subjected to the following two conditions:

$$[R(c, \boldsymbol{w}_1, \ldots, \boldsymbol{w}_n, a, b, \ldots) = 0] \Leftrightarrow [R(c, \boldsymbol{w}'_1, \ldots, \boldsymbol{w}'_n, a', b', \ldots) = 0] \quad (1')$$

(*Relativity Principle*)

$$R(c, 0, \ldots, 0, a, b, \ldots) = K(c, 0, \ldots, 0, a, b, \ldots) \quad (2)$$

(*Requirement that the classical law be a limiting case of the new law*)

We recall that

$$R(c, \boldsymbol{w}_1, \ldots, \boldsymbol{w}_n, a, b, \ldots) = R(c, \boldsymbol{v}_1/c, \ldots, \boldsymbol{v}_n/c, a, b, \ldots) = 0$$

is the relativistic law which is to replace the classical equation

$$K(c, \boldsymbol{w}_1, \ldots, \boldsymbol{w}_n, a, b, \ldots) = K(c, \boldsymbol{v}_1/c, \ldots, \boldsymbol{v}_n/c, a, b, \ldots) = 0.$$

If the relativistic law holds good in general, it will in particular be true for vanishing velocities: i.e. for $v_1 = \ldots = v_n = 0$ or equivalently for $w_1 = \ldots = w_n = 0$. By (2) it follows that:

$$K(c, 0, \ldots, 0, a, b, \ldots) = 0. \quad (3)$$

This last equation means that, when $v_1, \ldots, v_n$ all vanish, the relativistic law collapses into the classical one, which must therefore hold good in this particular case.[117]

There is a second way of making $v_1/c, \ldots, v_n/c$ tend to zero, namely by treating c as a variable parameter, fixing the velocities $\boldsymbol{v}_1, \ldots, \boldsymbol{v}_n$ and then letting c tend to infinity.[118] Putting $c = 1/\gamma$, we can write:

$$R = R_0(\gamma, \boldsymbol{v}_1, \ldots, \boldsymbol{v}_n, a, b, \ldots) \quad \text{and} \quad K = K_0(\gamma, \boldsymbol{v}_1, \ldots, \boldsymbol{v}_n, a, b, \ldots).$$

We now require that:

$$[R_0(\gamma, \boldsymbol{v}_1, \ldots, \boldsymbol{v}_n, a, b, \ldots) - K_0(\gamma, \boldsymbol{v}_1, \ldots, \boldsymbol{v}_n, a, b, \ldots)] \to 0,$$

as $c \to \infty$, i.e. as $\gamma = 1/c \to 0$.

[117] Both Einstein and Planck assumed that Newton's second law of motion holds good when the velocity vanishes (cf. below, p. 260).

[118] Note that, as $c \to \infty$, the Lorentz transformation collapses into the Galilean one.

Assuming that R_0 and K_0 are continuous, we obtain:

$$R_0(0, \boldsymbol{v}_1, \ldots, \boldsymbol{v}_n, a, b, \ldots) = K_0(0, \boldsymbol{v}_1, \ldots, \boldsymbol{v}_n, a, b, \ldots)^{119} \qquad (3')$$

One last way of meeting the requirement that R should tend to K as $(v_1/c, \ldots, v/c_n)$ tends to zero is to assume that R is a function of certain relativistic quantities and that K is the *same* function of the corresponding classical quantities. Then, if R and K are continuous and if each relativistic quantity tends to the corresponding classical one, it follows that:

$$R \rightarrow K \quad \text{as} \quad (v_1/c, \ldots, v_n/c) \rightarrow (0, \ldots, 0).^{120}$$

Having formulated the heuristic of the relativity programme in general terms, let me now give concrete examples illustrating the power of this heuristic.

The first example is concerned with Planck's modification of Newton's second law of motion. Following Einstein, Planck considered a slowly accelerated electron in an inertial frame I. By substituting Lorentz's expression for the ponderomotive force in Newton's second law, it is found that the motion of the electron is governed by the equation

$$e\left(\boldsymbol{E}+\frac{\boldsymbol{v}}{c}\wedge \boldsymbol{H}\right)-m\boldsymbol{a} = 0$$

where e is the charge of the electron, $\boldsymbol{v}$ is its velocity, $\boldsymbol{a}$ its acceleration and m its mass; as usual $\boldsymbol{E}$ and $\boldsymbol{H}$ are the electric and magnetic fields respectively. It is easily verified that this classical law is not Lorentz-covariant and thus has to be modified. Let us denote $e\left(\boldsymbol{E}+\frac{\boldsymbol{v}}{c}\wedge \boldsymbol{H}\right)-m\boldsymbol{a}$ by $K(\boldsymbol{v}/c, \boldsymbol{a}, \boldsymbol{E}, \boldsymbol{H})$. Planck implicitly assumed that the new relativistic law would involve the same variables as the classical one. Thus let $R(\boldsymbol{v}/c, \boldsymbol{a}, \boldsymbol{E}, \boldsymbol{H}) = 0$ be the relativistic equation which is to replace $K(\boldsymbol{v}/c, \boldsymbol{a}, \boldsymbol{E}, \boldsymbol{H}) = 0$.

Consider the electron at the time t when its velocity is $\boldsymbol{v}$, and choose an inertial frame I' which moves with the same velocity $\boldsymbol{v}$ with respect to I. In I' the electron is instantaneously at rest. If we denote by $\boldsymbol{v}'$, $\boldsymbol{a}'$, $\boldsymbol{E}'$ and $\boldsymbol{H}'$ the quantities in I' which correspond to $\boldsymbol{v}$, $\boldsymbol{a}$, $\boldsymbol{E}$ and $\boldsymbol{H}$ respectively, then $\boldsymbol{v}' = 0$. By the Relativity Principle, i.e. by the equivalence (1′) above:

$$[R(\boldsymbol{v}/c, \boldsymbol{a}, \boldsymbol{E}, \boldsymbol{H}) = 0] \Leftrightarrow [R(\boldsymbol{v}'/c, \boldsymbol{a}', \boldsymbol{E}', \boldsymbol{H}') = 0]$$
$$\Leftrightarrow [R(0, \boldsymbol{a}', \boldsymbol{E}', \boldsymbol{H}') = 0] \qquad (4)$$

119 The relativistic law of the conservation of momentum $\Sigma m_i \boldsymbol{v}_i/(1-v_i^2/c^2)^{\frac{1}{2}} = 0$ is a good illustration of equation (3′). $\Sigma m_i \boldsymbol{v}_i/(1-v_i^2/c^2)^{\frac{1}{2}} \rightarrow \Sigma m_i \boldsymbol{v}_i$ as $c \rightarrow \infty$. Letting $(v_1, \ldots, v_n)$ tend to zero serves no purpose in this case; since, if we start from an arbitrary function $f(v_i)$ and consider $\Sigma f(v_i)\, \boldsymbol{v}_i$, then: as $(v_1, \ldots, v_n) \rightarrow (0, \ldots, 0)$, $\Sigma f(v_i)\, \boldsymbol{v}_i \rightarrow 0 =$ value of $\Sigma m_i \boldsymbol{v}_i$ for $v_1 = v_2 = \ldots = v_n = 0$. This does not help us towards determining f.

120 Denote by μ_i the relativistic mass $m_i/(1-v_i^2/c^2)^{\frac{1}{2}}$. The relativistic momentum $\Sigma \mu_i \boldsymbol{v}_i$ and the classical momentum $\Sigma m_i \boldsymbol{v}_i$ are the same functions of the masses and of the velocities. Lewis and Tolman (tacitly) assumed that $\mu_i \rightarrow m_i$ as $v_i \rightarrow 0$. (Cf. Lewis [1908] and Lewis and Tolman [1909].)

We have already explained that, for vanishing velocities, a relativistic equation must coincide with its classical counterpart. (This is a direct consequence of the requirement (ii) above that the relativistic law should yield the classical one as a limiting case.)

Therefore:

$$[R(0, \boldsymbol{a}', \boldsymbol{E}', \boldsymbol{H}') = 0] \Leftrightarrow [K(0, \boldsymbol{a}', \boldsymbol{E}', \boldsymbol{H}') = 0]. \tag{5}$$

But:

$$K(0, \boldsymbol{a}', \boldsymbol{E}', \boldsymbol{H}') = e\left(\boldsymbol{E}' + \frac{0}{c} \wedge \boldsymbol{H}'\right) - m\boldsymbol{a}' = e\boldsymbol{E}' - m\boldsymbol{a}' \tag{6}$$

By (4), (5) and (6) it follows that:

$$[R(\boldsymbol{v}/c, \boldsymbol{a}, \boldsymbol{E}, \boldsymbol{H}) = 0] \Leftrightarrow [e\boldsymbol{E}' - m\boldsymbol{a}' = 0] \tag{7}$$

Using known transformation equations, we can express $\boldsymbol{E}'$ and $\boldsymbol{a}'$ in terms of $\boldsymbol{E}$, $\boldsymbol{H}$, $\boldsymbol{v}$ and $\boldsymbol{a}$, and thus obtain:

$$[e\boldsymbol{E}' - m\boldsymbol{a}' = 0] \Leftrightarrow \left[\frac{\mathrm{d}}{\mathrm{d}t}\left(\frac{m\boldsymbol{v}}{(1-v^2/c^2)^{\frac{1}{2}}}\right) = e\left(\boldsymbol{E} + \frac{\boldsymbol{v}}{c} \wedge \boldsymbol{H}\right)\right] \tag{8}$$

Thus $\frac{\mathrm{d}}{\mathrm{d}t}\left(\frac{m\boldsymbol{v}}{(1-v^2/c^2)^{\frac{1}{2}}}\right) = e\left(\boldsymbol{E} + \frac{\boldsymbol{v}}{c} \wedge \boldsymbol{H}\right)$ is the relativistic equation of motion for an electron moving in an electromagnetic field. Planck took the Lorentz force $e\left(\boldsymbol{E} + \frac{\boldsymbol{v}}{c} \wedge \boldsymbol{H}\right)$ to be the very paradigm of force and generalised the last equation as follows:

$$\frac{\mathrm{d}}{\mathrm{d}t}\left(\frac{mv}{(1-v^2/c^2)^{\frac{1}{2}}}\right) = \text{force} = \boldsymbol{f} \tag{9}$$

Equation (9) is the relativistic law which replaced Newton's second law of motion. By using (9), the expression of the relativistic kinetic energy $k(v)$ can be determined. It is:

$$k(v) = \int \boldsymbol{f}.\boldsymbol{v}\,\mathrm{d}t = mc^2\left(\frac{1}{(1-v^2/c^2)^{\frac{1}{2}}} - 1\right) \tag{10}$$

Thus, by using Einstein's heuristic together with the simple device of choosing an inertial frame I' in which the electron is instantaneously at rest, Planck modified the law $\boldsymbol{f} = m\boldsymbol{a}$ which had been considered an unshakeable convention of theoretical physics.

Let us now examine how Einstein, by using the same heuristic, arrived at his famour equation relating mass and energy:

$$E = mc^2/(1-v^2/c^2)^{\frac{1}{2}}$$

Einstein assumed that there had to be a relativistic law corresponding to the classical law of conservation of energy. By the Relativity Principle this new conservation law must hold in all inertial frames. Einstein considered an inertial frame I in which a stationary body B emits light and thereby loses a certain amount of energy Q. Since energy is conserved:

$$E_1 = E_2 + Q \tag{11}$$

where E_1 is the total energy of B before radiation and E_2 its total energy after radiation.

Einstein considers a second inertial frame I' moving with velocity $-\boldsymbol{v}$ with respect to I. The body B, which is at rest in I, moves with velocity $\boldsymbol{v}$ relatively to I'. By the Relativity Principle, energy must be conserved both in I and in I'. Hence:

$$E_1' = E_2' + Q' \tag{12}$$

where E_1', E_2' and Q' are the quantities in I' which correspond to E_1, E_2 and Q respectively.

Subtracting (11) from (12):

$$E_1' - E_1 = (E_2' - E_2) + (Q' - Q). \tag{13}$$

Einstein interpreted $(E_1' - E_1)$ as follows. E_1 is the energy of B in its rest-frame I (before light is emitted). E_1' is the energy of the same body B as seen from the moving frame I'. In I' the body B moves with velocity $\boldsymbol{v}$ but in I it is at rest. Hence $(E_1' - E_1)$ is the energy which accrues to the body B solely in virtue of its motion; i.e. $(E_1' - E_1)$ is the kinetic energy of B to within an additive constant. By (10)

$$E_1' - E_1 = (Mc^2/1 - v^2/c^2)^{\frac{1}{2}} - Mc^2 + h, \tag{14}$$

where h is a constant and M is the rest mass of B before radiation. Similarly:

$$E_2' - E_2 = (mc^2/1 - v^2/c^2)^{\frac{1}{2}} - mc^2 + h, \tag{15}$$

where m is the rest mass of B after radiation.

Note that Q is the energy lost through radiation in I. Einstein was in possession both of Maxwell's equations and of the tranformation laws for the field. From these he calculated that:

$$Q' = Q/(1 - v^2/c^2)^{\frac{1}{2}} \tag{16}$$

Substituting from (14)–(16) into (13):

$$(M - m)\, c^2 = Q; \quad \text{i.e.} \quad c^2 \Delta M = Q, \quad \text{where} \quad \Delta M = M - m. \tag{17}$$

Thus the *rest* mass of B has decreased by the amount $\Delta M = Q/c^2$. A supposedly immutable substance, namely the rest mass ΔM, can vanish and thereby give rise to the equivalent amount of energy $c^2 \Delta M$. This revolutionary result is a consequence of the Relativity Principle applied to the law of conservation of energy. True, Lorentz had shown that the electron possesses an electromagnetic inertia which *varies* with the speed. He had also found that the electromagnetic *rest* mass is a multiple of the electrostatic *energy*. But neither in Lorentz's [1904] nor in his [1909] is there any indication that the rest mass is a variable quantity![121]

The extraordinary power of the Relativity Principle is further displayed by the following fact. Given the Relativity Principle, the law of conservation of energy both implies and is implied by the law of conservation of momentum; where the momentum of a particle of rest mass m and velocity $\boldsymbol{v}$ is the vector $m\boldsymbol{v}/(1-v^2/c^2)^{\frac{1}{2}}$, and the energy of the particle is $mc^2/(1-v^2/c^2)^{\frac{1}{2}}$.

These examples show that *the revolutionary relativistic laws were not arrived at in a sudden flash of intuition or through some kind of mystical insight. The new laws were mathematically derived from assumptions like the Relativity Principle which seem so 'formal' and innocuous as to be devoid of empirical content.*

3 *Einstein's programme supersedes Lorentz's*

Einstein invented not a theory but a research programme with an immensely powerful heuristic. But research programmes are ultimately judged on their empirical rather than on their heuristic power. No matter how fruitful its heuristic guidelines for the construction of new theories are, the programme will not be successful if these theories are not empirically corroborated. In my view Einstein's relativity programme superseded Lorentz's in the empirical sense in 1915 with its explanation of the precession of Mercury's perihelion. This explanation requires the *general* theory. There were of course special relativistic results (e.g. $E = mc^2$) which could in principle be tested, but even by 1915 such tests seemed to be only a remote possibility.

My claim that Einstein's programme superseded Lorentz's with the explanation of the perihelion of Mercury raises two difficulties. *First*, since I wish to claim this as a success for the *whole* relativistic programme, I have to establish a continuity between the special and the general theories. *Secondly*, since the behaviour of Mercury was well-known, I shall have to show, in line with my definition of empirical support,[122] that the Mercury prediction was an unexpected consequence of the general theory. It may seem that this preoccupation with Mercury's perihelion is unnecessary in view of the fact that General Relativity made predictions which were novel in the *temporal* sense, e.g. the bending of light rays. However, the Mercury prediction, in contradistinction to the bending of the light rays, was in close agreement with observation and *also* depended on the full field equations.[123]

[121] The rest mass of an electron is a function of the charge and of the radius. Lorentz took both the charge and the radius to be constant.

[122] Cf. above, pp. 216–19.

[123] Cf. Adler, Bazin and Schiffer [1965], p. 194.

3.1 *The continuity between the Special and General Theories of Relativity*

The explanation of gravitation by General Relativity appears to involve a major shift in the methods used by Einstein. But I propose to show that after 1908 Einstein merely *strengthened* the methods already in use in Special Relativity;[124] the only new addition to the programme was in the heuristic function of the Principle of Equivalence (i.e. of the equality of gravitational and inertial masses).[125]

One might think that the General Theory constitutes simply a generalisation to the case of accelerating frames. But this is only a small part of the answer to the continuity problem. After 1905 Einstein's main problem was to devise a Relativistic Theory of Gravitation. He found it impossible to reconcile his equation:[126]

$$E = mc^2 = m_0c^2/(1-v^2/c^2)^{\frac{1}{2}}$$

with the Principle of Equivalence. This principle implies that, since the gravitational and the inertial masses of any material body are equal, all

[124] My views are in sharp contrast with those of Lanczos. In his [1972] Lanczos distinguishes between a young Einstein who was supposedly a strict empiricist and an older Einstein who indulged in speculation. Lanczos writes: '[Einstein] was at that time still a convinced empiricist who would not have dared to argue that perhaps nature is based on rational and universal principles, which cannot be found by experimentation but only by inspired and imaginative speculation;...In fact Einstein in the beginning of his career distrusted mathematics and considered the mathematical formulation of a physical event as the mere form in which a phenomenon is described, which does not touch on its substance...To think in experimental terms was Einstein's basic attitude in the prime of his career in marked contrast to his ideas in the last phase of his life when in search for the ultimate unification of nature he often fell victim to mere formalism.' Against this view I maintain that the roots both of General Relativity and of the various unified field theories go back to 1905. Had Einstein really been 'a strict empiricist and a follower of Mach who saw the task of theoretical physics purely in the more or less accurate description of experimental observations', then why did he insist that the Relativity Principle should hold not only at the *observational* but also at the *highest theoretical* level? An empiricist would have been perfectly satisfied with a solution such as Lorentz's which explains why no *experiment* could detect absolute motion. To insist that the Relativity Principle obtains at the level of *laws* presupposes a realistic interpretation of these laws beyond their functions as mere tools for the description of experimental observations. For Einstein observational symmetries are nothing but the mere reflection of a deeper symmetry at the ontological level. The empirical success of Einstein's earlier work and the empirical failure of Einstein's later work (unified field theory) cannot be explained by Lanczos's claim that Einstein degenerated from empiricism to speculation. High level speculation paid off handsomely in the case of General Relativity Theory; the programme achieved stunning empirical success in 1915 and in 1919. It is hardly surprising that Einstein 'overestimated' the speculative (or so-called *aprioristic*) method which enabled him to construct the General Theory and tried to apply the same methods to the construction of unified field theories. He was unlucky, but the lack of – empirical – success cannot be attributed, as Lanczos claims, to a change in *methodological attitude*. The *same* attitude helped Einstein in the case of the General Theory, but it failed him later in life. Einstein proposed and Nature disposed.

[125] For a more detailed account see Zahar [1973].

[126] *Warning to the reader:* In this secion m_0 denotes the rest mass and m the inertial mass i.e. $m = m_0/(1-v^2/c^2)^{\frac{1}{2}}$.

bodies fall with the same acceleration in the same gravitational field $\boldsymbol{g}$. In classical physics this result follows from Newton's second law

$$\boldsymbol{f} = \frac{\mathrm{d}}{\mathrm{d}t}(m_i \mathbf{v}) = m_i \boldsymbol{a}$$

and from the equality of the inertial and gravitational masses. The masses cancel out on both sides of $m_i \boldsymbol{a} = m_g \boldsymbol{g}$, leaving $\boldsymbol{a} = \boldsymbol{g}$. Einstein assumed that in Special Relativity the corresponding equation would be some relation of the form:

(inertial mass) $\boldsymbol{g}$ = rate of change of momentum;

i.e. $$m_0 \boldsymbol{g}/(1-v^2/c^2)^{\frac{1}{2}} = \frac{\mathrm{d}}{\mathrm{d}t}(m_0 \boldsymbol{v}/(1-v^2/c^2)^{\frac{1}{2}}).$$

In view of the equation:

$$E = \text{energy} = m_0 c^2/(1-v^2/c^2)^{\frac{1}{2}},$$

the rest mass m_0 may vary with the time; in other words we may have $(\mathrm{d}m_0/\mathrm{d}t) \neq 0$.

Dividing through by m_0, we obtain on the right hand side of

$$m_0 \boldsymbol{g}/(1-v^2/c^2)^{\frac{1}{2}} = \frac{\mathrm{d}}{\mathrm{d}t}(m_0 \boldsymbol{v}/(1-v^2/c^2)^{\frac{1}{2}})$$

a term $$\frac{1}{m_0}\frac{\mathrm{d}m_0}{\mathrm{d}t}\frac{\boldsymbol{v}}{(1-v^2/c^2)^{\frac{1}{2}}}$$

which may be different from zero.

Thus the motion of a material body under the effect of the gravitational field will generally depend on its rest mass m_0. The Principle of Equivalence is violated.

Rather than give up the Principle of Equivalence, Einstein gave up the hope of giving a special relativistic theory of gravitation. He changed his tactics and launched a two-pronged attack on the problem of gravitation:

(i) Einstein only now remembered his Machian scruples concerning the so-called myth of the inertial frame. If one purges Newtonian theory of the Absolute Space Hypothesis, one finds that Special Relativity is as 'absolute' as classical mechanics; both theories postulate a set of privileged inertial frames. From a Machian viewpoint this assumption is unacceptable.[127] Einstein decided to treat *all* coordinate systems on a par and to impose a condition of *general* covariance on *all* physical laws. This condition, which is a *strengthening* of the requirement of Lorentz-covariance (general covariance of course implies Lorentz-covariance), is an important element of continuity between the Special and the General theories of Relativity.

[127] This may have been a reason for Mach's rejection of Special Relativity. (Cf. Mach [1913], Preface.)

(ii) Einstein decided to go back to his original heuristic, in particular to the heuristic device which consists in scrutinising known empirical results and in isolating certain features in them which are 'unsatisfactorily' explained by current theories.[128] As I have shown, Einstein analysed the well-known 'fact' that all bodies fall with the same acceleration in the same way that he had analysed the result of the induction experiment.[129] Let me recall that Einstein reached the conclusion that all gravitational fields can be regarded as caused by a local acceleration of the frame of reference. It is thus obvious why the introduction of accelerated frames holds out the promise of solving the problem of gravitation.

The two prongs of the attack can now be seen to converge to the same result. We have here a second element of continuity between the Special and the General Theories: each involves an application of prescription II. However, Einstein now faced a new difficulty. In Newtonian mechanics there exist so-called inertial fields which arise if one chooses an accelerating frame of reference. These inertial fields are artificial in that they can be transformed away by one global change of coordinates, namely by a change which refers everything back to an inertial frame. This is not the case with 'real' gravitational fields; for example the field at a point near the earth's surface can be transformed away only at the cost of piling it up at the antipodes. How could one ever hope to be able to deal with two such dissimilar fields in the same way? By a tremendous stroke of genius Einstein turned this seeming impasse into a powerful heuristic device. Consider an accelerating frame of reference S in which there exists both a 'real' gravitational field $\boldsymbol{g}$ and an inertial field $\boldsymbol{i}$. Every particle P is acted upon by a force $\boldsymbol{F} = m_i \boldsymbol{i} + m_g \boldsymbol{g}$. Since m_i = inertial mass = gravitational mass $= m_g = m$ (say), it follows that $\boldsymbol{F} = m\boldsymbol{i} + m\boldsymbol{g} = m(\boldsymbol{i} + \boldsymbol{g}) = m\boldsymbol{G}$, where $\boldsymbol{G} = \boldsymbol{i} + \boldsymbol{g}$. Thus $\boldsymbol{i}$ and $\boldsymbol{g}$ always occur indissolubly fused into one global field $\boldsymbol{G}$; Einstein typically refuses to consider this 'fact' a mere accident and argues that there exists only one total field $\boldsymbol{G}$ which a new theory of gravitation would have to explain. In view of $\boldsymbol{G} = \boldsymbol{i} + \boldsymbol{g}$, $\boldsymbol{i}$ is now a special case of an Einsteinian gravitational field; $\boldsymbol{G}$ reduces to $\mathbf{i}$ if all matter in the universe is either annihilated or removed to an infinitely distant point. The field $\boldsymbol{i}$ can be globally transformed away through a single change of coordinates; in other words $\boldsymbol{i}$ is a reducible gravitational field. Reducible fields offer the advantage that they can be generated at will through an arbitrary acceleration of the frame of reference. One can heuristically exploit the Principle of Equivalence by generating a reducible field $\boldsymbol{i}$ through an acceleration of the frame, by studying the properties of $\mathbf{i}$ and finally by extending these properties to non-reducible fields $\boldsymbol{G}$.

[128] This is prescription II; cf. above, p. 239. The heuristic of Special Relativity is only part of Einstein's general heuristic as expressed in prescriptions I and II.
[129] Cf. above, pp. 239–41.

But how is this generalisation to be carried out? This is where the absolute differential calculus proved extremely helpful. The method is as follows: start from an inertial frame $\bar{S}(\bar{x}^0, \bar{x}^1, \bar{x}^2, \bar{x}^3)$ in which Special Relativity applies in its usual form; hence:

$$ds^2 = (d\bar{x}^0)^2 - (d\bar{x}^1)^2 - (d\bar{x}^2)^2 - (d\bar{x}^3)^2 = \bar{g}_{mn} d\bar{x}^m d\bar{x}^n$$

(we take the velocity of light to be unity) where:

$$(\bar{g}_{mn}) = \begin{pmatrix} 1 & & & O \\ & -1 & & \\ & & -1 & \\ O & & & -1 \end{pmatrix}.$$

Accelerate the frame $\bar{S}$; in other words carry out a non-linear transformation of coordinates. In the accelerating frame

$$S(x^0, x^1, x^2, x^3): ds^2 = g_{mn} dx^m dx^n,$$

where (g_{mn}) varies from one point to the next. The matrix(g_{mn}) is not arbitrary, since it satisfies the following condition which we denote by K: through a global transformation of coordinates, namely through the transformation $(x^m) \rightarrow (\bar{x}^m)$, (g_{mn}) is reducible to the constant matrix $(\bar{g}_{mn})$. There exists in S a reducible gravitational field generated by the acceleration of the frame S with respect to $\bar{S}$. We study the behaviour of this field in S, then we generalise our results by abstracting from, i.e. by lifting, the condition K. In doing this we have to study the same processes in two different frames $\bar{S}$ and S; so we need a method for translating the results obtained in $\bar{S}$ into results applying in S; the absolute differential calculus provides such a method.

Using such methods, Einstein determined the path of a particle moving freely in a gravitational field. In the frame $\bar{S}$ the trajectory of the particle is a straight line whose equations are: $d^2x^i/(dx^0)^2 = 0$ $(i = 0, 1, 2, 3)$. Since our aim is to 'look' at the same particle from the accelerated frame S, it proves useful to characterise the path in an invariant way, i.e. in a way which does not depend on any particular coordinate system. Using the intrinsic parameter s instead of $\bar{x}^0$ (i.e. t), we obtain:

$$d^2\bar{x}^i/ds^2 = 0 \tag{1}$$

These equations are easily seen to give the integral $\int ds$ a stationary value. In other words:

$$\delta\left(\int ds\right) = 0 \quad \text{when} \quad d^2\bar{x}/ds^2 = 0 \quad \text{in} \quad \bar{S}. \tag{2}$$

In view of the fact that ds is an invariant, this last equation means that, in $\bar{S}$ and in any other frame of reference, the trajectory of the particle is a geodesic. From the absolute differential calculus we know that a geodesic in S satisfies the following equations:

$$\frac{d^2x^i}{ds^2}+\begin{Bmatrix} i \\ mn \end{Bmatrix}\frac{dx^m}{ds}\cdot\frac{dx^n}{ds}=0 \tag{3}$$

where

$$\begin{Bmatrix} i \\ mn \end{Bmatrix} = g^{iu}[mn, u] = \tfrac{1}{2}g^{iu}\left(\frac{\partial g_{mu}}{\partial x^n}+\frac{\partial g_{nu}}{\partial x^m}-\frac{\partial g_{mn}}{\partial x^u}\right)$$

Comparing (1) and (3), we conclude that in the accelerating frame S the path of the particles is no longer a 'straight' line in the ordinary sense of the word: the quantity which deflects the particle from a straight trajectory is the quantity represented by the Christoffel symbol $\begin{Bmatrix} i \\ mn \end{Bmatrix}$. Since $\begin{Bmatrix} i \\ mn \end{Bmatrix}$ is a function of the g_{ij}'s and since we ascribe the deviation from a straight path to the action of a gravitational field, the latter is represented by the g_{ij}'s or rather by the partial derivatives $\partial g_{ij}/\partial x^m$; note that the coefficients $\begin{Bmatrix} i \\ mn \end{Bmatrix}$ vanish if the g_{ij}'s are constant. The g_{ij}'s can therefore be looked upon as the gravitational potentials.

So far we have implicitly assumed that the gravitational field is reducible, i.e. 'inertial' in the old terminology. We now generalise our results by extending them to the case where there need not exist a global transformation which makes all the g_{ij}'s constant. Hence, even if the field is irreducible, the path of a free particle is still a geodesic and the metric tensor (g_{ij}) still represents the gravitational potential.

This method also enabled Einstein to determine the effect of gravitation on other physical phenomena. He wrote the laws governing these phenomena in a generally covariant form, which generally involves the g_{ij}'s; he then abstracted from the condition K which was originally imposed on the metric tensor. The presence of gravitation manifests itself through the irreducible g_{ij}'s which occur in the new law.

There remained for Einstein the problem of finding the field equations which are satisfied by the g_{ij}'s. His attack on the problem was again two-pronged:[130]

(i) The new law would have to yield Poisson's equation $\nabla^2\phi = k\rho$ as a limiting case. This requirement is identical with the one made in the case of Special Relativistic laws. Because of Poisson's equation Einstein expected his own law to consist of second-order partial differential equations linear in the second derivatives $\partial^2 g_{ij}/\partial x^m\, \partial x^n$.

(ii) We have seen that g_{ij}'s have a dual function: on the one hand they represent the physical gravitational potentials and on the other they are coefficients in the expression of ds^2; this second function is a geometrical one. It can be said that Einstein geometrised gravitation, or alternatively

[130] For a more detailed account see Zahar [1973].

that he physicalised geometry. Since the field equations describe a geometrical state of affairs, they ought to be independent of any particular frame of reference; i.e. they ought to be generally covariant.

Thus Einstein and Grossmann[131] started from the rather vague assumption that the gravitational field is a geometrical entity ('geometry' is to be understood as synonymous with 'kinematics' or 'space–time geometry'). It had of course long been known that gravity was caused by the presence of massive bodies. Also, by the Special Relativistic equation $E = mc^2$, inertial mass and energy are interchangeable. Finally, by the Principle of Equivalence, inertial mass and gravitational mass are identical (*wesensgleich*). Putting all these assumptions together, Einstein guessed that gravitation is a geometrical phenomenon related to the energy content of space. Grossmann expected the field equations to be of the form $A = B$, where A and B respectively represent the geometry and the energy content of space. He took for granted that the geometry in question is Riemannian and not some more general, i.e. less structured, geometry. In the case of free space, B vanishes, so we are left with the equation $A = 0$. In order to determine A, Grossmann considered the tensor $B^{\sigma}_{.\lambda\mu\nu}$ which Riemann and Christoffel had shown to be essentially relevant to the geometrical properties of space. Grossmann knew that the equations $B^{\sigma}_{.\lambda\mu\nu} = 0$ imply that the space is flat and hence that the field can be globally transformed away. He also knew that the gravitational field is generally irreducible. So Grossmann weakened the relations $B^{\sigma}_{.\lambda\mu\nu} = 0$ by using a standard mathematical technique, namely the contraction of a contravariant index with a covariant one. Note that there exists essentially one way of contracting two such indices in $B^{\sigma}_{.\lambda\mu\nu}$, because $-B^{\sigma}_{.\lambda\sigma\mu} = B^{\sigma}_{.\lambda\mu\sigma}$ and $B^{\sigma}_{.\sigma\lambda\mu}$ vanishes identically. By using this type of reasoning, Grossmann finally obtained the equations $B^{\sigma}_{.\lambda\mu\sigma} = R_{\lambda\mu} = 0$, which are accepted today as the correct field equations for free space. Thus, through giving a precise mathematical formulation of his initial assumption that gravitation is a geometrical phenomenon linked to the energy content of space, Grossman had obtained the much stronger proposition: $R_{\lambda\mu} = 0$. In fact, as I have already said above, Grossmann's initial assumption was so weak as not even to imply that the geometry to be used is Riemannian. The reason for resorting to Riemannian geometry was its availability at the time as a fully developed mathematical system. This illustrates the point made earlier about the first heuristic role of mathematics in physical discovery.[132]

The solution $R_{ij} = 0$ for free space was rejected by its authors for two reasons, both of which turned out to be unfounded. First, Grossmann believed that $R_{ij} = 0$ would not yield the classical equation $\nabla^2\phi = 0$ as a limiting case for weak static fields. This was a relatively simple mathematical error. Secondly, both Einstein and Grossmann thought that, given the appropriate boundary conditions, the ten equations $R_{ji} = 0$ would

[131] Einstein and Grossmann [1913]. [132] Cf. above, pp. 224–6.

uniquely determine the ten functions g_{ij}. This means that we are not at liberty to choose an arbitrary frame of reference because the functions g_{ij} are generally altered by a change of coordinates. Thus it seems that the Relativity Principle is violated. Hilbert saved the situation by showing that the equations $R_{ij} = 0$ are not all independent; the left hand sides satisfy four identities, which give the exact degree of arbitrariness necessary for the free choice of a frame of reference (four identities corresponding to four coordinates).[133]

3.2 *The successful explanation of the perihelion of Mercury and its role in the further development of the General Theory*

In 1915 Einstein went back to the equations $R_{ij} = 0$ for free space; for the case where non-gravitational energy in the form of a symmetric tensor T_{ij} is present, Einstein found it natural to generalise the equations $R_{ij} = 0$ to $R_{ij} = kT_{ij}$. This generalisation turned out to be untenable. However, using only the equations $R_{ij} = 0$ for the field created by the sun, Einstein explained the precession of Mercury's perihelion. He published this result on November 22, 1915. This explanation of a well-known fact was tremendously important for the following reasons: the predicted fact is completely novel in the sense that I have previously explained;[134] that is *Einstein did not use the known behaviour of Mercury's perihelion in constructing his theory*. In fact this empirical prediction is all the more dramatic because it flows from a hypothesis which is so speculative, so 'metaphysical', that one may wonder whether it belongs to physics or to pure mathematics. *Thus, through explaining the 'anomalous' motion of Mercury's perihelion, Relativity Theory superseded its rivals from a strictly empirical point of view.*[135] *This empirical success also proved crucial for the further development of the Relativity Programme.*

Einstein realised that the equations $R_{ij} = kT_{ij}$ are untenable because the right hand side is divergenceless whereas the left hand side is not. Other things being equal, it would have been natural for Einstein to abandon this whole approach, i.e. to reject both the general equations $R_{ij} = kT_{ij}$ and the special case $R_{ij} = 0$. The fact that the equations $R_{ij} = 0$ had enabled him to explain the motion of Mercury convinced him that the fault lay not with his overall approach but with the method of generalising the equations $R_{ij} = 0$. In other words Einstein kept the equations $R_{ij} = 0$ for free space and looked for a new method of generalising these relations.[136]

[133] Cf. Einstein [1915e].

[134] Cf. above, pp. 217–19.

[135] Lorentz in his [1900] had produced a theory of gravitation which, however, explained only a small fraction (one tenth) of the residual angle of precession of Mercury's perihelion.

[136] Had the General Theory's prediction of Mercury's behaviour *not* been novel, e.g. had Einstein 'adjusted parameters' in order to obtain the correct experimental results, he would surely not have had such confidence in the equations $R_{ij} = 0$, in the face of the breakdown of the more general form $R_{ij} = kT_{ij}$.

At this critical stage Einstein was helped by the mathematical machinery of his system and by the special relativistic law about the interchangeability of inertial mass and energy. By using variational methods, he extracted from the relations $R_{ij} = 0$ a matrix t_i^j which obeys a formal conservation law: $t^j_{i|j} = 0$, i.e. $\partial t_i^j/\partial x^j = 0$. However, t_i^j is not a tensor and hence appears not to be susceptible of any physical interpretation. It seemed as if t_i^j ought to be treated as a mere mathematical entity which may be used for purposes of convenience but is otherwise devoid of physical meaning. Realising that he had reached an impasse with the equations $R_{ij} = kT_{ij}$, Einstein insisted against all odds on interpreting t_i^j as a gravitational energy matrix; but energy represents inertial mass and hence also gravitational mass (Principle of Equivalence); thus we reach the surprising physical result that gravitational energy acts as one source of the gravitational field. In passing from $R_{ij} = 0$ to $R_{ij} = kT_{ij}$ Einstein had 'mistakenly' supposed that he was going from a case in which energy was totally absent to a case in which it was not. He had forgotten that even when $R_{ij} = 0$, *gravitational* energy may be present. His solution consisted in rewriting the equation $R_{ij} = 0$ so as to bring out this dependence of the field on its own energy and then generalising by adding the non-gravitational energy kT_i^j to the gravitational energy t_i^j. In other words Einstein rewrote $R_{ij} = 0$ in the form $F(t_i^j, t) = 0$, where: $t = t_i^i$; he then generalised these equations by adding kT_i^j to t_i^j and kT to t. Einstein obtained

$$F(t_i^j + kT_i^j, t + kT) = 0$$

which turned out to be equivalent to the currently accepted field equations:

$$R_{ij} = -k(T_{ij} - \tfrac{1}{2}g_{ij}T)$$

This illustrates what I called the second role of mathematics in physical discovery.[137]

Let me end by reviewing the assumptions and methods which are common to the Special and to the General Theories of Relativity. These common assumptions and methods are the bridges connecting the two stages of the programme. Let me make the obvious point that in both theories Special Relativity holds locally, i.e. it holds in infinitely small domains about each point. Einstein starts the two theories by analysing two well-known 'facts' from the same point of view: he analyses the result of the induction experiment and the 'fact' that all bodies fall with the same acceleration. Common to both theories is the law concerning the interchangeability of mass and energy. The equation $E = mc^2$ was a dramatic new result implied by Special Relativity; moreover it was precisely this result which led Einstein to transcend the Special Theory and resort to General Relativity as a framework which would embody gravitation. We have also seen that the

[137] Cf. above, pp. 225–6

law about the interchangeability of mass and energy played a crucial role in enabling Einstein to modify his field equations $R_{ij} = +kT_{ij}$. Both the Special and the General Theories make use of the Covariance Principle: Lorentz-covariance in the case of Special Relativity and general covariance in the case of General Relativity. In both stages of the programme scientists exploited the assumption that classical theories ought to be limiting cases of the new relativistic laws. In Special Relativity the law of inertia, Maxwell's equations, Newton's second law and the laws of the conservation of energy and momentum were used in order to determine new and different Lorentz-covariant equations. In General Relativity Poisson's equation was exploited: since Poisson's equation was to be a limiting case of the law of gravitation, the latter was expected to consist of a system of second-order partial differential equations which would be linear in the second-order derivatives. The only essentially new method peculiar to General Relativity was the heuristic use which was made of the Principle of Equivalence and which has no analogue in the Special Theory.

Thus, the question asked in the title of this paper, '*Why did Einstein's Programme supersede Lorentz's?*', has now been answered. Already in 1905 the Relativity programme proved *heuristically superior* to its classical rival; at a time when the notion of a quasi-material ether was becoming heuristically barren, Einstein provided a powerful new tool for the construction of Lorentz-covariant laws yielding the corresponding classical theories as limiting cases. However, heuristic power gives a measure only of *intellectual* achievement and not of *scientific* progress. After all, science is empirical. Special Relativity by itself did not empirically supersede Lorentz's programme. Bucherer's experiment[138] confirmed *both* Lorentz's and Einstein's hypotheses and Kaufmann's experiment[139] disconfirmed them both. Indeed, before the advent of General Relativity the scientific community (e.g. Plank, Poincaré, Bucherer, Kaufmann and Ritz) spoke of the Lorentz–Einstein theory and contrasted it with the more classical theories of Abraham and Ritz: they regarded the theories of Lorentz and Einstein as observationally equivalent.[140]

It was only when Einstein's programme yielded General Relativity that it superseded Lorentz's *empirically* by successfully explaining the 'anomalous' precession of Mercury's perihelion.[141] This explanation constitutes empirical progress because, according to my amended definition of 'novel fact', the behaviour of Mercury, although well known, is nonetheless a novel fact predicted by General Relativity.

This new (General Relativistic) phase in which *empirical* success was achieved was, as it happened, more *speculative* than the previous (Special Relativistic) phase. In this later phase Einstein strengthened his earlier heuristic and thus arrived at a covariant theory of gravitation (he had

[138] Cf. Bucherer [1909].
[139] Cf. Kaufmann [1905].
[140] Cf. Ehrenfest [1913], p. 321.
[141] Cf. Einstein [1915*c*].

been unable to accommodate gravitation within the confines of Special Relativity[142]).

Nevertheless there is a strong continuity between the Special Theory and the General Theory. The latter can be regarded as a more powerful realisation of essentially the same outlook and the same heuristic which had previously led Einstein to Special Relativity. During the earlier phase the deep differences between Lorentz and Einstein remained primarily *heuristic* (and of course metaphysical) ones. It was only with the development of the General Theory that the underlying conflict between the two programmes was reflected at the *empirical* level: with regard to Mercury's perihelion, the bending of the light rays and the red shift, General Relativity made predictions which were never matched by Lorentz's (or Ritz's) theories.

References

Adler, R., Bazin, M. and Schiffer, M. [1965]: *Introduction to General Relativity.*

Born, M. [1956]: 'Physics and Relativity', *Helvetica Physica Acta,* Supp. IX, p. 248.

Born, M. [1962]: *Einstein's Theory of Relativity.*

Bridgman, P. W. [1936]: *The Nature of Physical Theory.*

Bucherer, A. H. (1909): 'Die experimentelle Bestätigung des Relativitätsprinzips', *Annalen der Physik,* **28**, pp. 513–36.

Duhem. P. [1906]: *La Théorie Physique: Son Objet, Sa Structure.*

Eddington, A. S. [1920]: *Space, Time and Gravitation.*

Eddington, A. S. [1923]: *The Mathematical Theory of Relativity.*

Eddington, A. S. [1939]: *The Philosophy of Physical Science.*

Ehrenfest, P. [1913]: *Zur Krise der Lichtäther-Hypothese.* Page reference is to the reprint in M. J. Klein (ed.): *Paul Ehrenfest, Collected Scientific Papers,* pp. 306–28.

Ehrenfest, P. [1923]: *Professor H. A. Lorentz as Researcher,* in M. J. Klein (ed.): *Paul Ehrenfest, Collected Scientific Papers,* pp. 471–8.

Einstein, A. [1905]: 'Zur Elektrodynamik bewegter Körper', *Annalen der Physik,* **17**, pp. 891–921. Page reference is to the English translation, 'On the Electrodynamics of Moving Bodies', in A. Einstein and others: *The Principle of Relativity,* pp. 35–65.

Einstein, A. [1907]: 'Über die vom Relativitätsprinzip geforderte Trägheit der Energie', *Annalen der Physik,* **23**, pp. 371–84.

Einstein, A. [1912]: 'Relativität und Gravitation. Erwiderung auf eine Bemerkung von M. Abraham', *Annalen der Physik,* **38**, pp. 1059–64.

Einstein, A. [1915*a*]: 'Zur allgemeinen Relativitätstheorie', *Sitzungsberichte der preussichen Akademie der Wissenschaften,* 1915, Part 2, pp. 778–86.

Einstein, A. [1915*b*]: 'Zur allgemeinen Relativitätstheorie (Nachtrag)', *Sitzungsberichte der preussischen Akademie der Wissenschaften,* 1915, Part 2, pp. 799–801.

Einstein, A. [1915*c*]: 'Erklärung der Perihelbewegung des Merkurs aus der allgemeinen Relativitätstheorie', *Sitzungsberichte der preussischen Akodemie der Wissenschaften,* 1915, Part 2, pp 831–9.

Einstein, A. [1915*d*]: 'Die Feldgleichungen der Gravitation', *Sitzungsberichte der preussichen Akademie der Wissenschaften,* 1915, Part 2, pp. 844–7.

[142] Cf. Einstein [1912].

Einstein, A. [1915*e*]: Letter to Sommerfeld, 28. 11. 1915; in A. Hermann (ed.): *Albert Einstein und Arnold Sommerfeld: Briefwechsel*, 1968, pp. 32–6.

Einstein, A. [1916]: 'Grundlage der allgemeinen Relativitätstheorie', *Annalen der Physik*, **49**, pp. 768–822. Page reference is to the English translation 'The Foundation of the General Theory of Relativity', in A. Einstein and others: *The Principle of Relativity*, pp. 111–64.

Einstein, A. [1918]: 'Prinzipielles zur allgemeinen Relativitätstheorie', *Annalen der Physik*, **55**, pp. 241–4.

Einstein, A. and others [1923]: *The Principle of Relativity.*

Einstein, A. [1934]: *Mein Weltbild.*

Einstein, A. [1949]: 'Autobiographical Notes', in P. A. Schilpp (ed.): *Albert Einstein: Philosopher–Scientist*, pp. 1–95.

Einstein, A. [1950]: *Out of my Later Years.*

Einstein, A. and Grossmann, M. [1913]: 'Entwurf einer verallgemeinerten Relativitätstheorie und einer Theorie der Gravitation', *Zeitschrift für Mathematik und Physik*, **62**, pp. 225–61.

Einstein, A., Born, M. and Born, H. [1969]: *Briefwechsel 1916–1955.*

Grünbaum, A. [1959]: 'The Falsifiability of the Lorentz–Fitzgerald Contraction Hypothesis', *The British Journal for the Philosophy of Science*, **10**, p. 48.

Grünbaum, A. [1961]: 'The Genesis of the Special Theory of Relativity', in H. Feigl and G. Maxwell (eds.): *Current Issues in the Philosophy of Science*, pp. 43–50.

Grünbaum, A. [1963]: 'The Bearing of Philosophy on the History of Science', *Science*, **143**, pp. 1406–12.

Haas-Lorentz, G. L. de (ed.) [1957]: *H. A. Lorentz: Impressions of his Life and Work.*

Holton, G. [1969]: 'Einstein, Michelson, and the "Crucial" Experiment', *Isis*, **60**, pp. 133–97.

Kaufmann, W. [1905]: 'Über die Konstitution des Elektrons', *Sitzungsberichte der preussischen Akademie der Wissenschaften*, Part **2**, pp. 949–56.

Kennedy, R. J. and Thorndike, E. M. [1932]: 'Experimental Establishment of the Relativity of Time', *Physical Review*, **42**, pp. 400–18.

Kilmister, C. W. [1970]: *Special Theory of Relativity.*

Kompaneyets, A. S. [1962]: *Theoretical Physics.*

Kretschmann, E. [1917]: 'Über den physikalischen Sinn der Relativitätspostulate, A. Einstein's neue und seine ursprüngliche Relativitätstheorie', *Annalen der Physik*, **53**, pp. 575–614.

Kuhn, T. S. [1962]: *The Structure of Scientific Revolutions.*

Lakatos, I. [1968*a*]: 'Changes in the Problem of Inductive Logic', in I. Lakatos (ed.): *The Problem of Inductive Logic*, pp. 315–417.

Lakatos, I. [1968*b*]: 'Criticism and the Methodology of Scientific Research Programmes', *Proceedings of the Aristotelian Society*, **69**, pp. 149–86.

Lakatos, I. [1970]: 'Falsification and the Methodology of Scientific Research Programmes', in I. Lakatos and A. Musgrave (eds.): *Criticism and the Growth of Knowledge*, pp. 91–195.

Lakatos, I. [1971*a*]: 'History of Science and its Rational Reconstructions' in R. C. Buck and R. S. Cohen (eds.): *Boston Studies in the Philosophy of Science*, **8**, pp. 91–136.

Lakatos, I. [1971*b*]: 'Popper zum Abgrenzungs- und Induktionsproblem' in H. Lenk (ed.): *Neue Aspekte der Wissenschaftstheorie*, pp. 75–110; translated into English as 'Popper on Demarcation and Induction' in P. A. Schilpp (ed.): *The Philosophy of Sir Karl Popper*, 1973.

Lakatos, I. and Zahar, E. G. [1976]: 'Why did Copernicus's Programme supersede Ptolemy's?', in R. Westman (ed.): *The Copernican Achievement.*

Lanczos, C. [1972]: 'Einstein's Path from Special to General Relativity', in L. O'Raifeartaigh (ed.): *General Relativity*, pp. 5–18.

Laue, M. von [1911]: *Das Relativitätsprinzip.*

Laue, M. von [1952]: *Die Relativitätstheorie, I.*

Lewis, G. N. [1908]: 'A Revision of the Fundamental Laws of Matter and Energy', *Philosophical Magazine*, Series 5, **16**, pp. 705–17.

Lewis, G. N. and Tolman, R. C. [1909]: 'The Principle of Relativity and Non-Newtonian Mechanics', *Philosophical Magazine*, Series 5, **18**, pp. 510–23.

Lorentz, H. A. [1886]: 'De l'Influence du Mouvement de la Terre sur les Phénomènes Lumineux', *Versl. Kon. Akad. Wetensch. Amsterdam*, **2**, pp. 297–358. Reprinted in *Collected Papers*, **4**, pp. 153–218.

Lorentz, H. A. [1892*a*]: 'La Théorie Electromagnétique de Maxwell et son Application aux Corps Mouvants', *Arch Néerl.*, **25**, pp. 363–642; reprinted in *Collected Papers*, **2**, pp. 164–343.

Lorentz, H. A. [1892*b*]: 'The Relative Motion of the Earth and the Ether', *Versl. Kon. Akad. Wetensch. Amsterdam*, **1**, pp. 74–8; reprinted in *Collected Papers*, **4**, pp. 219–23.

Lorentz, H. A. [1895]: *Versuch einer Theorie der electrischen und optischen Erscheinungen in bewegten Körpern*; reprinted in *Collected Papers*, **5**, pp. 1–138.

Lorentz, H. A. [1899]: 'Théorie simplifiée des Phénomènes Electriques et Optiques dans des Corps en Mouvement', *Versl. Kom. Akad. Wetensch. Amsterdam*, **7**, pp. 507–23; reprinted in *Collected Papers*, **5**, pp. 139–55.

Lorentz, H. A. [1900]: 'La Gravitation', *Scientia*, **16**, pp. 21–51; reprinted in *Collected Papers*, **7**, pp. 116–46.

Lorentz, H. A. [1904]: 'Electromagnetic Phenomena in a System Moving with any Velocity less than that of Light', *Proceedings of the Royal Academy of Amsterdam*, **6**, pp. 809–34; reprinted in *Collected Papers*, **5**, pp. 172–97.

Lorentz, H. A. [1909]: *The Theory of Electrons.*

McCormmach, R. [1970]: 'Einstein, Lorentz and the Electron Theory', in R. McCormmach (ed.): *Historical Studies in the Physical Sciences*, **2**, pp. 41–87.

Mach, E. [1883]: *Die Mechanik in ihrer Entwicklung, historisch-kritisch dargestellt.*

Mach, E. [1913]: *The Principles of Physical Optics.*

Newton, I. [1686]: *Principia Mathematica*; page reference is to the English edition of F. Cajori: *Sir Isaac Newton's Mathematical Principles*, 1934.

O'Rahilly, A. [1965]: *Electromagnetic Theory*, **2**.

Pearce Williams, L. [1968] (ed.): *Relativity Theory: its Origins and Impact on Modern Thought.*

Peierls, R. E. [1963]: 'Field Theory since Maxwell', in C. Domb (ed.): *Clerk Maxwell and Modern Science*, pp. 26–43.

Planck, M. [1906]: 'Das Prinzip der Relativität und die Grundgleichungen der Mechanik', *Verhandlungen der Deutschen Physikalischen Gesellschaft*, pp. 136–41.

Poincaré, H. [1902]: *La Science et l'Hypothèse.* (References are to English translation: *Science and Hypothesis*, 1905.)

Poincaré, H. [1905]: 'Sur la Dynamique de l'Electron', *Comptes Rendus de l'Académie des Sciences*, **140**, pp. 1504–8.

Poincaré, H. [1906]: 'Sur la Dynamique de l'Electron', *Rendiconti del Circolo Matematico di Palermo*, **21**, pp. 129–76.

Polanyi, M. [1958]: *Personal Knowledge.*

Popper, K. R. [1935]: *Logik der Forschung.*

Popper, K. R. [1963]: *Conjectures and Refutations.*

Popper, K. R. [1969]: *Logik der Forschung (Dritte, vermehrte Auflage).*

Reichenbach, H. [1958]: *The Philosophy of Space and Time.*

Reichenbach, H. [1965]: *Axiomatik der relativistischen Raum-Zeit-Lehre.*

Ritz, W. [1911]: *Gesammelte Werke.*
Schaffner, K. [1969]: 'The Lorentz Electron Theory and Relativity', *American Journal of Physics*, **37**, pp. 498–513.
Shankland, R. S. [1963]: 'Conversations with Albert Einstein', *American Journal of Physics*, **31**, pp. 47–57.
Stephenson, G. and Kilmister, C. W. [1958]: *Special Relativity for Physicists.*
Watkins, J. W. N. [1958]: 'Confirmable and Influential Metaphysics', *Mind*, **67**, pp. 344–65.
Whittaker, E. T. [1951]: *A History of the Theories of Aether and Electricity, I.*
Whittaker, E. T. [1953]: *A History of the Theories of Aether and Electricity, II.*
Zahar, E. G. [1973]: *The Development of Relativity Theory: a Case Study in the Methodology of Scientific Research Programmes.* (Doctoral dissertation, University London.)

The rejection of Avogadro's hypotheses*

MARTIN FRICKÉ
UNIVERSITY OF OTAGO

1 *Introduction*

Many historians tell us that the development of the atomic theory in the early nineteenth century was due primarily to three men: Dalton, Gay-Lussac, and Avogadro. They state that Dalton revived the atomic theory at the beginning of the nineteenth century, and that in 1808 Gay-Lussac proposed the experimental law that reacting gases combine in such a way that the constituents and the resultant, if gaseous, are all in simple proportions by volume to each other.[1] They then point out that it was common knowledge that this law is a simple consequence of the atomic theory and the hypothesis, which I shall call the *equal numbers thesis*, that equal volumes of all gases, at a given temperature and pressure, contain the same number of particles. But, they observe, the equal numbers thesis seemed to be refuted by at least two independent sets of experimental results.[2] And they

* The first drafts of this paper were criticised and improved by Imre Lakatos and John Worrall and especially by Peter Clark. A very early, and somewhat different, version was submitted to *Studies in History and Philosophy of Science*, and I have benefited from the perceptive remarks and suggestions of Gerd Buchdahl and the Referee for his journal.

[1] See Gay-Lussac [1808].

[2] There were many negative results exemplified by the combination of nitrogen and oxygen: one volume of nitrogen reacts with one volume of oxygen to form two volumes of 'nitrous gas'. Gaseous elements were assumed to consist of single atoms; and atoms, by their very nature, could not be split; and so the reaction was interpreted as:

1 atom nitrogen + 1 atom oxygen → 1 particle 'nitrous gas'. Consequently, the 'nitrous gas' cannot have more than half as many particles in a unit volume as the oxygen or the nitrogen. (See Dalton [1808], pp. 70, 71.)

The other category involved those gases of unusual density; for instance, oxygen is more dense than steam, yet the particles of steam, since they contain an atom of oxygen, must be the heavier of the two, and so a smaller number of particles of steam than of oxygen can exist in a unit volume. (See *Dalton's Notebooks* (*M/S*), vol. 1, 246 – quoted in Roscoe and Harden [1896], p. 27.)

conclude their story by arguing that in 1811 Avogadro exposed these refutations as being merely apparent by showing that they depended upon the false assumption that all gaseous elements consist of single atoms.[3]

This tale is common. It poses two well-known and much discussed historical problems. The first one is: *why did Dalton explicitly reject Gay-Lussac's law, despite the fact that it seemed to support his atomic theory?* And the second problem is: *why did nearly all working chemists ignore Avogadro's hypotheses for fifty years?*

The usual explanations of these two strange, apparently irrational pieces of behaviour invoke non-intellectual factors, factors external to the intellectual problem-situation at the time. I will discuss these standard external solutions to the two historical problems in the final section of my paper. But first I am going to argue that we can solve them without invoking 'external factors', *provided we reconstruct the intellectual problem-situation correctly*. Given an accurate reconstruction, our two problems can be given a purely 'internal' solution – that is, the behaviour of the nineteenth-century chemists will be seen to be rational.

The structure of my solution to the two historical problems consists of three theses:

(i) Dalton and Gay-Lussac were not working within the same research programme; and the law of combining volumes, as interpreted by Dalton, was specifically denied by Dalton's programme: in order for Dalton to have accepted Gay-Lussac's law he would have had to have given up his own programme; but Dalton's programme, having predicted a whole range of novel facts, was progressive; and so we can defend Dalton's strategy of shelving Gay-Lussac's law as an anomaly in the hope that he would be able eventually either to show that the law was simply false or to successfully incorporate it into his programme as an approximation.

(ii) Avogadro's hypotheses – the equal numbers thesis and the second hypothesis – solved no problems other than the ones that they were designed to answer, that is, they were *ad hoc*; and Berzelius had a different and better solution to the same problems; no wonder, then, that Avogadro was ignored.

(iii) The theories of Cannizzaro, dated from 1858, were progressive and had Avogadro's second hypothesis as an unexpected consequence; and so it was Cannizzaro and not Avogadro who played the major role in this aspect of the development of the atomic theory.

[3] Avogadro resolved the inconsistency between the equal numbers thesis and experiment by means of a *second hypothesis that some gaseous elements consist of polyatomic molecules rather than of single atoms*. If the nitrous gas reaction was reinterpreted as one involving the splitting of complex molecules, thus: $N_2+O_2 \rightarrow 2NO$, then the equal numbers view would not be contradicted. And if the steam was formed by the replacement of one of the two oxygen atoms by two hydrogen atoms: $O_2+2H_2 \rightarrow 2H_2O$, then the oxygen gas would be more dense than steam, and the unusual densities would also be innocuous. (See Avogadro [1811].)

2 *The birth of the Chemical Atomic Programme and Dalton's rejection of Gay-Lussac's law*

Dalton rejected the law of combining volumes because he held the view that there was no simple relation between the numbers of particles in a set volume of different gases.[4] I will call Dalton's view the *unrelated numbers thesis*. No atomist of the period – and, for that matter, no atomist since – would assent both to the unrelated numbers thesis and to Gay-Lussac's law: in practice, either the law or the thesis had to be given up. For Dalton the choice was easy. The unrelated numbers thesis was as essential part of his atomic theories; indeed, it was responsible for the success of the theories. So he gave up the law.

I intend in this section to show that the unrelated numbers thesis was a central tenet of Dalton's atomic theories and to argue that his theories were extremely good ones. The method that I will adopt to achieve this end is to trace the sequence of Dalton's problems and his answers to them. The sequence is an excellent example of a progressive shift of problems; each member of the sequence is intimately and organically connected with its predecessor, and each solution explains or predicts something independently of its own problem; in short, the sequence is not *ad hoc*. Moreover, the unrelated numbers thesis is the central thread connecting the theories in this sequence.

Dalton's first interest was meteorology and it was his answers to several of the physical problems in this field that were eventually to lead to the chemical atomic programme. His major problem concerned the atmosphere; he asked:

> ...why the oxygen gas being specifically the heaviest, should not form a distinct *stratum* of air at the bottom of the atmosphere, and the azotic gas [nitrogen] one at the top of the atmosphere.[5]

Dalton had certain background assumptions about gases with which any solution to the atmosphere problem would have to conform. He looked at gases in a modified Newtonian way. In Proposition 23, Book 2 of *Principia Mathematica*, Newton had derived Boyle's law from the hypothesis that a gas consists of small particles which repel each other. Dalton took

[4] Dalton states this in the Appendix to his [1810]. I see no reason to doubt or to ignore this passage as it ties in well with Dalton's subsequent writings on the subject. See, for instance, the letter to Berzelius of 1812 (quoted in Roscoe and Harden [1896] p. 47) and p. 349 of Dalton's [1827].

[5] J. Dalton 'On Heat', Lecture 15, *Lectures to the Royal Institution* (1810). The problem was also given an alternative formulation in his [1808], p. 150: '...[why] two elastic fluids, having apparently no affinity for each other, should not arrange themselves according to their specific gravities as liquids do in like circumstances.'

Dalton was almost the only scientist working on this problem for other researchers generally believed that the atmosphere was made up of a mild chemical combination of gases and consequently, for them, the atmosphere's homogeneous constitution raised no problem at all. Dalton's reasons for rejecting the orthodox view are to be found in his [1793], pp. 125ff.

this derivation to be a firm proof of the nature of gases and, in consequence, adopted a type of Newtonian physical atomism. The change that Dalton introduced was that of using caloric to provide the repulsive force. Thus, gases were held to consist of small particles; each particle was enclosed within a caloric shell of about one or two thousand times its size; the shells were in contact and, since caloric repels other caloric, this created the repulsive force responsible for the pressure exerted by a gas on the walls of a closed container.[6]

Dalton's first answer to the atmosphere problem was the hypothesis that:

> ...the atoms of one kind did not repel the atoms of another kind, but only those of their own kind.[7]

This hypothesis is known as the *1st Theory of Mixed Gases* (*1801*). It certainly explains the constitution of the atmosphere; if atoms repel only similar atoms and have no effect whatever on the atoms of another gas then, with a mixture of gases, each individual gas will distribute itself more or less uniformly throughout the available space and a homogeneous mixture will be formed. Not only does the 1st Theory solve the problem, but it also predicts something new: namely, that the pressure exerted by a fixed amount of gas is independent of the presence of other gases (the law of partial pressures). This prediction comes about because the atoms of a given type repel only each other and not atoms of another type, and pressure is produced by this repulsion. Dalton turned to his favourite topic of vapours to confirm the prediction and thus he was able to point out that:

> it was perfectly obvious why neither more nor less vapour could exist in air of extreme moisture than in a vacuum of the same temperature...[8]

And the result that the saturated vapour pressure in air exactly matched the vapour pressure in a vacuum applied to other vapours.

The 1st Theory, as well as predicting novel facts, also withstood severe criticism.[9] The opponents of the theory were those scientists, for example Berthollet, Thomson, Murray, Gough and William Henry, who were convinced that there was an affinity between the different gases in a gaseous mixture. The question of whether it was Dalton or whether it was his critics who were right was settled by the phenomena of the solution of gases in water. Dalton hoped to extend his theory to account for the solution of gases in water but he was thwarted because he was unable to find the laws governing this aspect of the behaviour of gases. Henry took up the challenge of discovering the laws. He did so because he was sure that the law covering the solution of gases in water, when actually found,

[6] See Dalton [1808], pp. 147, 163.

[7] J. Dalton 'On Heat', *Lectures to the Royal Institution* (1810). This is a rewritten version of the passage appearing in Dalton [1808], pp. 154–5.

[8] Dalton [1808], pp. 154–5.

[9] This point is brought out by A. W. Thackray in his [1966].

would show that the manner in which any particular gas dissolved would be affected by the presence or absence of other gases. In other words, Henry expected that the weak affinity that was thought to exist between mixed gases would influence their manner of forming solutions. But Henry found a law which was inconsistent with his views about affinities. The law states that, at a given temperature, the volume of a gas dissolved by a unit volume of water is independent of the pressure of that gas; and it entails that the volume of a given gas that dissolves in water is unaffected by the presence of other gases. This unexpected outcome convinced Henry of the merit of Dalton's views; in 1803 he wrote:

> ...[The 1st Theory of Mixed Gases]...is better adapted than any former one, for explaining the relation of mixed gases to each other, and especially the connection between gases and water.[10]

Dalton, encouraged by the result of Henry's researches, no longer hesitated in applying the 1st Theory to solubility problems. The next question he considered was why gases varied in their degrees of solubility, and he tried to answer it by postulating that the particles of the various gases had different weights.[11] This is Dalton's first suggestion that the particles might have different relative weights and thus it marks the point at which the physical atomic theories become less important and the Chemical Atomic Research Programme begins.[12]

We have seen that the *1st Theory of Mixed Gases* was a good solution to the atmosphere problem. But the attentive reader must have noticed that it was not an adequate one. The 1st Theory invoked a selective repulsive force which only operated between atoms of the same gas. But the caloric shells which gave rise to the force certainly did not have this property; caloric repelled caloric without regard to the type of central atom that was within the shell. And so, throughout the period of the development of the 1st Theory, and *independently of any empirical results*, Dalton was searching for a replacement for it which would be consistent with his gas model.

Its successor was the hypothesis that the size of the atoms (that is, the diameter of their caloric shells) of any specific gas was the same and yet no two gases had particles of the same size (at a given temperature and pressure). The idea embodied in the *2nd Theory of Mixed Gases* (*1804–5*) was two

[10] *Phil. Trans.* **93** (1803); quoted from Thackray [1966].

[11] Dalton wrote in a paper entitled 'On the Absorption of Gases by Water and Other Liquids': 'The greatest difficulty attending the mechanical hypothesis arises from different gases observing different laws. Why does not water admit of its bulk of every gas alike?... I am nearly persuaded that this circumstance depends upon the weight and number of the ultimate particles...' (*Mem. of the Lit. and Phil. Society of Manchester*, Ser. II, volume 1, London, 1805. Read October 21, 1803.)

[12] The idea that atoms of different substances might have different weights was not new. Indeed, the homogeneous constitution of the atmosphere only becomes a problem if it is assumed that the ultimate particles of the mixed gases have different weights. Even so, this puzzle over dissolving gases has a special significance because it emphasises the importance of actually determining the relative particle weights.

fold. First, that a single gas, having all its particles the same size and any two mutually repulsive, would distribute itself throughout any available space so that its particles formed a symmetrical pattern rather like a pile of shot in a box. Second, that if another gas was introduced with different-sized particles, again a symmetrical pattern, or homogenous mixture, would result.

The 2nd Theory was expressed entirely in terms of the modified Newtonian model and needed only caloric to provide the repulsive force. It was therefore a heuristic advance on the 1st Theory. Moreover, it explained one of the anomalies to the 1st Theory. Under the 1st Theory, diffusion would be a fast, if not instantaneous, process because, as Henry neatly paraphrased it, 'every gas is a vacuum to every other gas'. The theory did not square with the facts over this point; but, with the 2nd Theory, because the particles had to slither and slide over each other, diffusion should be a slow process, in agreement with observation.

The 1st Theory led to the problem of determining relative particle weights and the 2nd Theory also led to the same problem. To give a more specific solution to the atmosphere problem, the 2nd Theory had to be augmented by information on the relative sizes of the gaseous particles. Gases were made up of touching spheres of caloric, and it is the diameter of these spheres that has to be found. Since the diameter of a sphere is proportional to the cube root of its volume, and the volume of one of these spheres is proportional to its weight divided by the density of the gas, and finally the weight of one of the spheres is the sum of the weight of the caloric shell (which is zero) and the weight of the atom or particle at the centre of the sphere – the following formula applies:[13]

$$\text{The diameter of the caloric shell of a gaseous particle} \propto \left(\frac{\text{atomic weight}}{\text{density of gas}}\right)^{\frac{1}{3}}$$

The densities of gases could be easily measured in the laboratory and so there remained one task: to find atomic weights. This was the core problem of the *Chemical Atomic Research Programme.*

Before discussing the Chemical Programme, let me summarise my reconstructions. I maintain that the Chemical Programme was prefaced by the following two intertwined sequences of problems:

(i) Atmosphere problem → 1st Theory of Mixed Gases → solubility problem → problem of relative particle weights;

(ii) Atmosphere problem → 2nd Theory of Mixed Gases → problem of relative particle sizes → problem of relative particle weights.

This shift in the physical problems, from the atmosphere problem to the problem of relative particle weights, enabled Dalton to give a statement of the plan of the Chemical Atomic Research Programme:

The different *sizes* of the particles of elastic fluids...being once established, it became an object to determine the relative *sizes* and *weights*, together with the

[13] Dalton [1810], p. 226.

relative *number* of atoms in a given volume. This led the way to the combination of gases, and to the number of atoms entering into such combinations... Other bodies besides elastic fluids, namely liquids and solids, were subject to investigation, in consequence of their combining with elastic fluids. Thus a train of investigation was laid down for determining the *number* and *weight* of all chemical elementary principles which enter into any sort of combination one with another.[14]

This project presupposed the programme's metaphysical hard core. This included the assumptions that each chemical element consists of its own characteristic type of indivisible, indestructible atom, that the property which characterises these atoms is their weight, that compounds are definite combinations of these different atoms, and that, during chemical reactions between compounds, atoms are neither created nor destroyed but only rearranged to form new compounds. These views do not, of course, originate with Dalton. Even his idea to use size and, from this, weight as the identifying characteristic of an atom was an old one; but Dalton's contribution was that he was the first scientist to develop these ideas far enough to be able to attempt the determination of these weights.

At first sight, it seems easy to ascertain relative particle weights. The gravimetric composition of a substance can be measured; together with the formula of the substance this yields relative particle weights. For example, water consists of $11\frac{1}{9}$ percent hydrogen and $88\frac{8}{9}$ percent oxygen by weight; its formula, we assume, is H_2O; this means that oxygen atoms must weigh sixteen times as much as hydrogen atoms. The calculation goes through as follows. Let one hydrogen atom weigh h, and an oxygen atom weigh o, then the water molecule weighs $(2h+o)$; but the water molecule consists of $88\frac{8}{9}$ percent oxygen by weight, and so:

$$\frac{o}{(2h+o)} = \frac{88\frac{8}{9}}{100}$$

$$9o = 8(2h+o)$$

$$o = 16h.$$

Therefore, oxygen atoms weigh sixteen times as much as hydrogen atoms; or, if the weight of one hydrogen atom is conventionally set as 1, the atomic weight of oxygen is 16 and, further, the molecular weight of water is 18.

The Chemical Atomic Programme was immediately successful. It gave a rationale to Prout's law of constant composition, it explained Richter's law of equivalents, and it predicted the law of multiple proportions by weight. Finally, it possessed enormous power as a tool for analysing substances – Dalton himself identified marsh gas and allied compounds and resolved the confusion over the many oxides of nitrogen, and Thomson and Wollaston startled the scientific world with their studies on oxalates.

[14] J. Dalton, 'On Chemical Elements', Lecture 17, Lectures to the Royal Institution (1810); paraphrased in Dalton [1810], p. 213.

Let me now link up Dalton's various theories with the unrelated numbers thesis. The connection is simple: from the point of view of his theories, *the central tenet that the ultimate particles varied in their weight was equivalent to the unrelated numbers thesis.* For if particle weight varied then, given the actual densities of the gases, so did particle size;[15] in turn, since the particle shells were touching, this meant that unlike gases had different numbers of particles in a unit volume.[16]

The unrelated numbers view was thus central to an advancing series of problems and to the existence of a progressive research programme.[17]

This, of course, means that Gay-Lussac just could not have been working along the same lines as Dalton. A look at Gay-Lussac's [1808] reveals that it was argued within a framework of combining volumes with barely a mention of atoms. That the law can be explained by the equal numbers thesis was pointed out only by Dalton. As, in the opinion of many historians, the value of the law is that it suggested the equal numbers view, it is puzzling on their account why Gay-Lussac does not claim all the credit due to him. It is interesting that weak philosophies of science do not attach any significance to Gay-Lussac's apparent reticence. For instance, in the Preface to [A.C.R. 4], J.W. gives a typical inductivist account:

> Gay Lassac's important experimental work on the combining volumes of gases then shows the necessity of a simple relationship between the ultimate particles of gases and their volumes, *although he does not point this out in his paper.* [My italics]

J.W. makes no further comment.

I suggest that the reason why it took Dalton to bring out the strength in Gay-Lussac's law is that these two scientists were not working within the same research programme. Gay-Lussac was not an atomist: he did not accept the hard core of Dalton's Chemical Atomic Research Programme. Gay-Lussac shared with his Arcueil colleagues a positivistic view of chemistry. He regarded gases as 'ideal' and solids and liquids as mere approximations to the perfection of gases. And he measured gases

[15] This conclusion is derived using the formula referred to in note 13, above.

[16] There were additional arguments for the unrelated numbers thesis. The theory that the particles were touching opened the way to a line of reasoning from specific heat to unrelated numbers; and the unrelated numbers thesis could also provide qualitative solutions to other physical problems. Dalton's [1808] is rich in arguments of one kind or another.

[17] The physical theories, after their initial successes, ran into difficulties from about 1812. Dalton, when he had obtained atomic weights by means of the chemical programme, used the formula to calculate particle sizes. The fundamental trouble was that the size of the particles turned out to be fairly uniform; in fact, several had the same size! This made it look as though the 2nd Theory could not solve the atmosphere problem, and in the meantime Dalton had admitted that it would not solve the solubility problem; further, the law of partial pressures could not be derived from it. Therefore, on the one hand the 1st Theory answered the atmosphere problem but conflicted with his standards; and on the other hand, the 2nd Theory was above board but was unable to answer the problem that it was designed to solve. As as result, Dalton vacillated between the two theories – in his [1827] he again favoured the 1st Theory – and the major part of his time was wisely given to the more promising chemical programme.

volumetrically, as opposed to gravimetrically, for practical reasons.[18] In short, chemistry, in Gay-Lussac's eyes, was about the volumetrically 'ideal' ways in which gases combine. He even takes great care in his [1808] to disassociate himself from Dalton's programme; he wrote about Dalton's law of multiple proportions:

> ...Dalton has been led to this idea by systematic considerations; and one can see from his work that his researches have no connection with mine.

3 *The central problem of atomism*

Gay-Lussac was not alone in rejecting atomism. There was a widespread tendency to replace the theoretical concept of 'atom' by the measurable notions of 'volume', 'equivalent', or 'measure'. Dalton's empirical laws, such as the law of multiple proportions, were considered to be of great scientific value, but his theories were discarded as speculations. This scepticism was certainly justifiable, for the problems which were later to show the superiority of an atomic view of chemistry were slow to arise. For example, questions concerning isomerism (the problem which converted Gay-Lussac into an atomist) and polymerism became important only with the development of organic chemistry from 1820 onwards. The primary reason for the prevalent atomic scepticism of the time was the central problem of atomism which I shall call the *independence problem.*

Chemical atomism uses three items: formulas, to represent the number and type of atoms in a compound; atomic weights, to measure the relative weight of an atom (perhaps using 1 as the conventional weight of a hydrogen atom); and gravimetric combining proportions. This last quantity, which can be measured in the laboratory, is the composition by weight of the substance; for example, a compound of iron and oxygen can be analysed and found to consist of, say, forty percent oxygen and sixty percent iron. In Dalton's day, *two of the three* above items were unknown. If one of the unknowns is fixed then, given the gravimetric combining proportions, the other is also determined. I have already shown how the two quantities are interrelated with the example of water. The independence problem is that of settling one of the unknowns in a non-arbitrary way. Now, the sceptics argued that it was unnecessary to invoke two unknowns to account for the experimental results. And their line of reasoning ended with the conclusion that the superfluous atomic weights and formulas should be replaced by combining 'equivalents', 'measures', or whatever.[19]

[18] See Gay-Lussac [1808] and Crosland [1961].

[19] This idea has its roots in Richter's work on acid–base reactions. In 1793, Richter published his conclusion that it was possible to set out a table of equivalents or standard quantities of acids and bases such that an equivalent of any of the acids would exactly neutralise an equivalent of any of the bases, even if the reaction had never been tried before. Positivists thought that this could be extended to other types of chemical reaction.

Therefore, the first task for the atomist was to solve the independence problem by finding some satisfactory way of determining either atomic weights or formulas.

4 *Early solutions to the independence problem: Dalton, Berzelius and Avogadro*

The first attempt at solving the independence problem was made by Dalton. He advocated a rule of greatest simplicity which tentatively dictated a compound's formula. The rule stated that if two elements A and B formed several compounds then 'unless some cause appeared to the contrary' it should be assumed as a 'guide'[20] that the first compound had the formula AB, the second AB_2, the third A_2B, and so on. If subsequent factors indicated a fault with the assigned formula of a compound then it could be revised. Some such rule was needed to permit research to proceed. But Dalton's rule did not silence the sceptics. Even atomists, such as Berzelius and Avogadro, criticised it on the grounds that it was arbitrary and difficult to apply. Dalton was well aware of the shortcomings of the rule and tried to give it a physical foundation by arguing that it was based on stability considerations arising from the caloric-shelled models.[21]

An opportunity to give a better solution to the independence problem arose with Gay-Lussac's discovery of the law of combining volumes. It was suspected that some connection between the numbers of particles in a unit volume of the different gases could be established, and consequently that the combination of gases by volume might indicate how the particles themselves combined. But first the relation between volumetric combination and particle combination had to be discovered.

Historians often assert that the equal numbers thesis provided the only possible connection between the two, and that the 'refuting evidence' mentioned in footnote 2, above, then showed the need for the splitting of complex or polyatomic molecules. In other words, Avogadro's hypotheses were a necessary consequence of Gay-Lussac's law and its 'refuting evidence'.[22]

However, that this view is clearly mistaken is established by the existence of a real alternative which was discussed at the time, namely that the numbers of particles in a given volume of the different gases were

[20] Dalton [1808], p. 213.

[21] See Dalton's paper in *Nicholson's Journal*, **29** (1811), page 147, Meldrum [1920?], p. 10, and Thackray [1970].

[22] For example, Ida Freund maintains that Gay-Lussac's law is incompatible with the chemical indivisibility of particles and so two orders of finite particles (atoms and molecules) of elements have to be assumed. See Freund [1904]. Lowry refers to Avogadro's suggestion as 'the most obvious, perhaps even the only possible method of explaining Gay-Lussac's law...' Lowry [1915], p. 338.

in a simple ratio. Avogadro mentions the alternative in his [1811], and Dalton had already written:

In fact, [Gay-Lussac's] notion of measures is analogous to mine of atoms; and if it could be proved that all elastic fluids have the same number of atoms in the same volume, *or numbers that are as 1, 2, 3 &c.* the two hypotheses would be the same.[23]

One assumption is a special case of the other. The hypothesis that, in a given volume of the different gases, the numbers of particles are in some simple ratio has as a limiting case, with the ratios 1:1:1 and so on, the equal numbers thesis.

Now let us see how these rival explanations of Gay-Lussac's law face up to the 'refuting evidence' of the nitrous gas type reactions and the unusual densities.

On the one hand, the 'refuting evidence' can be accepted and used to decide between the two explanations. If this is done, then the experiments show that the equal numbers thesis is false and therefore that the viable explanation is provided by the hypothesis that the numbers of particles in a given volume of the different gases are in some simple ratio. This is a very general hypothesis and consequently it is better for exact discussion to restrict it to the minimal case consistent with the facts: that the ratio of the numbers of gaseous particles is 1:1 for gaseous elements and left undecided for compound gases. *This is the explanation of the law that was championed by Berzelius.*[24]

On the other hand, the equal numbers thesis can be assumed to be correct and the inconsistency between theory and the 'refuting evidence' resolved by rejecting, or reinterpreting, the experiments. This, as I have shown in footnote 3, requires the further conjecture that the particles split. *This is Avogadro's explanation of the law.*

Thus, there were two contenders as explanations of Gay-Lussac's law. Berzelius held that elementary gases each had the same number of particles in a unit volume, but that compound gases did not. And Avogadro maintained that all gases had the same number of particles in a unit volume and that, as a result, some elementary gases consisted of polyatomic molecules which split up into their constituent atoms during a reaction.

The law of combining volumes, in itself, was never a problem, and the scientists of the time were not interested solely in explaining it. The

[23] Dalton [1810], p. 555; my italics.

[24] Berzelius thought that Gay-Lussac's law proved that atoms and volumes were identical and, initially, he preferred the positivistic volume account to the atomic one. A muddled argument then saw him through the 'refuting evidence' – the nitrous gas result virtually demands the conclusion that the nitrous gas is made of half an atom of nitrogen and half an atom of oxygen – Berzelius believed that if one said that nitrous gas was made of half a volume of nitrogen and half a volume of oxygen then the difficulties would disappear. He later favoured the atomic version of his 'volume' theory.

independence problem was their preoccupation and Gay-Lussac's law was merely an intermediate step in some of the attempted solutions of it. *We should therefore appraise the rival researches of Berzelius and Avogadro to answer the independence problem*, and not simply consider their explanations of Gay-Lussac's law.

I shall argue that, for a chemist convinced of the merits of atomism, Berzelius's work was a more attractive proposition than that of Avogadro. Berzelius and Avogadro worked within the Chemical Atomic Research Programme, and both of them were trying to solve the independence problem. In fact, neither of them did solve it and consequently the atomic programme as a whole was stagnating. But, I suggest, Berzelius's attempt at a solution was the better one of the two. I shall conclude that:

(i) Berzelius's account of Gay-Lussac's law was better than Avogadro's. These two widely different explanations were observationally equivalent until the mid-1820s, but Avogadro's views involved *ad hoc* manoeuvres.

(ii) Berzelius's attempt to solve the independence problem was superior to Avogadro's.

(iii) After 1820 it was possible to distinguish by experiment the two explanations of the law mentioned in (i). The experiments, however, went against Avogadro's theories.

Let us first consider conclusion (i).

Basically, Avogadro's hypotheses dispelled the 'refuting evidence', formed a guide to some formulas, and provided a technique for calculating molecular weights. Berzelius's explanation of Gay-Lussac's law did much the same.

We have already seen how both explanations can account for the 'refuting evidence'.

Avogadro's hypotheses constituted a guide to formulae as follows. Gay-Lussac's law describes the volumetric ways in which gases combine; since, by Avogadro's hypotheses, a given volume of any gas whatever contains a set number of molecules, Gay-Lussac's law can be interpreted as describing the way in which molecules combine. Take, for example, the formation of water. Two volumes of hydrogen combine with one of oxygen to form two of water (steam). This means that two molecules of hydrogen combine with one molecule of oxygen to form two molecules of water, and so the formulae might reasonably be guessed to be H_2, O_2, and H_2O respectively. But Berzelius arrived at exactly the same formula for water. For him, the elementary gases were monatomic – that is, hydrogen was H and oxygen O – and so, since the elementary gases each had the same number of atoms in a given volume, two Hs combine with one O to form H_2O. In a similar way Berzelius concluded that the formula of ammonia was NH_3. And so the two scientists described the reactions as follows.

	2 volumes hydrogen	+ 1 volume oxygen		→ 2 volumes steam
Avogadro:	$2H_2$	+	O_2	→ $2H_2O$
Berzelius:	$2H$	+	O	→ H_2O

and

	3 volumes hydrogen	+ 1 volume nitrogen		→ 2 volumes ammonia
Avogadro:	$3H_2$	+	N_2	→ $2NH_3$
Berzelius:	$3H$	+	N	→ NH_3

One point to note is that, for Berzelius, ammonia and steam had half as many particles per unit volume as the elementary gases.[25]

Finally, Avogadro's hypotheses provided a method for determining relative molecular weights by density comparison. Since all gases had the same number of molecules as each other in a set volume, the density of a gas was directly proportional to the weight of one of its molecules; and if the density of a gas was compared with the density of hydrogen, which was conventionally set as 1 (or 2), then the molecular weight of the gas would be obtained. Berzelius, because he did not accept that all gases had the same number of particles in a unit volume as each other, denied the validity of this technique. Hence there was a direct test that could be made between the two explanations. If any compound gas were chosen and its molecular weight found by standard means, such as summing the appropriate multiples of the atomic weights of its constituents, then Berzelius's theory predicts that this value will not equal the density of the gas relative to hydrogen, whereas Avogadro's hypotheses predict that the relative density and the molecular weight will be identical. But, with the compound gases available at the time, this test fails to distinguish the rivals. This is because Avogadro regards the elementary gases as being diatomic and Berzelius takes them to be monatomic, and this results in the differences between them being systematically cancelled out. Take the formation of water, mentioned previously. According to Berzelius, steam has half as many particles in a unit volume as hydrogen, but hydrogen has the formula H; whereas, on Avogadro's account, steam has the same number of particles in a unit volume as hydrogen, but hydrogen has the formula

[25] There is an interesting problem here. Some writers like to use modern views as a criterion of the value of historical theories. In this vein, Linus Pauling wrote: 'Avogadro was the first man in the world to know that water is H_2O... [he]... was always far ahead of his contemporaries. There was never a time, after 1811, when any other scientist proposed a system of atomic weights of elements that contained fewer errors than the system contemporarily used by Avogadro.' (Pauling [1956], p. 708.) The difficulty with this criterion is that it does not separate the sheep from the goats because both Avogadro and Berzelius 'knew' that water is H_2O. And so Coley modified Pauling's statement to: 'Avogadro's formulae were not merely "correct" in fact but were also *derived* by means of the only correct method based on the Molecular Hypothesis. Cf.... Berzelius whose "correct" formulae (e.g. H_2O for water), only happened to be so by a fortunate set of circumstances.' (Coley [1964*a*], p. 11, footnote 1.) My question is: when, and how, could one tell that the circumstances were fortunate?

H_2. So, both Avogadro and Berzelius predict that steam has the same density relative to hydrogen; one makes the forecast in terms of H_2O/H_2 and the other in terms of $\frac{1}{2}H_2O/H$. For the test to be effective, it has to be applied to gases or vapours which have not been formed from the elementary gases. (This extension was carried out in the mid-1820s and its outcome will be discussed later in my paper.) The remaining question of whether or not the assertions that the elementary gases were monatomic and that they were diatomic were observationally equivalent will be answered affirmatively when we come to evaluate Avogadro's research.

At this point it will probably be objected that there is a clear advantage to Avogadro's account, namely that it explains the volume of the resulting gas. In other words, one can understand, in the light of Avogadro's theories, why there should be *two* volumes of steam and *two* volumes of ammonia produced in the preceding examples; whereas for Berzelius's theory this remains a mystery. But this objection is groundless. There is no experimental outcome that Avogadro's hypotheses could not 'explain'. For instance, if six volumes of steam had been produced, Avogadro would have described the reaction as:

$$2H_6 + O_6 \rightarrow 6H_2O$$

For the explanation to be satisfactory there must be good reasons for holding that hydrogen and oxygen are H_2 and O_2, or H_6 and O_6, or whatever. The degree of submolecularity should be decidable. It is often said that Avogadro was responsible for the great theoretical advance of considering the elementary gases as being diatomic. This is not true, for the molecules of gases were permitted to have any degree of submolecularity, provided that there was either one atom or an even number of atoms in a molecule.[26] Indeed, the actual number of atoms in a molecule was 'exactly what is necessary to satisfy the volume of the resulting gas'.[27] In fact, the 'great theoretical advance' of diatomic molecules was merely an *ad hoc* device designed solely to square the formulae for the reaction with the volume of the resulting gas. For a new reaction, Avogadro's hypotheses made no predictions about either the degree of submolecularity or the volume of the resulting gas. In this respect, Avogadro's explanation, just like Berzelius's, lagged behind the facts.

Now let us turn to conclusion (ii).

Avogadro tried to solve the independence problem by using the reactions of gases as a guide to formulae. This technique, which I have explained, relies upon the equal numbers view. But this view was contradicted by the 'refuting evidence' and, as a result, Avogadro introduced his second hypothesis. Now, anybody can change a theory so as to absorb counterevidence. Some such modifications will be conventionalist stratagems, *ad hoc* manoeuvres, or unscientific steps; others will have inde-

[26] See Mundy [1967].

[27] Avogadro [1811], p. 32.

pendently testable consequences. If Avogadro's second hypothesis is to constitute progress, it must have testable implications external to its own problem. It did not. There was no way of checking the conjecture that there existed polyatomic molecules which could divide up during reactions. And the best argument that Avogadro himself managed to produce in favour of the splitting of complex molecules was the weakest possible one – from simplicity.[28, 29]

Furthermore, Avogadro failed to develop his views into a testable theory. The hypotheses were first published in his 1811 paper and in this paper they were applied to elementary gases. Three years later the ideas were extended to cover simple compound molecules.[30] After this second paper, Avogadro stopped working on the independence problem and never again published on it. His inactivity is surprising, for there were opportunities for him to show the wider application of his hypotheses. One key opening was provided by Gay-Lussac. As I have already pointed out, Gay-Lussac thought that chemistry was about the combination of gases. This view made chemistry a very restricted subject because there

[28] 'The possibility of this division of compound molecules might have been conjecture *a priori*; for otherwise the integral molecules of bodies composed of several substances with relatively large numbers of molecules, would come to have a mass excessive in comparison with the molecules of simple substances. We might therefore imagine that nature had some means of bringing them back to the order of the latter...' (Avogadro [1811], p. 32.)

[29] It later became possible to test the theory that there were polyatomic molecules of elements, and the theory was also introduced to answer problems outside those involving combinations of gases. I shall sketch these developments. In the 1830s, it was suggested that polyatomic molecules might explain the isomeric modifications of elements; for instance, it was widely thought that the different varieties of sulphur might be S, S_2, S_3. Again, there was no test of this idea. During the 1840's ozone was discovered and, despite the confusion about its exact properties, at least one scientist – Soret – held that the ratio of the density of ozone to that of oxygen was 3:2. With hindsight we can see that this ratio fits in neatly with Avogadro's hypotheses and the view that ozone is O_3 and oxygen O_2; I think that no one argued this at the time.

There were also problems arising from odd heats of reaction. In 1843, Dulong found that there was less heat produced by the combustion of carbon in oxygen than from burning it in nitrous oxide. In the latter case, some heat is needed to split up the nitrous oxide molecule to yield available oxygen. With this in mind, Favre and Silbermann put forward the idea in 1846 that oxygen gas consisted of polyatomic molecules and that a large amount of heat was required to divide up the oxygen molecule to obtain usable oxygen. Furthermore, their conjecture accounted for the well-known fact that oxygen and hydrogen will *not* react to form water unless they are heated. In the 1850s Brodie and Laurent used arguments from the existence of the phenomenon of nascent action to justify the conclusion that some elements existed in polyatomic form. And, as we shall see in the next section, in 1857 Cannizzaro gave excellent reasons for supposing that there were polyatomic molecules. Finally, during the 1860s, the kinetic theory of gases and experiments involving the absorption of radiant heat gave independent confirmation of Cannizzaro's work.

[30] The 1814 paper also contained some of Avogadro's eccentric ideas. The scientific community must have been startled to read in it that the first hypothesis was intended to also apply to vapours of *non-volatile* compounds – a particular oxide of iron was supposed to consist of one volume of iron 'gas' with two volumes of oxygen, and the molecular weight of iron 'gas' was $106\frac{1}{2}$ – and that some solids had polyatomic molecules: for example, carbon and sulphur were said to be diatomic.

were only four elementary gases known at the time. But Gay-Lussac was not a man to be hemmed in. If gases were needed, then he would make some. As a result he started to investigate vapours of volatile substances, produced the experimental techniques and apparatus that were needed to work at high temperatures, and showed that his law applied to vapours. Avogadro, despite the fact that this new domain was ideal for his hypotheses, paid scant attention to Gay-Lussac's work. It was left to Dumas to use the hypotheses on vapours of volatile substances. And again Avogadro took no interest. Instead he turned his attentions to fields that were far removed from the independence problem. Avogadro's later work is completely without value. It consists largely of vacuous systems of definitions, wild generalisations, obscure and counter-intuitive conclusions and – whenever there was a link with the world – inconsistency with the facts.[31] The only substantial arguments that Avogadro ever offered for his hypotheses were contained in the first two papers.

Berzelius, on the other hand, made great progress in determining the atomic weights of elements. By 1819 he had three unrelated techniques to help him with this task, one arising from his explanation of Gay-Lussac's law, another from Dulong and Petit's law of atomic heat, and a third from Mitscherlich's law of isomorphism.[32] He was remarkably successful in assigning atomic weights. Indeed, it was universally acknowledged by the scientists of the day that his tables of atomic weights and formulae were the best available. However, like most scientific theories, the laws of Dulong and Petit and of Mitscherlich had many exceptions, anomalies, and counter-instances.[33] Berzelius resolved some of the inconsistencies that arose. But, as a matter of fact, his explanation of the law of combining volumes was never involved in any of the corrections that he made to atomic weights or formulae.

[31] One conclusion was that there could be fractions of atoms, that water for example was made of a quarter of an atom of hydrogen and one eighth of an atom of oxygen. Even commentators favourable to Avogadro's work do not know quite what to make of his utterances on this subject. Crosland wrote: 'His 1838 memoir... probably shows Avogadro at his very worst. Fractional atoms were introduced in the most irresponsible way...' (Crosland [1970] page 349).

The lesser-known writings of Avogadro are described in Coley [1964*a*], [1964*b*], and Crosland [1970].

[32] The atomic heat law states that the product of the atomic weight of an element and its specific heat is a constant. And the law of isomorphism states that, as a mechanical consequence of similarities in atomic construction, compounds which crystallise out into isomorphic structures have similar formulae.

[33] The atomic heat of carbon was only thirty percent of the other values, and this led to the Dulong and Petit law being restricted to the metallic elements. Moreover, it was well known that the law needed modification because the specific heat of an element is a function of temperature and this variation is not uniform among the elements. Mitscherlich's law also had exceptions, for example compounds of the ammonium radical NH_4 were isomorphous to potassium ones. And the law was 'refuted' by dimorphism – some compounds crystallise out in two alternative ways, and therefore crystalline form cannot be determined solely by the number and arrangement of the compound's atoms.

In short, Berzelius made a better attempt than Avogadro to solve the independence problem. And the few mistakes and errors made by Berzelius did not seem attributable to his explanation of Gay-Lussac's law.

To reach conclusion (iii), let us look at the series of experiments that contradicted Avogadro's hypotheses.

Avogadro's and Berzelius's explanations of the law of combining volumes did diverge over one point. Avogadro asserted that *all* gases had the same number of particles in a given volume and Berzelius denied this. And so for Avogadro, but not for Berzelius, the determination of molecular weights by density comparison had universal application.

One of the few advocates of Avogadro's hypotheses was Dumas.[34] His explicit intention was to argue, using vapours, against the 'arbitrary data' of Berzelius's explanation of the law of combining volumes and for the 'definite conceptions' of Avogadro's account. In other words, he planned to put into practice Avogadro's proposal for finding molecular weights.

The standard way of determining molecular weights was to add up the appropriate multiples of the atomic weights of the constituents of the compound concerned. This means that atomic weights would always be more accurate than molecular weights. And so, Avogadro's hypotheses predict that, in the case of the most demanding set of experiments, the atomic weights of volatile elements should exactly match their respective vapour densities. There were four volatile elements readily available at the time: phosphorus, sulphur, arsenic, and mercury.

During the period 1826–32 Dumas performed the required series of experiments. It had the following outcome. The vapour density of phosphorus was roughly twice that of its atomic weight, sulphur's vapour density was three times its anticipated value, and so on![35]

[34] Dumas's theories were not identical to Avogadro's. What we now call an 'atom' Dumas called a 'molecule', while our 'molecule' was labelled an 'atom'. Hence, for Dumas, water consisted of an atom of hydrogen and half an atom of oxygen. (This representation is exactly the same as Avogadro's view that water was made up of a molecule of hydrogen together with half a molecule of oxygen.) Many thought that Dumas's conclusions were ridiculous. For instance, Berzelius said of them: 'Till now it has been usual to discard a hypothesis as soon as it leads to absurdities but to some modern investigators this course seems too inconvenient.' (Quoted from Freund [1904], p. 339.) (Berzelius was never fond of Dumas; Meyer, in his [1898], quotes Berzelius describing Dumas as a 'charlatan, French wind-bag, chemical dancing master...'.)

The essential difference between Dumas's and Avogadro's theories is that Dumas restricted the second hypothesis to diatomic molecules only.

[35] No experimental result can demolish a theory with a single blow. A theoretician can always adapt a hypothesis to any counter-evidence. There were two such attempts to nullify the setback caused by Dumas's test. Dumas himself quickly realised that the results were roughly integral multiples of their expected values; and so he guessed that sulphur, for instance, somehow existed as S_6 in the vapour state (relative to hydrogen as H_2). He followed this guess by the conjecture that the heat necessary to convert the sulphur from a solid into a vapour, had not quite broken the sulphur down into the form it takes in chemical reactions. He therefore tried to explain the failure of the prediction by stating that heat was less effective than chemical action for splitting up an element into its smallest constituent parts. This hypothesis was quickly shown to be valueless by the

It soon became apparent that there were examples of an even more perplexing nature than these. After all, these vapour densities, although anomalous to Avogadro's hypotheses, did at least bear some relation to their atomic weights. However, Dumas actually found that sulphur's vapour density was only three times its expected value at certain temperatures; at other temperatures it had different values; *moreover, within certain limits, the variation was continuous*. And there were many compounds, for example ammonium chloride, which had this property of a variable vapour density.

Dumas capitulated. In 1832 he wrote:

it must be clearly stated that gases, even when they are simple, do not contain in equal volumes the same number of molecules or at least chemical molecules.[36]

It is worth remembering that this is the statement of a man who had been determined to show that Avogadro was right and Berzelius wrong.

Dumas's defeat was not the end of the matter. The flow of counter-instances to Avogadro's hypotheses continued unabated. In 1838, Bineau showed that Avogadro's theories were inconsistent with his own results on the vapour densities of ammonium and phosphonium salts.[37] Further instances of odd vapour densities were observed by Cahours in 1844 and 1847.[38] By 1850, Avogadro's hypotheses had been 'refuted' by the vapour densities of sulphur, phosphorus, mercury, iodine, arsenic, ammonium chloride, ammonium carbamate, phosphorus pentachloride, and many more elements and compounds.

To sum up this section. Avogadro's hypotheses were not widely accepted by atomists because (*a*) his research was degenerate, involving the *ad hoc* second hypothesis and failing to pass Dumas's test, and (*b*) there was a better rival theory available – that of Berzelius. In turn, atomism itself was not popular amongst the scientific community because the Chemical Atomic Research Programme was degenerating; true, it had started well by predicting the law of multiple proportions, but then it was realised that almost all of the positive results of the programme could be obtained, without recourse to such theoretical entities as atoms, by using measurable quantities such as equivalents. In short, the atomists were held up by the independence problem.

vapour of mercury. The vapour density of mercury is half that of its normal 'atomic weight' and so it contradicted both the original prediction and Dumas's protective modification. (See Buchdahl [1959].) Gaudin also tried to account for the failure of Dumas's research. He postulated that the molecules of the vapours varied in their sub-molecular constitution – that sulphur was S_6, phosphorus P_4, and so on. This suggestion was *ad hoc* and was ignored. (Gaudin's work is discussed in Mauskopf [1969].) Both of the rescue operations failed.

[36] Quoted from Partington [1964], p. 219. To make Dumas's terminology more modern, I have substituted the word 'molecule' for 'atom'.

[37] *Ann. Chim. Phys.* **68** (1838), pp. 416–41. To have accounted for them by modifying Avogadro's second hypothesis, as Gaudin had done for Dumas's oddities, Bineau would have had to assume that nitrogen was N_{12}, carbon C_6, chlorine Cl_4, bromine Br_4, sulphur S_{12} and so on.

[38] *Comptes Rendus*, **19** (1844), pp. 771–3 and *Ann. Chim. Phys.* **20** (1847), pp. 369–78.

5 *Cannizzaro's solution to the independence problem*

The eventual success of Avogadro's theories is easy to explain. They were simple consequences of a creative shift in the atomic programme which was instituted by Cannizzaro in 1858. *Cannizzaro produced a completely new and original solution to the independence problem. His work cannot be classified, as it usually is, as a mere extension of Avogadro's because it did not in any way rely upon reactions between gases. Cannizzaro, unlike Avogadro, had no use for Gay-Lussac's law of combining volumes.*

Cannizzaro's research, as I will show, predicted Avogadro's second hypothesis and the known anomalies to Avogadro's theories as novel 'facts'. This is why Cannizzaro was successful in convincing the scientific world of the value of his theories. To recapitulate, for Avogadro the postulation of polyatomic molecules was *ad hoc*; for Cannizzaro, *their existence was a startling and unexpected new consequence which was unrelated to his original problem.*

Physicists were intrigued by the variable vapour densities of certain substances. In 1857, St Clair Deville attempted to explain them by the hypothesis that these vapours were mixtures of dissociation products. (Dissociation was eventually defined as the (possibly reversible) spontaneous decomposition of substances by heat, without interference by chemical agents.) This hypothesis was difficult to test because the substances concerned re-associated when cooled, and thus only the original compounds were available for laboratory inspection. Even so, there were certain facts in its favour. It had been noted in 1849 that the vapour of phosphorus pentabromide showed the striking orange colour of bromine. And Marignac had found that the quantity of heat required to volatilise ammonium chloride was altogether of the wrong magnitude as that expected from the mean heats of volatilization, and yet was equal to that produced by the combination of ammonia and hydrogen chloride. Conversely, it was known that if ammonia is added to hydrogen chloride at high temperatures no heat is yielded. Cannizzaro thought that Deville's tentative explanation was substantially correct and in 1857 published a paper on the subject. Cannizzaro assumed, as a physical hypothesis, that there were no exceptions to the universal law that equal volumes of all gases contain equal numbers of particles (*the equal numbers thesis*) and that dissociation products caused certain substances to have a variable vapour density. Thus far, Cannizzaro's theories have nothing to do with chemistry.

However, if it were true that equal volumes of all gases contain equal numbers of particles then comparison of the density of gases or vapours could be used to compare the weight of their particles. And thus Cannizzaro could find relative particular (or molecular) weights.[39]

[39] There was nothing new in this. Dalton was well aware of the relationship between the density of a gas, the number of particles it had in a unit volume, and its molecular

And he did set about determining the relative particular weights of all volatile elements and compounds. It is worth remarking that even at this stage Cannizzaro was on his own. Earlier chemists just did not see any point in finding molecular weights because they knew that a solution to the independence problem would give molecular weights simply by the addition of the appropriate multiples of the atomic weights of the constituents of the compounds. *But Cannizzaro inverted previous ideas by using the novel approach of ascertaining molecular weights in order to solve the independence problem.*

At this juncture I must explain Cannizzaro's big idea. Cannizzaro found out the molecular weights of volatile compounds and, like everyone else, he also knew the gravimetric composition of these compounds. By dividing the molecular weight of a compound up in the proportions of its composition he could obtain figures which expressed the quantities of the elements in the compound in the same units as those in which atomic and molecular weights were expressed.[40] To return to my example, Cannizzaro knew that water had a molecular weight of 18 and that it was composed of $11\frac{1}{9}$ percent hydrogen and $88\frac{8}{9}$ percent oxygen; he therefore was able to calculate that the quantity of hydrogen, expressed in the same units as atomic and molecular weights, was $11\frac{1}{9}/100 \times 18 = 2$; and the quantity of oxygen was $88\frac{8}{9}/100 \times 18 = 16$. He also realised that the figures that he had obtained, namely 2 for hydrogen and 16 for oxygen, must be some integral multiple of these elements' respective atomic weights. This last conclusion comes about because there have to be integral numbers of atoms in a compound. In other words, he realised that 2 must either be the atomic weight of hydrogen or be some integral multiple of it, and similarly that 16 must be some integral multiple of oxygen's atomic weight. Of course, this information is not much use on its own because we do not know what multiples are involved. However, if the same calculation is performed on a second compound of hydrogen (or oxygen) then another figure is obtained which is also an integral multiple of the atomic weight of hydrogen (or oxygen). Clearly, if enough compounds are subjected to this treatment then a set of figures is obtained, all of which figures are integral multiples of the atomic weight of hydrogen (or oxygen), and so it should be possible to guess what the atomic weight is; and also, from this, deduce what the multiples are.

weight; but he could not accept the equal numbers thesis and so denied the validity of Cannizzaro's technique. Avogadro used the method but did not think it important – the title of his first paper was 'Essay on a manner of determing the relative masses of the *atoms* of bodies...' [My italics, modern terminology].

[40] This simple idea is difficult to explain. The reader might fare better with Cannizzaro's own account: 'If the body is composite, its elementary analysis is made, and thus we discover the constant relations between the weights of its components; then the weight of the molecule is divided into parts proportional to the numbers expressing the relative weights of the components, and thus we obtain the quantities of these components contained in the molecule of the compound, referred to the same unit as that to which we refer the weights of all the molecules.' (Cannizzaro [1858], p. 8.)

So Cannizzaro's idea was to take as many compounds of a given element as possible; determine their molecular weights by comparison of vapour densities; use the gravimetric composition of these compounds together with their molecular weights to calculate the various quantities of the element that are in each of the compounds; tabulate these quantities; and finally, infer from these tables the atomic weight of the element and also the number of atoms of it that are in each of the compounds.

Hydrogen was the first element on which Cannizzaro tried his idea. He conventionally set the molecular weight of hydrogen gas as 1 and produced the following table for the quantities of hydrogen in its various volatile compounds:

In hydrogen gas	= 1	In formic acid	= 1
In hydrochloric acid	= $\frac{1}{2}$	In phosphine	= $1\frac{1}{2}$
In hydrobromic acid	= $\frac{1}{2}$	In ammonia	= $1\frac{1}{2}$
In hydrocyanic acid	= $\frac{1}{2}$	In acetic acid	= 2
In hydriodic acid	= $\frac{1}{2}$	In ethylene	= 2
In water	= 1	In alcohol	= 3
In hydrogen sulphide	= 1	In ether	= 5

There are some dramatic conclusions to be drawn from this table. We can read off that there is half as much hydrogen in hydrochloric acid as there is in hydrogen gas. Therefore, hydrogen gas cannot be monatomic! All the figures in the table are multiples of $\frac{1}{2}$ and so the easiest assumption to make is that $\frac{1}{2}$ is the atomic weight of a hydrogen atom.[41] This makes hydrogen gas diatomic. The fractions can then be dispensed with by rewriting the table. This time the atomic weight of hydrogen is conventionally set as 1 or, equivalently, the molecular weight of hydrogen gas is conventionally set as 2. I shall also insert in the adjusted table the information as to the number of hydrogen atoms in each of its compounds:

In hydrogen gas	= 2 which is (2×1), therefore it contains two hydrogen atoms
In hydrochloric acid	= 1 which is (1×1), therefore 1 atom
In hydrobromic acid	= 1 which is (1×1), therefore 1 atom
Etc.	
In water	= 2 which is (2×1), therefore 2 atoms
Etc.	
In phosphine	= 3 which is (3×1), therefore 3 atoms
Etc.	
In acetic acid	= 4 which is (4×1), therefore 4 atoms
In alcohol	= 6 which is (6×1), therefore 6 atoms
In ether	= 10 which is (10×1), therefore 10 atoms

[41] Cannizzaro wrote: 'The different quantities of the same element contained in the different molecules are all whole multiples of one and the same quantity, which, always being entire, has the right to be called an atom.' (Cannizzaro [1858], p. 11.)

The tables for the other elements, which are now expressed relative to hydrogen gas having the molecular weight of 2, are just as revealing. An abbreviated table for oxygen reads:

In oxygen gas	= 32	(2 × 16)
In water	= 16	(1 × 16)
In ether	= 16	(1 × 16)
In acetic acid	= 32	(2 × 16)

and so the atomic weight of oxygen is 16 and oxygen gas is diatomic. And a shortened list for chlorine is:

In chlorine gas	= 71	(2 × 35.5)
In hydrochloric acid	= 35.5	(1 × 35.5)
In chloride of arsenic	= 106.5	(3 × 35.5)
In chloride of tin	= 142	(4 × 35.5)

and consequently the atomic weight of chorine is 35.5 and chlorine gas is diatomic. More information is yielded by cross reference between tables. For instance, given that hydrochloric acid is a compound of hydrogen and chlorine only, the hydrogen table implies that its molecule contains only one atom of hydrogen and the chlorine table implies that it has one atom of chlorine, and therefore its formula has to be HCl.

Other conclusions of relevance include the ones that sulphur vapour exists as S_6 below 1000 °C and that mercury vapour is monatomic.[42]

To sum up. Cannizzaro's problem was the independence problem, and he solved it by means of the tables. Avogadro's second hypothesis, Gay-Lussac's law and the 'refuting evidence' to the law of combining volumes were all completely independent of the problem. However, Cannizzaro's tables led to the unexpected conclusion that oxygen gas, hydrogen gas, and nitrogen gas were all diatomic: they predicted that Avogadro was substantially correct with his second hypothesis and with his explanation of the law of combining volumes and its 'refuting evidence'. And, unlike Avogadro, Cannizzaro was able to determine the degree of submolecularity of polyatomic molecules. Not only that but Cannizzaro could also explain why Dumas's experiments went awry. If sulphur was S_6 as a low temperature vapour, then its vapour density would be three times the value that Dumas expected; and similar explanations applied to the other anomalous non-variable vapour densities, like that of mercury.

Cannizzaro's research and his conclusions were all published in his 1858 paper. The paper did not have a stunning impact. I think that this was probably because it relied upon the comparison of vapour density technique which was still contradicted by the substances with variable vapour densities. Although it should be noted that the paper itself con-

[42] Not all compounds and elements are volatile, and so Cannizzaro occasionally had to use standard methods for settling the chemical parameters.

stituted good evidence for the soundness of the technique. As Wurtz perceptively wrote:

Let us not forget that the [molecular hypothesis] has been verified in so large a number of cases that the contrary facts undoubtably assume the character of exceptions...[43]

In 1860 Cannizzaro attended the Karlsruhe conference.[44] He went with the intention of convincing the chemists who were present of the value of his ideas. He was the only person there who had prepared for the conference: only he had written out an address in advance, and also he had brought with him duplicated copies of his [1858]. The outcome was that a few chemists, notably Meyer, were converted although the remainder readily acknowledged the depth of Cannizzaro's theories.

Of course, there still remained the question of whether or not Deville and Cannizzaro had hit upon the correct explanation of the substances with variable vapour densities. In fact, as was established in the mid-1860s, they had. Avogadro's hypotheses, now supplemented by Deville's theory of dissociation, no longer had any anomalies. In 1869, Wurtz was confident enough to say of the 'large number of exceptions to the law of Avogadro' (the substances which dissociated as vapours) that:

They have often been cited as arguments tending to upset this law. And they would indeed be capable of embarrassing its advocates, if they were not susceptible to a very simple interpretation, which deprives them of all demonstrative force in the question. There is, indeed, nothing to prove that the compounds in question can really exist in the state of a vapour without undergoing more or less complete decomposition; their boiling points are generally high enough to render such a supposition very probable.[45]

But it does not follow from this that Avogadro had been unjustly ignored. For this 'simple interpretation' could not have been made before 1857, and was not independently tested and confirmed until the mid-1860s.

6 *Postscript: some historical myths*

My paper has provided an internal account of the growth of atomism and, in particular, of Dalton's rejection of Gay-Lussac's law and the fifty year delay in the acceptance of Avogadro's hypotheses. This has never been done before. It is true that the delay has often been explained; but, as a result of the principles of rationality of the philosophies of science that have been used, the explanations have had to rely upon external factors. The standards of these philosophies are severe and unrealistic. No scientist can be rational under an inductivist code; consequently myths are needed to explain why scientists deviate from inductivist rationality. Just as austere

[43] Wurtz [1869], p. 179.

[44] Cannizzaro's propaganda is described in de Milt [1948] and [1951], and Green [1960].

[45] Wurtz [1869], p. 175.

moral standards result in lies, deceit, and hypocrisy, intolerant standards of rationality lead to historical myths.[46] In this section I intend to show how some of these myths have been generated and to criticise them.

Most historians are unwitting inductivists. And most histories of atomism have described it as a case of cumulative growth: first Dalton, second Gay-Lussac, and third Avogadro.[47] Such a viewpoint can result in extreme statements like:

> The object of Gay-Lussac's paper was to confirm and establish the new atomic theory by exhibiting it in a new point of view.[48]

and

> [The hypothesis of atoms] was confirmed by the great discoveries of Gay-Lussac.[49]

In its strongest form, inductivism will discard from the history of science those theories which do not appear in the modern textbook or which do not approximate, within wider observational limits, to recent knowledge. (Although these theories may be retained in supplementary histories of metaphysics, myths, magic and the like.) This strong inductivism omits Dalton's *physical* theories, which depend upon the caloric theory, and starts the history of atomism with the law of multiple proportions. As a result, the problem of Dalton's rejection of Gay-Lussac's law is thrown into sharp relief. Roscoe and Harden wrote:

> So long as we hold the view that the atomic theory was inspired by the discovery of the law of combination in multiple proportions, it remains almost incredible that...[Dalton]...should have denied his adherence to such a brilliant extension of the same principle. Gay-Lussac's law, interpreted by the atomic theory, leads at once to the conclusion that the number of particles in equal volumes of the different gases are either equal or stand in some simple ratio to each other.[50]

Now, how is the inductivist, weak or strong, to explain why Dalton rejected the law? To all intents and purposes, the law of combining volumes was a simple experimental law. How could Dalton fail to accept the 'facts'? For an inductivist, the answer seems to be that Dalton was just unable to determine the 'facts'. This suspicion can quickly be made to bear fruit. Dalton comments on the law in the Appendix to his [1810]; he states:

> The truth is, I believe, that gases do not unite in equal or exact measures in any one instance; when they appear to do so, it is owing to the inaccuracy of our experiments. In no case, perhaps, is there a nearer approach to mathematical exactness than that of one measure of oxygen to two of hydrogen; but here, the most exact experiments I have ever made gave 1.97 hydrogen to 1 oxygen...

[46] This thesis and most of the other methodological points that I make are to be found in the works of Imre Lakatos. See his [1970] and [1971].

[47] The best examples are old ones: Thomson [1830–1], Meyer [1898], Muir [1907], Holmyard [1931], and J. W.'s Preface to [A.C.R. 4].

[48] Thomson [1830–1], vol. 2, p. 229.

[49] Wurtz [1898], p. 1.

[50] Roscoe and Harden [1896], p. 47.

Two points should be noted: Dalton's other values do not even approximate to those required under the law, and, in the Preface to the same volume, he explicitly takes responsibility for the experiments.[51] Therefore it was Dalton who was at fault. He performed the experiments and yet was unable to obtain the correct results. The obvious conclusion is that Dalton's failure to measure properly was the major factor in his ignoring the law of combining volumes. Roscoe, for instance, stated in this context:

> Dalton, it must be admitted, was not great in exact experimental chemistry. Although it may be urged that he was self-taught, and began his work when the resources of the experimentalist were scanty and imperfect, yet it is evident that there must have been some inherent deficiency either in his mind or in his hands, which disqualified him for accuracy in experimentation.[52]

The inductivist does not have to rely upon Dalton's description alone. Berzelius, often said to be the best chemist of the period, wrote to Dalton:

> ...and moreover I must warn you that I do not think that Gay-Lussac is easily mistaken, *especially in a matter where it is only a question of good or bad measurement.*[53]

To sum up. Inductivist standards compel the historian to regard the argument between Dalton and Gay-Lussac as one solely about the measurable facts of a simple laboratory experiment. Dalton was in the wrong; therefore he was unable to determine the facts. Hence, *there is the myth that Dalton was a poor experimenter*. Even non-inductivist authors have related this story. Nash, for example, wrote that 'Dalton never became a very able laboratory worker.'[54]

However, this myth generates yet more historical problems. After all, Dalton did initiate the modern atomic theory and so must have had some positive qualities as an experimenter. Obviously the myth leads to an inconsistency. One popular attempt to dissolve this contradiction is to temper the criticism of Dalton by adding the thesis that he was handicapped by crude apparatus. W. C. Henry wrote, describing Dalton's apparatus which he had inherited:

> His instruments of research, chiefly made by his own hands, were incapable of affording accurate results.[55]

and Roscoe asserted:

> The apparatus employed by J. Dalton in his chemical researches...was of the simplest and rudest character.[56]

Thus, *the myth requires the embellishment that Dalton used unreliable instruments.*

[51] 'Having been in my progress so often misled, by taking for granted the results of others, I have determined to write as little as possible but what I can attest by my own experience...' (Dalton [1810], Preface.)

[52] Roscoe [1901], p. 158.

[53] Letter from Berzelius to Dalton, October 16, 1812; quoted from Roscoe and Harden [1896], p. 162, my italics.

[54] Nash [1950], p. 32.

[55] Quoted from Farrar [1968].

[56] Quoted from Farrar [1968].

In short, inductivists assent to the following four statements. Atomism developed in three stages: Dalton, then Gay-Lussac, then Avogadro. The disagreement between Dalton and Gay-Lussac concerned only measurable facts. Dalton was a poor experimenter. He was handicapped by his laboratory equipment.

I shall now criticise these four statements, taking them in reverse order from the last to the first.

Dalton had exactly the same equipment as his contemporaries. W. C. Henry, who later had the apparatus in his possession, probably never even looked at it! I refer you to the more detailed arguments to this end by Farrar in her very interesting [1968]. Her thesis is that this myth originated from a desire to show that it is possible for a poor country boy to make good without any initial advantages of training or financial backing. This may have been the first motive behind the story but I would maintain that subsequently its use was much more sinister. Patterson, in her [1970], also observes that Dalton's posture as a 'simple rustic' was hardly credible: 'Dalton, from the first, invested a relatively large part of his earnings in excellent apparatus and books.' (p. 203).

It is doubtful that, as a matter of fact, Dalton was a poor experimenter.[57] Partington, amongst others, has argued forcibly that Dalton was sound in laboratory techniques.[58] Similarly, Trengrove has concluded:

> We cannot prove that he [Dalton] was a very great experimenter – only that reports of his inadequacies have been greatly exaggerated.[59]

Partington has also maintained that Dalton had a good sense as to the reliability of results:

> He [Dalton] had the true investigator's gift of finding the right result almost by intuition, which is worth much more than tedious refinement and elaboration without inspiration.[60]

Therefore, unanimous agreement has yet to be reached as to the truth or falsehood of the statement that Dalton was a poor experimenter. Moreover, it is not even important to settle the statement's truth value for, as I have argued and will emphasise further, the disagreement between Dalton and Gay-Lussac went far beyond the measurable facts and consequently Dalton's skill, or lack of it, is irrelevant to the problem of why he rejected the law.[61]

[57] True, some of Dalton's results do not correspond with modern ones; but the use of recent values as a criterion of experimental adequacy, because it would make the assessment a function of time, is not acceptable. By the way, if this criterion is employed then Gay-Lussac is also censured. The law of combining volumes is not exact: two volumes of hydrogen, for instance, do *not* combine with one of oxygen.

[58] Partington [1962].

[59] Trengrove [1968].

[60] Partington [1962], p. 758.

[61] Although historians have tried to answer the question of whether or not Dalton was a poor experimenter, they have not raised the problem of why there should be such a widespread myth that he could not measure.

Gay-Lussac rarely left results untouched. He approximated some, and rejected others without explicit statement or explanation as to why they were unsatisfactory. In his [1808] he employs both practices. Take Davy's three gravimetric results which Gay-Lussac cited as strong antedated evidence for the law. Two of the three were rounded off by well over one percent. (Remember that Dalton was happy to reject the law with some values that were within one and a half precent of exactly confirming it.) Gay-Lussac discards the third one, which was nine precent out, and asserts unconvincingly that:

The difference, however, is not very great, and is such as we might expect in experiments of this sort; and I have assured myself that it is actually nil.[62]

It was his expectation that there would be an ideal volumetric combination between gases that enabled him to make these bold approximations. His opponent, Dalton, when it actually comes to producing objective arguments against the law, invokes several authorities like Berthollet, Davy, and Henry. He is not himself responsible for all the experimental results that he employs. Both Dalton and Gay-Lussac select their basic data from a variety of sources; and so, Gay-Lussac's skill and Dalton's clumsiness, even if real, were not relevant to their decisions. I think that it is misleading to say, as Partington does, that 'Dalton did not accept Gay-Lussac's law since it did not agree with his own experiments' (Partington [1962], p. 81). Let me conclude this paragraph by interpreting the passages that the inductivists are fond of quoting. Dalton's comments on the law, mentioned on p. 300 above, mean the following. He has rejected the law on theoretical grounds and is consequently faced with explaining why Gay-Lussac should think it true. To this end he suggests experimental inadequacy on Gay-Lussac's part. Historians often make smug remarks on this. Nash wrote:

It is a little amusing to observe Dalton, whose experiments were generally very crude, quoting his own values as a refutation of those of Gay-Lussac, who was an acknowledged experimental virtuoso.[63]

Knight, another well-known authority, does not commit himself here:

Dalton was no great experimentalist, and the prospect of him upbraiding one of the best observers in Europe for his alleged inaccuracies has entertained or outraged many commentators.[64]

The merriment of many commentators might be dampened by this thought. Dalton, as part of his argument against the law which appears in the Appendix to his [1810], also upbraided 'one of the best observers in Europe' for his results on the oxides of nitrogen. This time, Dalton was right. In 1816, Gay-Lussac concurred and revised his findings as

[62] Gay-Lussac [1808], p. 13.
[63] Nash [1950], p. 68.
[64] Knight [1967], p. 85.

directed.[65] Therefore, there should be no precept that the 'crude and clumsy' Dalton should not try to correct anybody. Finally, Berzelius was wrong in the implication of his letter cited in note 53 above. He saw the situation as a case of inductive growth.[66]

Finally, as I have argued in the main body of the paper, Dalton and Gay-Lussac were not working within the same research programme. And so there was no cumulative growth from Dalton to Gay-Lussac. There is a sophisticated variation of the one-two-three story of the development of atomism. Some historians report that Dalton was faced with the independence problem, and Avogadro, by correctly reading the implications of Gay-Lussac's law, solved the problem for Dalton. Nash wrote:

> In 1808 Gay-Lussac announced his discovery... It was this observation that, when correctly interpreted, led to a sounder criterion for the establishment of molecular formulas (and, thence, of atomic weights) than was provided by Dalton's 'rule of greatest simplicity'.[67]

And Knight has stated:

> The solution to the problem of how to avoid being arbitrary in assigning formulae to compounds was soon solved [by Gay-Lussac and Avogadro].[68]

and

> [Avogadro's] hypotheses not only made it possible to readily explain Gay-Lussac's results, but also provided a method of finding true molecular formulae without arbitrary simplicity rules.[69]

Now, atomic interpretations of Gay-Lussac's law have never been widely used to determine formulae. Avogadro originated the technique of using reacting gases to make inferences of the type: 1 volume of gas A combines with 2 volumes of gas B, therefore 1 molecule of A combines with 2 molecules of B. This method has limited application. There are few gases and few reactions between gases; yet the independence problem was that of determining the molecular formulae of *all* substances, whether they be solids, liquids, or gases. It was Cannizzaro who solved the independence problem.

The second problem for the inductivist is to explain the almost universal rejection of Avogadro's hypotheses. It seems unlikely that all other scientists, misled by preconceived ideas, were blind to the truth.[70] It is

[65] This point is dealt with at greater length in Crosland [1961].

[66] The previous sentence of the letter reads 'I would have thought the [the law] was the best proof of the probability of the atomic theory.'

[67] Nash [1950], p. 44.

[68] Knight [1967], p. 83.

[69] Knight [1967], p. 88.

[70] Some, however, do prefer to seek an explanation in these terms. Hartley wrote that 'Cannizzaro saw so clearly that this confused state of chemical theory was due to the reluctance of chemists to accept wholeheartedly the logical conclusions from the work of Gay-Lussac and Avogadro owing to their preconceived ideas on one aspect or another.' (Hartley [1966], p. 57.) And Ihde, although in his other publications holding a similar position to my own, stated: '[Cannizzaro] avoids earlier uncertainties [by] refusing to entertain preconceived notions regarding molecular composition.' (Ihde [1961].)

more probable that there was a breakdown in communication. Avogadro's contemporaries could not see the wisdom of his ideas not because of some fault with their eyes, but because the proposals were not there to be seen clearly. Thus the myth arose that Avogadro's exposition lacked clarity.[71]

At first glance, there seems to be good evidence for this explanation. Avogadro does formulate his ideas obscurely: most of his sentences are a paragraph long and he uses difficult terminology. For example, to understand his [1811], one has to make careful distinctions between the words 'molécule', 'molécule intégrante', 'molécule constituente', and 'molécule élémentaire'. On closer inspection, however, the value of this evidence is dubious. Avogadro was perfectly consistent in his thinking and in his application of the terms; and the terms themselves were standard amongst the chemists of the period.

Several modern authorities, suspecting that Avogadro's exposition was sound, have still sought a partial explanation in terms of a failure of communication. If Avogadro stated his case clearly and distinctly and yet still his ideas did not become widely accepted, this could only mean that he was not widely read. In turn, this must be because his papers were not printed in the best journals. This is why Crosland asserts that Avogadro had to struggle against an editorial policy that only permitted the one non-Italian outlet: the French 'comparatively obscure and second-rate Bulletin...de Ferussac', and therefore that:

> Some part of the blame for the lack of attention given to Avogadro's Hypothesis by his contemporaries must attach to the editors of the influential British, French, and German Scientific periodicals.[72]

Crosland's explanation does not apply to the first two papers, for they were published in the very influential *Journal de Physique*. Mellor comes neatly to the rescue. The *Journal de Physique* stopped publication eleven years after the printing of Avogadro's first paper and so:

> It is just possible that the impact of Avogadro's work was to some extent prejudiced by the nature of the journal in which he published it and by the fact that it went out of existence soon afterwards.[73]

All this is unnecessary as well as false. For once, the various editors were exactly right. The papers did not receive decent publication because they did not deserve it. The original two papers, both of which had some merit, were published well. Most of the others are just vacuous. It would have been a disgrace for any of them to have been printed in the better journals.

Again, it just does not matter whether Avogadro was clear or not, or whether he was published in the best journals or not. One cannot adequately explain, except on a partial, trivial, and eclectic level, the rejection of

[71] Even the non-inductivist Nash has stated '[Avogadro's] exposition was anything but lucid' (Nash [1950], p. 87).

[72] Crosland [1970], p. 346.

[73] Mellor [1971], p. 80.

Avogadro's work on the grounds of its lack of lucidity if it *still* would not have been generally acknowledged as valuable even if it had been perfectly understood. This is the essence of my paper. I have assumed, probably against the facts, that Avogadro's work was perfectly well known and comprehended by every chemist. Then I have argued that I can explain the scientists' instinctive recognition of the work as unsatisfactory and their subsequent lack of interest in it by showing that, on clearly articulable public standards, the work was poor.

References

[A.C.R. 4]: *Alembic Club Reprints*, **4** (Edinburgh, 1923).

Avogadro, A. [1811]: 'Essay on a manner of determining the relative masses of the elementary molecules of bodies, and the proportions in which they enter into these compounds', in [A.C.R. 4], pp. 28–51. Originally published in *J. de Phys.*, **73** (1811), pp. 58–76.

Buchdahl, G. [1959]: 'Sources of Scepticism in Atomic Theory', *Brit. J. Phil. Sci.*, **10**, pp. 120–34.

Cannizzaro, S. [1858]: 'Sketch of a Course of Chemical Philosophy', *Alembic Club Reprints*, **18** (Edinburgh, 1910). First published in *Il Nuovo Cimento*, **7** (1858), pp. 321–66.

Cardwell, D. S. L. (ed.) [1968]: *John Dalton and the Progress of Science*.

Coley, N. G. [1964*a*]: 'Avogadro – The Molecular Hypothesis' (M.Sc. Thesis, University of Leicester, 1964).

Coley, N. G. [1964*b*]: 'The Physico-Chemical Studies of Amedeo Avogadro', *Annals of Science*, **20**, pp. 195–210.

Crosland, M. P. [1961]: 'The Origins of Gay-Lussac's Law of Combining Volumes of Gases', *Annals of Science*, **17**, pp. 1–26.

Crosland, M. P. [1970]: 'Amedeo Avogadro', *Dictionary of Scientific Biography*, **1**, pp. 343–50.

Dalton, J. [1793]: *Meteorological Observations and Essays*.

Dalton, J. [1808]: *A New System of Chemical Philosophy*, vol. 1, Part 1.

Dalton, J. [1810]: *A New System of Chemical Philosophy*, vol. 1, Part 2.

Dalton, J. [1827]: *A New System of Chemical Philosophy*, vol. 2, Part 1.

Farrar, K. R. [1968]: 'Dalton's Scientific Apparatus', in Cardwell [1968], pp. 159–186.

Freund, I. [1904]: *The Study of Chemical Composition*.

Gay-Lussac, J.-L. [1808]: 'Memoir on the combination of gaseous substances with each other', in [A.C.R. 4]. The paper was read on December 31, 1808 and first published in *Mémoires de la Société d'Arcueil*, **2**, pp. 207–34.

Green, J. H. S. [1960]: The Conference at Karlsruhe, 1860, and the Development of Chemical Theory', *Proc. Chem. Soc.*, pp. 329–32.

Hartley, Sir H. [1966]: 'Stanislao Cannizzaro &c.', *Notes and Records of the Royal Society*, **21**, pp. 56–63.

Holmyard, E. J. [1931]: *Makers of Chemistry*.

Ihde, A. J. [1961]: 'The Karlsruhe Congress: A Centennial Retrospect', *Journal of Chemical Education*, **38**, pp. 83–6.

Knight, D. M. [1967]: *Atoms and Elements*.

Lakatos, I. [1970]: 'Falsification and the Methodology of Scientific Research Programmes', in I. Lakatos and A. Musgrave (eds.): *Criticism and the Growth of Knowledge*, pp. 91–195.

Lakatos, I. [1971]: 'History of Science and its Rational Reconstructions', in R. C. Buck and R. S. Cohen (eds.): *Boston Studies in the Philosophy of Science*, **8**, pp. 91–136.
Lowry, T. M. [1915]: *Historical Introduction to Chemistry.*
Mauskopf, S. H. [1969]: 'The Atomic Structural Theories of Ampère and Gaudin: Molecular Speculation and Avogadro's Hypotheses', *Isis*, **60**, pp. 61–75.
Meldrum, A. N. [1920?]: *The Development of the Atomic Theory* (London: no date, catalogued B.M. 1920).
Mellor, D. P. [1971]: *The Evolution of the Atomic Theory.*
Meyer, E. von [1898]: *A History of Chemistry*, 2nd edition, translated by G. McGowan.
Milt, C. de [1948]: 'Carl Waltzein and the Congress at Karlsruhe', *Chymia*, **1**, pp. 153–69.
Mitt, C. de [1951]: 'The Congress at Karlsruhe', *J. Chem. Educ.*, **28**, pp. 421–5.
Muir, M. M. P. [1907]: *History of Chemical Theories and Laws.*
Mundy, B. W. [1967]: 'Avogadro on the Degree of Submolecularity of Molecules', *Chymia*, **12**, pp. 151–5.
Nash, L. K. [1950]: *The Atomic–Molecular Theory.*
Partington, J. R. [1962]: *A History of Chemistry*, volume 3.
Partington, J. R. [1964]: *A History of Chemistry*, volume 4.
Patterson, E. C. [1970]: *John Dalton and the Atomic Theory.*
Pauling, L. [1956]: 'Amedeo Avogadro', *Science*, **124**, pp. 708–13.
Roscoe, H. E. [1901]: *John Dalton and the Rise of Modern Chemistry.*
Roscoe, H. E. and Harden, A. [1896]: *A New View of the Origin of Dalton's Atomic Theory.*
Thackray, A. W. [1966]: 'The Emergence of Dalton's Chemical Atomic Theory; 1801–8', *Brit. J. Hist. Sci.*, **3** (1966–7), pp. 1–23.
Thackray, A. W. [1970]: *Atoms and Powers.*
Thomson, T. [1830–1]: *History of Chemistry* (2 vols).
Trengrove, L. [1968]: 'Dalton as an Experimenter', *Brit. J. Hist. Sci.*, **4**, pp. 394–8.
Wurtz, A. [1869]: *A History of Chemical Theory*, translated and edited by Henry Watts.
Wurtz, A. [1898]: *The Atomic Theory*, translated by E. Cleminshaw.

On the critique of scientific reason

PAUL FEYERABEND

UNIVERSITY OF CALIFORNIA, BERKELEY

Summary

The historical studies on which the following essay comments use the idea of a research programme (explained in the first essay by Lakatos) and define two types of relation between research programmes and the evidence. Let me call these types type A and type L respectively. They examine episodes where one research programme, R'', replaces another research programme, R' (or fails to be replaced by it), i.e. R'' is made the basis of research, argument, metaphysical speculation by the great majority of competent scientists. The authors find that the relation of R'' to the evidence is always of type L while that of R' is of type A (other circumstances being present when this is not the case). Assuming the historical analysis to be correct this is an interesting *sociological law*. The authors do not present their results in such terms, however. Making A and L part of a *normative methodology*, they claim to have shown that the acceptance of R'' was *rational* while the continued defence of R' would have been *irrational*, and they express this belief of theirs by calling research programmes exhibiting relation L to the evidence *progressive* research programmes, while research programmes which stand in relation A to the evidence are called *degenerating*. They also claim that such judgements are *objective*, independent of the whims and subjective convictions of the thinkers who make them. Using such a normative interpretation of their sociological results they also claim to possess *arguments* for and against research programmes. For example, they would say that today most versions of environmentalism degenerate and that it is irrational to continue working on them. Fortunately this puritanical superstructure of the otherwise excellent sociological studies need not be taken seriously. The reason is that the superstructure is arbitrary, or 'subjective' in at least five different ways. (i) The basic philosophy behind the normative appraisals makes modern science the source of the standards without giving reasons for this choice; (ii) despite all its praise for modern science it uses a streamlined version of it without (*a*) making the principles of streamlining explicit and without (*b*) arguing for them; (iii) the standards that are obtained via the arbitrary steps (i) and (ii) are not strong enough to praise any action as 'rational' or condemn it as 'irrational', which means that such judgements are without any support from the arbitrarily accepted methodology; (iv) in some of the essays research programmes are selected in an idiosyncratic manner, the purpose being to make the general philosophy appear true (not that such truth would be of any use – see item (iii) above); (v) the attempt to show that competent scientists always act 'rationally' is not applied to all scientists but only to those whose actions seem to fit into the general methodology (for 'seem' see again (iii), above). The superstructure which is subjective in the five ways just enumerated is supposed to guide scientists, while the case studies are to show that the guide has some substance – he is not merely a philosopher indulging in abstract dreams of law and order. I argue that the alleged substantiality is moonshine and that one can reject the standards just as arbitrarily as they have been introduced. To sum up: in the enclosed essays we have (*a*) the discussion of certain sociological regularities; (*b*) the proposal of arbitrary standards which have no practical force; (*c*) the insinuation that the regularities are not merely factual, but are features of rationality and that they lend support to the standards, give them the authority they need to be accepted as guides of research. (*a*) may be accepted with the caution we extend to any new 'discovery' in sociology; (*c*) must be rejected (and with it the tendentious terminology used in the presentation of the results of (*a*)); (*b*) may be accepted, or rejected, depending on mood, the weather, etc. Environmentalists, however, may continue on their path, for no argument has been raised against their enterprise.

1 *Two fundamental questions; only one of them examined by the methodology of research programmes*

There are two questions that arise in the course of any critique of scientific reason. They are:

(i) *What is science?* – how does it proceed, what are its results, how do its procedures, standards and results differ from the procedures, standards and results of other enterprises?

(ii) *What's so great about science?* – what makes science preferable to other forms of life, using different standards and getting different kinds of results as a consequence? What makes modern science preferable to the science of the Aristotelians, or to the ideology of the Azande?

Note that in trying to answer question (ii) we are not permitted to judge the alternatives to science by scientific standards. We are now *examining* these standards, we are *comparing* them with other standards rather than making them the *basis* of our judgements. Azande results must be judged by Azande standards and the question is: are these results and these standards preferable to science, or are they not? And if they are not, then, what are the reasons for the deficiency?

Questions (i) and (ii) arise with all abstract concepts. We can ask them about truth, knowledge, beauty, goodness, and so on. In the history of thought answers to question (ii) are often taken for granted. For example, it is taken for granted that Truth is something quite excellent and that all we need to know are the detailed features of this Excellent Thing. This means that one starts with a *word* and uses the enthusiasm created by its *sound* for the support of questionable ideologies (cf. the Nazis on Freedom). A rare exception is the 'long standing quarrel between poetry and philosophy' (Plato *Republic*, 607b6f). The philosophers' case was stated as follows: 'If you...admit the honeyed muse in epic or in lyrical verse, then pleasure and pain will usurp the sovereignty of law and of the principles always recognised by common consent as the best' – i.e. poetry will drive out knowledge. The case of the poets was stated, in a very weak form, in the following story in Plutarch's *Life of Solon:* 'When the company of Thespis began to exhibit tragedy, and its novelty was attracting the populace but had not yet gone as far as public competition, Solon being fond of listening and learning and being rather given in his old age to leisure and amusement, and indeed to drinking parties and music, went to see Thespis act in his own play, as was the practice in ancient times. Solon approached him after the performance and asked him if he was not ashamed to tell so many lies to so many people. When Thespis said *there was nothing dreadful in representing such works and actions in fun*, Solon struck the ground violently with his walking stick: "If we applaud these things in fun", he said, "we shall soon find ourselves honouring them in earnest".' The italicised passage gives the weak defence of Thespis. (The story seems historically impossible, yet it elucidates a widespread attitude in early Greece; cf. also chapter 8 of Forsdyke [1964].) Plato challenges the champions of poetry 'to plead for her...that she is no mere source of pleasure but a benefit to society and to human life' (607d5). The challenge is taken up by Aristotle in his *Poetics* where it is shown that 'poetry tends to give general truths while history gives particular facts': poetry, dealing with the *nature* of things, is more philosophical than history (1451b2ff). Note that the champions of poetry are already on the defensive: they defend poetry by trying to show that it *aids* knowledge. They do not defend it as an *alternative* to the kind of knowledge sold by the Presocratics and their followers. In the course of their defence they emphasise historical *tendencies* over individual idiosyncracies (primacy of the plot in Aristotle) and thus inhibit the advance of freedom. (Cf. my [1976].)

There are at least two ways of dealing with question (i). We may use the method of an *anthropologist* who examines the behaviour and the ideology of an interesting and peculiar tribe. In this case statements such as 'science proceeds by induction' are *factual* statements of the same kind as statement describing how a particular tribe builds houses, how the foundation is laid, how the walls are erected, what rites accompany the procedure, and so on. On the other hand we may consider *ideal demands* and examine their consequences. Such a procedure is only loosely connected with actual (scientific) practice, and it may be entirely divorced from it. This applies to many investigations dealing with the 'logic' of science. Occasionally the difference is noticed, but emphasised as an advantage: actual science has not yet achieved the purity of an enterprise that agrees with the demands, of a so-called 'rational' enterprise; it must be 'reconstructed' and the reconstruction, obviously, will be different from the real thing. Of course, nobody can say whether a reconstruction, *when inserted into the historical surroundings that gave birth to actual science*, will produce comparable results. It may not give any results at all. (Who would expect that one can climb Mount Everest using the 'rational' steps of classical ballet?) We need the anthropological method to find out whether reconstruction improves science, or whether it turns it into a useless though perfect adornment of logic books. The procedure of the anthropologist therefore takes precedence over the procedure of the logician.

There is a third way in which science and, for that matter, any practice can be examined. Considering the standards and the aims of a certain form of life we may ask whether the practice agrees with the standards and whether it leads up to the aim. In this case we compare the results of an anthropological inquiry (the practice, the aims, the standards which have been found to constitute the form of life) with what we know, or think we know about the laws of nature and man's relation to them. For example, we may point out that a certain way of building houses is not very efficient and that houses built in this way cannot last very long. Or we may point out that induction does not get us very far, and does not provide certainty. The first criticism assumes (*a*) that the builder wants to build a solid house and (*b*) that our knowledge of building houses is at least as good as the knowledge of the culture we are examining. That (*a*) is not always satisfied can be seen from our own civilisation which relies on obsolescence. And as regards (*b*), it suffices to remind the reader of recent discoveries which show that 'primitive' procedures are often superior to their scientific rivals.[1] The second criticism, the criticism of induction,

[1] To understand an ideology we have to understand (*a*) its aims; (*b*) how the ideology fits into the society that accepts it; (*c*) how it fits into nature. Modern sociologists and anthropologists always make *science* the measure of (*c*). Thus, Merton explains the rain dance of the Hopi Indians as a kind of social glue assuming, as a matter of course, that rain dancing does not bring rain. Even a perceptive writer such as Frank Waters interprets the Hopi Genesis 'as a depth psychology' (Waters [1963], xxiii), and asserts that 'the basic meaning of the Hopi creation myth and the symbol which expresses it is subjective' (*op. cit.*, p. 31, following Jung's general point of view). Why this reluctance to assume that the Hopi were practical people who knew how to create rain? (We have to speak of earlier periods of the history of the tribes as these tribes no longer possess their former integrity.) The reason cannot be empirical. There are no statistical studies about the effectiveness of rain dances, *properly carried out* (remember that an experiment succeeds only if it is properly carried out). Nor can the reason lie in what is said by any *specific* scientific theory that is highly confirmed by the evidence. There is not a single respectable theory I know of that has been shown to forbid the success of rain dances (we can be pretty sure of that: nobody knows physics, and psychology, and rain dancing well enough to construct the needed arguments). The reason why rain dances are declared to be emotional performances unsuited for the intended effect is an *ideology* that is never spelled out in detail and for which one claims the authority of science. Now in the past decades this ideology has been shattered more than once. Respectable scientific institutions, such as the American Academy of Science, the American Association for the Advancement of Science and others have reacted accordingly and have incorporated organisations dedicated to the study of psychic phenomena (for example). They have accepted the reality of the phenomena examined. Now, so far these phenomena have been examined only in the case of individuals. Is it not possible that they get amplified in certain social contexts so that a society can produce rain by dancing *provided* the psychophysical atmosphere of the remaining planet does not produce too much background noise? (It is quite possible that the ceremony worked during the period of the first world but that the objective conditions of success disappeared with the disturbance of the harmony between living beings that introduced the second world and caused animals to draw away from people – cf. Waters, *op. cit.*, p. 15.) I repeat, the discussion of such possibilities is no longer frowned upon by institutionalised science (though there is a sufficient number of scientists who are not aware of such developments and who would not welcome them, were they informed). I know a follower of Lakatos will take this as a sign of the degeneration of late twentieth-century science and will try to return to more Victorian times. But he can be

assumes that its users want to get very far (they do not) and that the classes corresponding to universal properties have misleading subclasses[2] (if there are no misleading subclasses then induction will succeed despite the alleged invalidity of the inference from the particular to the general). We see: a true Critique of Scientific Reason cannot take anything for granted. It must examine the most obvious assumptions.

2 *The excellence of science is* assumed, *it is not argued for. The same is true of the standards proposed by the methodology of research programmes. These standards are obtained by an analysis of modern science. Their excellence is therefore again assumed, it is not argued for. There is not a single argument to show that they are better than the standards that underlie the practice of magic*

Such a critical attitude is only rarely found among philosophers of science. Scientists have by now gone very far in the revision of basic cosmological ideas and they have some up with some amazing suggestions (subject-dependence of the physical world; synchronicity in addition to causality; telekinesis; non-sensory information-gathering by plants and ability to recognise individuals; non-causal reaction of deep sea organisms to the position of the sun and of the moon; artificial character of the first satellite of Mars; existence of an international astronomy at 30000 B.C. and so on). There is no longer any antagonism between the most advanced parts of science and ancient points of view which have degenerated because of scientific warfare. Ancient myths are reconsidered, brought into testable form, examined. Surprising and revolutionary results have been obtained, in the Soviet Union,[3] in China,[4] in the United States.[5] Speculation on the frontiers of knowledge is often indistinguishable from mythmaking and does not follow any easily recognisable method. There may be law and order in some domains; there is absolutely no law and order in others. It is true that the great majority of scientists is still quite hostile towards such mobility (the National Science Foundation, for example, refuses to support the most interesting research on plant-communication that is being carried out by some members of the Stanford Research Institute in Menlo Park). This is a familiar phenomenon, explainable by prejudice and an anxious

easily tamed by pointing out *that his own method, as misunderstood by him* (for the 'mis-' cf. §§3 and 6 below) *forces him to take psychic research seriously*. For psychic research, unfortunately, *is a progressive research programme* (example: Vassiliev's confirmation of the hypothesis of telehypnosis).

Apart from such possibilities we have the discovery of real achievements of ancient ideologies that were forgotten, and even forbidden by science, and that had to be brought to the fore by force. An example is discussed in chapter 4 of my [1975].

[2] Cf. below, on the cosmological criticism of methodologies.

[3] Ostrander and Schroeder [1970] as well as the literature in Thelma Moss' contribution to Mitchell [1974].

[4] Cf. the immense literature on traditional medicine in China as well as my brief account in chapter 4 of my [1975].

[5] Cf. the reports and literature in Mitchell, *op. cit.*

commitment to the status quo. On the other hand, one would have thought that philosophers of science, being aware of such developments and being less impressed by scientific orthodoxy than their specialist colleagues, might have developed a suitable philosophy, providing additional stimuli for speculation.

This has not been the case. Quite the contrary – most philosophers of science still seem to be living in Victorian times when only a few clouds were dimly perceived on a distant horizon. Their craving for orderliness easily exceeds that of the most systematic scientist and approaches that of a catatonic. They have a strong faith in the basic orderliness of science, they have a strong faith in the excellence of (non-dialectical) logic (and this despite the many open problems one finds even in this discipline), and they spend their lives trying to find a point of view that enables them to uphold both kinds of faith. In this they often succeed, for 'science' is for them a particular logical system, or set of systems, rather than the historical process usually designated by that name, and 'logic' is a very simple and dull part of that discipline, a kind of pidgin logic. 'Problems of Science', however, are the internal problems of the chosen system, or set of systems, illustrated with the help of bowdlerised examples from science itself.[6] Kuhn has shown the dream-like character of the whole enterprise. The essays in the present book and the methodology of research programmes that forms their background are an attempt to move from dream to reality *without* any loss of logic and reconstruction. Let us inquire to what extent the attempt succeeds.

We see at once that question (ii) remains unanswered. It is of course assumed that science is vastly better than any other research programme of comparable scope and generality. But we do not find a single reason in favour of this assumption. Occasionally the assumption enters a detailed argument concerning some different matter, apparently lending it additional force when all we have is a dogmatic and ritual reassertion of the greatness of science.[7]

[6] According to Nelson Goodman, 'A World of Individuals', quoted from Benacerraf and Putnam [1964], p. 207, the inventions of the scientists 'become raw material for the philosopher whose task is to make sense of all this...in understandable terms'. Considering that logicians are only rarely capable of following scientists on their flights of fancy this would indeed seem to be 'the task' of 'the philosopher' – only Goodman is not that modest. If *he* does not comprehend a thing, then the thing uncomprehended is *intrinsically obscure* and must be 'clarified', i.e. it must be translated into a language which he understands (pidgin logic, in most cases). If *he* understands a language, then the language is intrinsically clear and must be spoken by everyone. This is also how the demand for reconstruction arises. Logicians cannot make sense of science – but they can make sense of logic and so they stipulate that science must be presented in terms of their favourite logical system. This would be excellent comedy material were it not the case that by now almost everyone has started taking the logicians seriously.

[7] It is surprising to realise how difficult it is to see science in perspective. Carl Sagan, surely one of the most imaginative scientists alive, warns us not to unduly restrict the possibilities of life, and he mentions various types of 'chauvinism' (oxygen chauvinism: if a planet has no oxygen, then it is uninhabitable; temperature chauvinism: low tem-

Thus John Worrall in a position paper on critical rationalism[8] compares the measures which Marxists use to get rid of *prima facie* refuting instances with the measures used by scientists, and he asserts that the former do not lead to any increase of content while the latter do. Had he examined the matter with the care he has spent on his story of Young and Fresnel he would have come to a different conclusion.[9] But let us assume he is right – what follows? We can infer that Marxism does not agree with the standards of science as reconstructed by critical rationalists. We cannot infer that Marxism is *inferior* to science unless we have reasons for the standards which are independent of the fact that they are part of science. No such reasons are found in the philosophy of research programmes. Quite the contrary: it is explicitly stated that science is the measure of method and that good method is the method practised by the 'scientific élite'.[10] This, at least, is the theory defended by Lakatos in the introductory eassay to this volume.

peratures such as those on Jupiter and high temperatures such as those on Venus make life impossible; carbon chauvinism: all biological systems are constructed of carbon compounds) which he regards as unwarranted (Sagan [1975], chapter 6). He writes (p. 179): 'It is not a question of whether we are emotionally prepared in the long run to confront a message from the stars. It is whether we can develop a sense that beings with quite different evolutionary histories, beings who may look far different from us, even "monstrous" may, nevertheless, be worthy of friendship and reverence, brotherhood and trust.' Still, in discussing the question whether the message on the plaque of *Pioneer 10* will be comprehensible to extraterrestrial beings he says that 'it is written in the only language we share with the recipients: science'. (Cf. p. 217: messages to extra-terrestial beings 'will be based upon communalities between the transmitting and the receiving civilization. Those communalities are, of course, not any spoken or written language or any common, instinctual encoding in our genetic materials, but rather what we truly share in common – the universe around us, science and mathematics.') In times of stress this belief in science and its temporary results may become a veritable mania making people disregard their lives for what they think to be the Truth. Cf. Medvedev's account of the Lysenko case in Medvedev [1969].

[8] Worrall [1975], pp. 2/21 ff.

[9] Critical rationalists take great care to show that *prima facie* disreputable procedures in science, when looked at in detail, turn out to be quite acceptable (cf. Zahar on the Lorentz–Fitzgerald contraction, or Worrall on the fate of Young's version of the wave theory). They also know that there are good scientists and bad scientists and that the procedures of the former are not discredited by the errors of the latter: no one would abandon science because it contains complementarity. The attitude towards Marxism, or astrology, or other traditional heresies is very different. Here the most superficial examination and the most shoddy arguments are deemed sufficient. Worrall uses some Marxist interpretations of events in Hungary to discredit the whole approach but without saying what the interpretations are, who has put them forth, and where they can be found. Popper ([1945], vol. 2, pp. 187ff) mentions the hypothesis of colonial exploitation as a perfect example of an *ad hoc* hypothesis although it is accompanied by a wealth of novel predictions (the arrival and structure of monopolies being one of them). And whoever has read Rosa Luxemburg's reply to Bernstein's criticism of Marx or Trotsky's account of why the Russian Revolution took place in a backward country (cf. also Lenin [1968], vol. 19, pp. 99ff.) will see that Marxists are pretty close to what Lakatos would like any upstanding rationalist to do, though there is absolutely no need for them to accept his rules. After all, all *he* can say in favour of these rules is that the elite of some enterprise he loves *sometimes* sticks to the rules. (See below.)

[10] This volume, p. 23.

According to Lakatos methodologies are tested, i.e. either defended or attacked, by reference to historical data. The historical data which Lakatos uses are '"basic" appraisals of the scientific elite'[11] or 'basic value judgements'[12] which are *value* judgements about *specific* achievements of science. Example: 'Einstein's theory of relativity of 1919 is superior to Newton's celestial mechanics in the form in which it occurs in Laplace.' For Lakatos such value judgements (which constitute what he calls a 'common scientific wisdom') are a suitable basis for methodological discussions because they are accepted by the great majority of scientists: 'while there has been little agreement concerning a *universal* criterion of the scientific character of theories, there has been considerable agreement over the last two centuries concerning *single* achievements.'[13] Basic value judgements can therefore be used for checking theories about science or *rational reconstructions* of science much in the same way in which 'basic' *statements* are used for checking theories about the world. The ways of checking depend on the particular methodology one has chosen to adopt: a falsificationist will reject methodological rules *inconsistent* with basic value judgements, a follower of Lakatos will accept methodological research programmes which 'represent a *progressive shift* in the sequence of research programmes of rational reconstructions'.[14] The standard of methodological criticism thus turns out to be the best methodological research programme that is available at a particular time. So far a first approximation of the procedure of Lakatos.

The approximation has omitted two important features of science.

On the one side basic value judgements are not as uniform as has been assumed. 'Science' is split into numerous disciplines, each of which may adopt a different attitude towards a given theory, and single disciplines are further split into schools, heresies, and so forth. The basic value judgements of an experimentalist will differ from those of a theoretician (cf. Rutherford, or Michelson, or Ehrenhaft on Einstein), a biologist will look at a theory differently from a cosmologist, the faithful Bohrian will regard modifications of the quantum theory with a different eye than will the faithful Einsteinian. Whatever unity remains is dissolved during revolutions, when no principle remains unchallenged, no method unviolated. In addition there are individual differences: Lorentz, Poincaré and Ehrenfest thought that Kaufmann's experiments had refuted the special theory of relativity and were prepared to abandon relativity in the form in which it had been introduced by Einstein, while Einstein himself retained it because of its comprehensiveness.

On the other hand, basic value judgements are only rarely made for good reasons. Everyone agrees now that Copernicus's hypothesis was a big step forward but hardly anyone can give a halfway decent account of it,

[11] This volume, p. 30.
[12] *Ibid.*
[13] This volume, p. 23.
[14] This volume, p. 30.

let alone enumerate the reasons for its excellence. Newton's theory of gravitation was 'highly regarded by the greatest scientists',[15] most of whom were unaware of its difficulties and some of whom believed that it could be derived from Kepler's laws. The quantum theory which suffers from quantitative and qualitative disagreements with the evidence and which is also quite clumsy in places is accepted not *despite* its difficulties, in a *conscious violation* of naive falsificationism, but because 'all evidence points with merciless definiteness in the ...direction...[that] all processes involving...unknown interactions conform to the fundamental quantum law'.[16] And so on. *These* are the reasons which produce the basic value judgements whose 'common scientific wisdom' Lakatos occasionally gives such great weight.[17] Add to this the fact that most scientists accept basic value judgements on trust, they do not examine them, they simply bow to the authority of their specialist colleagues, and one will see that *common scientific wisdom is not very common and it certainly is not very wise.*

Lakatos is aware of the difficulty. He realises that basic value judgements are not always reasonable[18] and he admits that 'the scientists' judgement [occasionally] fails'.[19] In such cases, he says, it is to be balanced and perhaps even overruled, by the 'philosophers' statute law'.[20] The 'rational reconstruction of science' which Lakatos uses as a measure of method is therefore not just the sum total of all basic value judgements; nor is it the best research programme trying to absorb (or so produce) them. It is a 'pluralistic system of authorities'[21] in which basic value judgements are a dominating influence as long as they are uniform *and* reasonable. But when the uniformity disappears, or when 'a tradition degenerates',[22] then general philosophical constraints come to the fore and enforce (restore) reason and uniformity.

Now I have the suspicion that Lakatos vastly underestimates the number of occasions when this is going to be the case. He believes that uniformity of basic value judgements prevailed 'over the last two centuries'[23] when it was actually a very rare event. (Here he is in the same predicament as Kuhn who assumes that a particular normal science may have lasted for decades when in fact it was a very rare event.) But if that is the case, then his rational reconstructions are dominated either by common sense,[24] or by the abstract standards of the 'philosopher's statute law'. Moreover, he accepts a uniformity only if it does not stray too much from his standards: 'When a scientific school degenerates into pseudoscience, it may be worthwhile to force a methodological debate.'[25] This means, of

[15] This volume, p. 24.

[16] Leon Rosenfeld in Körner [1957].

[17] 'Is it not...*hubris* to try to impose some *a priori* philosophy of science on the most advanced sciences?...I think it is.' This volume, p. 35.

[18] This volume, p. 23, footnote 8.

[19] This volume, p. 35.

[20] *Ibid.*

[21] *Ibid.*

[22] This volume, p. 36.

[23] This volume, p. 23.

[24] This volume, p. 16, footnote 58.

[25] This volume, p. 36.

course, that the judgements which Lakatos passes so freely are ultimately neither the results of research, nor parts of scientific practice; they are part of an *ideology* which he tries to lay on us in the guise of a 'common scientific wisdom'. We discover here a most interesting difference between the *wording* of Lakatos's proposals and their *cash value*. The methodology of research programmes is introduced with the purpose of aiding rationalism. It is supposed to find historical support for methodological standards. Such standards are to be grounded in history, not in the abstract discussion of abstract possibilities. But the reconstructions which are to provide the historical support are very close to the abstract methodologies supposedly aided and they merge with them at times of crisis. Despite the difference in rhetoric ('Is it not *hubris* to try to impose some apriori philosophy of science on the most advanced sciences?...I think it is' – a sentiment that is forgotton the moment Lakatos enters 'the most advanced' parts of atomic physics[26]), despite the decision to keep things concrete ('there has been considerable agreement...concerning single achievements'[27]) Lakatos does not really differ from the traditional epistemologists *except that they argue for their abstract principles while he does not but uses propaganda instead*: he announces that he is going to support his principles by historical research when the results of this research are overruled the moment they conflict with what he thinks a 'rationalist' should do. Here I prefer the procedure of Watkins who in the position paper I have already mentioned simply says, 'in a letter to Santa Claus', that the science described by critical rationalists is the science he 'would like to have'.[28] This is not exactly the most sophisticated answer to question (ii) but it is the answer that emerges whenever we examine the procedure of our most recent critical rationalists in somewhat greater detail. It is the answer implicit in the essays that have been collected in this book. Any charge of irrationalism or praise of rationalism which these papers contain is therefore purely subjective, unsupported by either abstract or historical reasons. This will become even clearer as our analysis proceeds.

Let us look at the matter from a slightly different point of view. A 'rational reconstruction' as described by Lakatos comprises concrete judgements about results in a certain domain as well as general standards (we have seen that it is the general standards that really run the reconstruction, and in an arbitrary manner, but let us forget this for the time being). A rational reconstruction as described by Lakatos is rational in the sense that it reflects *what is believed to be a valuable achievement* in the domain. It reflects what one might call the *professional ideology* of the domain. Now even if this professional ideology consisted of a uniform bulk of basic value judgements only, even if it had no abstract ingredients whatsoever, even then it would not guarantee that the corresponding field has worthwhile

[26] This volume, p. 36, footnote 131.

[27] This volume, p. 23.

[28] Watkins [1975], p. 1/3.

results, or that the results are not illusory. Every medicine man proceeds in accordance with complex rules, he compares his tricks and his results with the tricks and the results of other medicine men of the same tribe, he has a rich and coherent professional ideology – and yet no rationalist would be inclined to take him seriously. Astrological medicine employs strict standards and contains fairly uniform basic value judgements, and yet critical rationalists reject its entire professional ideology as 'irrational'. For example, they are not prepared even to consider the 'basic value judgement' that the tropical method of preparing a chart is preferable to the sidereal method or vice versa (the latter opinion being that of Kepler). This possibility of rejecting professional standards *tout court* shows that 'rational reconstructions' *alone* cannot solve the problem of method. To find the right method one must reconstruct the *right discipline*. But what is the right discipline?

Lakatos does not consider the question, and he need not consider it as long as his aim is to find out how post-seventeenth-century science proceeds and as long as he can take it for granted that this enterprise rests on a coherent and uniform professional ideology (we have seen that it does not). But Lakatos and his followers go much further. Having finished their 'reconstruction' of modern science they turn it against other fields *as if it had already been established* that modern science is superior to magic, or to Aristotelian science, and that it has no illusory results. *It is assumed* that question (ii) has already been answered, and that it has been answered in the affirmative. However, there is not a shred of an argument to support this assumption.[29] 'Rational reconstructions' take 'basic scientific wisdom' for granted; they do not show that it is better than the 'basic wisdom' of witches and warlocks. Nobody has shown that only science (of 'the last two centuries'[30]) has results that conform to its 'wisdom' while other fields have no corresponding results that conform to their 'wisdom'. What *has* been shown by more recent anthropological studies is that *all* sorts of ideologies and associated institutions produce, and have produced, results that conform to their standards and other results that do not conform to their standards. For example, Aristotelian science has been able to accommodate numerous facts without changing its basic notions and its basic principles, thus conforming to its own standard of *stability*. We obviously need further considerations for deciding what field to accept as a measure of method.[31]

Exactly the same problem arises when we consider *particular*

[29] At this point critical rationalists and followers of the methodology of research programmes usually introduce the criterion of content increase: Aristotle was defeated, and justly so, because he did not conform to this criterion. This assumes (*a*) that Aristotelians *wanted* to conform to the criterion (they did not – see below, p. 332) and (*b*) that the criterion is preferable to, say, the criterion of stability, or to the criterion that the best explanations are *post hoc* explanations. But (*b*) is the assumption under examination.

[30] This volume, p. 231.

[31] Watkins, in his 'letter to Santa Claus', points out that the ideal preferred by him is

methodological rules. It is hardly satisfactory to reject naive falsificationism because it conflicts with the basic value judgements of eminent scientists. Most of these eminent scientists retain refuted theories not because they have some insight into the limits of naive falsificationism, but because they do not realise that the theories are refuted. Besides, even a more rational practice would not be sufficient to reject the rule: universal leniency towards refuted theories may be nothing but a mistake. It certainly is a mistake in a world that contains well-defined species whose properties are only rarely misread by the senses. In such a world the basic laws are manifest and recalcitrant observations are rightly regarded as indicating an error in our *theories* rather than in our *methodologies*.[32] The situation changes when the disturbances become more insistent and assume the character of an everyday affair. A cosmological discovery of this kind forces us to make a choice: shall we retain naive falsificationism and conclude that knowledge is impossible; or shall we opt for a more abstract and recondite idea of knowledge and a correspondingly more liberal (and less 'empirical') type of methodology? Most scientists, unaware of the nomological–cosmological background of the problem, and even of the problem itself, retain theories that are incompatible with established observations and experiments and praise them for their excellence. One might say that they make the right choice *by instinct*[33] – but one will hardly regard the resulting behaviour as a measure of method, especially in view of the fact that the 'instinct' has gone wrong on more occasions than one. The cosmological criticism just outlined (omnipresence of disturbances) is to be preferred (and was, as a matter of fact, preferred by Aristotle: see his criticism of the Presocratics).[34]

'really an amalgam of ideas' found in Bacon and Descartes (*op. cit.*, p. 1/4). That may be so, but does not establish its superiority over, say, the ideology of the Aristotelians or of John Dee. Thus, the argument always becomes circular at the most decisive point. Among the participants of this volume it is only Elie Zahar who approaches the problem in a more rational manner. In his account of the Copernican Revolution he assumes that all the competitors *shared the same standards* and that the Aristotelians lost because their theories did not conform to these shared standards. This still does not give us an answer to question (ii) – all it does is to extend the domain from which the basic value judgements are taken. But it gives a rational account of the victory of the Copernicans *provided* the shared standards had some *force*. This problem will be discussed in the next section.

[32] In such a world the demand for depth (Watkins [1975], pp. 1/4ff) is unrealistic and cannot be satisfied.

[33] 'Up to the present day it has been the scientific standards, as applied 'instinctively' by the scientific elite in *particular cases*, which have constituted the main – although not the exclusive – yardstick of the philosopher's universal laws.' This volume, p. 35.

[34] This possibility of *choosing* a methodology on the basis of cosmological considerations shows that there can be different types of science: given fairly clear species with not too many disturbances we may decide to remain naive falsificationists and absorb the exceptions by methods such as monster-barring, or various means of adaptation, but we may also decide to use basic laws for the explanation of *all* events and so become research programmists. Aristotle made the first decision, Galileo as seen by some thinkers made the second. Thus the idea that there can be only one science – one physics, one biology, one chemistry – which is found even among so-called dialectical

To sum up: the methodology of research programmes does not argue for the superiority of science ('of the last two centuries'); it takes this superiority for granted and pretends to use it as a basis for the standards it employs. It does not use it as such a basis either because it implies 'a pluralistic system of authority' in which 'the philosophers' statute law' plays an important role, side by side with 'common scientific wisdom' ('the philosophers' statute law' – these are the abstract principles the methodology of research programmes was supposed to support by appeal to historical facts). Now: what is the content of this 'philosophers' statute law', what are the reasons for it, and when does it come to the fore and overrules 'common scientific wisdom'? It comes to the fore 'when the scientists' judgement fails'[35] and *that* occurs whenever there is massive support for degenerating research programmes.[36] Thus the standards, instead of being supported by history, are the *criterion* by which we decide when to accept historical trends and when to reject them. Moreover, the methodology of research programmes does not offer any abstract (philosophical) arguments in their favour (or against alternative standards). The standards are therefore arbitrary, subjective, and 'irrational'. They do not provide *objective* reasons for eliminating Marxism, or Aristotelianism, or Hermeticism, or for attacking new developments in the sciences. They merely indicate what critical rationalists would 'like to have' at this stage of the development of their ideology.

But the situation is even worse. So far I have argued that Imre Lakatos has *not* provided *any objective reasons for accepting the standards*, he has not shown them to be rational in any sense of the word he is prepared to accept. I shall now argue that *the standards have no force either*, they are too weak to condemn any action as 'irrational'. It follows that an author who uses Puritanical language of this kind – and the authors collected in this volume use it rather frequently – either subscribes to a rationality theory different from Lakatos's, or else he is content with rhetorical flourishes, unconnected with any argument. In the latter case he gives us an interesting sociological study and uses it as a club for forcing people to accept standards which are very different from those he pretends to defend. Let us now examine the assertion that the standards which Lakatos recommends have no force to condemn any action as irrational.

materialists (cf. Zhores A. Medvedev in Medvedev [1969], pp. 133, 247) is again but a result of insufficient analysis.

[35] This volume, p. 35.

[36] According to Lakatos it seems that modern particle physics represents a degeneration. He also speaks of the development of 'new bad traditions' such as modern sociology, psychology, social psychology. These traditions are indeed *new*. But they are *bad* only if the standards of science 'of the last two centuries' have been shown to be good. They are *assumed* to be good – this much is sure. But there is no argument to support this assumption and, besides, the standards are overruled whenever they seem to conflict with the house philosophy of the methodology of research programmes.

3 *Nor are the standards strong enough to praise individual actions as 'rational' or condemn them as 'irrational'. All that can be said is that the actions have taken place, and that they have certain features*

When a theory enters the scene, it is usually somewhat inarticulate, it contains contradictions, the relation to the facts is unclear, ambiguities abound, the theory is full of faults. However, it can be developed, and it may improve. The natural unit of methodological appraisals is therefore not a single theory, but a succession of theories, or a *research programme*; and we do not judge the state in which a research programme finds itself at a particular moment; we judge its history, preferably in comparison with the history of rival programmes.

According to Lakatos, the judgements are of the following kind: 'A research programme is said to be *progressing* as long as its theoretical growth anticipates its empirical growth, that is as long as it keeps predicting novel facts with some success...; it is *stagnating* if its theoretical growth lags behind its empirical growth, that is, as long as it gives only *post hoc* explanations of either chance discoveries or of facts anticipated by, and discovered in a rival programme.'[37] A stagnating programme may *degenerate* further until it contains nothing but 'solemn reassertions' of the original position coupled with a repetition, in its own terms, of (the successes of) rival programmes.[38] Judgements of this kind are central to the methodology Lakatos wishes to defend. They *describe* the situation in which a scientist may find himself. *They do not yet advise him how to proceed.*

Considering a research programme in an advanced state of degeneration one will feel the urge to abandon it, and to replace it by a more progressive rival. This is an entirely legitimate move. *But it is also* legitimate to do the opposite and to retain the programme. For any attempt to demand its removal on the basis of a rule can be criticised by arguments almost identical with the arguments that eliminate say, naive falsificationism: if it is unwise to reject faulty theories the moment they are born because they might grow and improve, then it is also unwise to reject research programmes on a downward trend because they might recover and attain unforeseen splendour (the butterfly emerges when the caterpillar has reached its lowest state of degeneration).[39] Hence, one cannot *rationally* criticise a scientist who sticks to a degenerating programme and there is no *rational* way of showing that his actions are unreasonable.

Lakatos agrees with this. He emphasises that one 'may rationally stick to a degenerating programme until it is overtaken by a rival *and even after*'[40] – 'programmes may get out of degenerating troughs'.[41] It is true that his

[37] This volume, p. 11. [38] This volume, p. 16.

[39] This remark shows that the methodology of research programmes, too, makes certain *cosmological* assumptions concerning the relationship between research programmes and the world.

[40] This volume, p. 15. [41] Lakatos [1970], p. 164.

rhetoric frequently carries him much further, showing that he has not yet become accustomed to his own liberal proposals.[42] But when the issue arises in explicit form, then the answer is clear: the methodology of research programmes provides standards that aid the scientist in *evaluating* the historical situation in which he makes his decisions; it does not contain *rules* that tell him what to do.

However, even this very modest formulation still goes much too far. Speaking of *risks*[43] it assumes that the progress initiated by progressive phases will be greater than the progress that follows a degenerating phase – after all, it is quite possible that progress is always followed by long-lasting degeneration while a short degeneration (say, 50 or 100 years) precedes overwhelming and long-lasting progress.[44] Speaking of evaluation and using evaluative terms such as 'progressive' and 'degenerating' it assumes that 'progress' is preferable to 'degeneration' both intrinsically and as regards consequences. The second case has just been dealt with. The first case (intrinsic advantage of 'progress') leads back to question (ii). Question (ii) is unanswered, hence the question of the intrinsic advantage of progress is unanswered, too.[45]

[42] 'I give...rules for the "elimination" of whole research programmes'; this volume, p. 11 – note the ambiguity introduced by the quotation marks. Occasionally, the restrictions are formulated in a different way, by denying the 'rationality' of certain procedures. 'It is perfectly rational to play a risky game', says Lakatos, this volume, p. 16 – 'what is irrational is to deceive oneself about the risk': one can do whatever one wants to do if occasionally one remembers (or merely recites?) the standards *which, incidentally, say nothing about risks or the size of risks*. Speaking about risks either involves a *cosmological* assumption (nature rarely permits research programmes to behave like caterpillars), or a *sociological* assumption (*institutions* only rarely permit degenerating research programmes to survive), and thus lead to exactly the same conundrum which Hume explained so nicely more than 200 years ago. Lakatos in passing (this volume, p. 12) admits the need for such additional assumptions: only they 'can turn science from a mere game into an epistemologically rational exercise'. But he does not *discuss* them in detail and those he takes for granted are very doubtful, to say the least. Take the cosmological assumption I have just mentioned. It is interesting, and it certainly deserves to be studied in greater detail. Such a study, I venture to suggest, would reveal that the research programme corresponding to it is now in a degenerating phase (to see this, one needs only consider anomalies such as the Copernican Revolution, the revival of the atomic theory, the revival of assumptions concerning celestial influences; and so on). The sociological assumption, on the other hand, is certainly true – which means that given a world in which the cosmological assumption is false we shall forever be prevented from finding the truth.

[43] Cf. the quotation in the previous footnote.

[44] This is, of course, again a version of Hume's problem. Hume turns up in all methodologies because all methodologies make cosmological assumptions. The naive falsificationist assumes that there are no oceans of anomalies. The conventionalist assumes that the world is built in a simple way. The research programmist assumes that progress, once realised, does not put an end to further progress and that it leads to the truth (after all, a progressive theory may lead us further and further away from the truth; cf. the life of Paphnutius as presented by Anatole France). And so on.

[45] John Worrall writes (p. 2/29, footnote 3 of Worrall [1975]): 'A scientist *would* be pronounced "irrational" (or rather mistaken) by the methodology of research programmes if he stuck to the old programme denying that the new programme had any merits not shared by the old one and thus denying that his own programme needed improvement in

Alan Musgrave has written an interesting paper in which he agrees with some of the criticisms voiced above.[46] Lakatos, he writes (p. 15) 'develops an elaborate account of what is good science and what bad, but he refuses (apart from "Thou shalt not lie") to give advice to the scientists'. '"Anything goes" is the position which Lakatos finally adopts' (footnote 53). But a methodology must 'provide advice or directives' (p. 22). This advice and these directives are to be addressed 'to science, ...to the community of scientists, as a whole' (p. 22). They would 'forbid wholesale persistence with degenerating programmes, or premature mass conversion to a budding one' but permit the individual scientist to go his way: 'we cannot condemn Priestley for his die-hard adherence to phlogistonism; but we could condemn the community of late nineteenth-century chemists had they all done the same'. Musgrave thinks he 'can provide a purely deductive argument' for such directives (p. 23). The argument proceeds from the premiss that 'science ought to devote energy to investigating unsolved scientific problems'. Now, a 'progressive research programme throws up more unsolved problems than a degenerating one' hence, 'science ought to devote more energy to' progressive programmes than to degenerating ones (p. 24). In reply one can point out, first, and still in accordance with the methodology of research programmes, that every success of a progressive programme is a problem for its degenerating rival, so in the end it will be a degenerating programme that 'throws up more unsolved problems'. Secondly, it is not only the *number* of problems

order to catch up with the new one. It is in such circumstances that we shall begin to suspect the operation of extra-rational motives.' The arguments in the last section, in the text above, and in footnote 42 show that it is rather this judgement of Worrall's which makes us 'suspect the operation of extra-rational motives' where by 'extra-rational' we mean motives either not in accordance with the standards, or not dictated by them. Assume I have a research programme which degenerates and I am told so by a research programmist. My reply might well be that I am interested in certainty and not in novelty and that I prefer a programme that can incorporate newly found facts without revision, to a programme that constantly upsets basic convictions. When being told that this means I am not being 'scientific' I can reply that the excellence of science is still a matter of debate, that it is *assumed* by my opponent (though discarded by him when it goes counter to his own pet ideology – see above), that it is not supported by argument. (Nor is there any argument to show that non-scientific ideologies are worse than science in addition to being different from it; of course, there is a general *belief* that this is so, and this belief may even be quite reasonable, but what I am now talking about is the ability of the methodology of research programmes to give a *reason* for the belief.) Adopting the point of view of science, I can add that degeneration when taken seriously may be followed by bigger progress than progress and that progress may lead away from the truth. Or is a scientist supposed to be satisfied with temporary spectacles only? Is it enough for him to impress everyone by first predicting, and then discovering, a new planet (e.g. Neptune) *without any implications for the quality of future research*? And, finally, one might comment on the futility of a point of view where a thief can steal as much as he wants, is praised as an honest man by the police and by the common folk alike provided he tells everyone that he is a thief. If *that* is the sense in which the methology of research programmes differs from anarchism (Worrall [1975], p. 2/30, footnote 1), then I am ready to become a research programmist. For who does not prefer being praised to being criticised if all he has to do is to describe his actions in the lingo of a particular school?

[46] Musgrave [1976].

that counts, but also their *quality*. Now it is certainly more difficult to find the right questions than to answer 'problems' that are already spelled out in detail. Again the directive advises us to pay attention to degenerating programmes. Thirdly, letting the individual scientist do the dirty work of improving a degenerating programme prejudges the issue in a very unfair way. Today an individual can only rarely attack, let along solve the problems that arise in the course of research. Without computers, without expensive equipment, without the help of colleagues and assistants he is doomed from the start. (Just consider the expenses involved in experiments such as those of Reines, or Weber; and where would General Relativity be today had Einstein had to carry the expenses of all the experiments carred out to test it?) Musgrave's directive, and that is my fourth point, is also uncomfortably close to the directive of some politicians who advise us to spend educational funds only on those who are already well educated and to let the less educated fight for themselves, the difference being that the advantages of an educated person are much more obvious than the advantages of a 'progressive' research programme. And my fifth point is (question): why should one prefer programmes which have successfully anticipated experimental discoveries to programmes which have no such record? Does such a preference not indicate an inductivistic prejudice? Musgrave thinks it should not and this is why he speaks of problems rather than of the successful anticipation of novel facts. But why should a research programme that creates lots of problems be preferable to a research programme that creates none? For an Aristotelian absence of problems is a sign that certainty and agreement with facts has been achieved. Popperians do not like certainty and they reject the moves that help us achieve it. They do not like certainty and they think they have also found arguments to support their dislike: certainty is not part of science, therefore it should be rejected. This is firstly not true (cf. the arguments of early Newtonians and Cartesians) and secondly not sufficient: why should we accept science as a measure of excellence? We have to conclude that Musgrave's rescue manoeuvre does not succeed. Its principles are arbitrary, and they lead to results very different from those envisaged by him.

To sum up this part of the argument: the standards which Lakatos has chosen neither issue abstract orders (such as 'abandon degenerating research programmes') nor do they support general judgements concerning the rationality, or irrationality of a certain course of action (such as 'it is irrational to support degenerating research programmes'). Such orders and such judgements give way to concrete decisions in complex historical situations. Hence, if the enterprise that contains the standards is to be different from the 'chaos' of anarchism, *then such decisions must be made to occur with a certain regularity*. Taken by themselves the standards cannot achieve the regularity, as we have seen. But psychological or sociological *pressures* may do the trick.

Thus, assume that the institutions which publicise the work and the results of the individual scientist, which provide him with an intellectual home where he can feel safe and wanted and which because of their eminence and their (intellectual, financial, political) pull can make him seem important, adopt a *conservative attitude* towards the standards; they refuse to support degenerating research programmes, they withdraw money from them, they ridicule their defenders, they do not publish their results, they make them feel bad in every possible way. The outcome can easily be foreseen: scientists, who are as much in need of emotional and financial security as anyone else, especially today when science has ceased to be a philosophical adventure and has become a business, will revise their 'decisions', and they will tend to reject research programmes on a downward trend.

This conservative attitude adopted by the institutions is not irrational, for it does not conflict with the standards. It is the result of collective policies of the kind encouraged by the standards. The attitude of the individual scientist who adapts so readily to the pressures is not irrational either, for he again decides in a way that is condoned by the standards. We have thus achieved law and order without reducing the liberalism of our methodology. And even the complex nature of the standards now receives a function. For while the standards do not prescribe, or forbid, any particular action, while they are perfectly compatible with the 'anything goes' of the anarchist who is therefore right in regarding them as mere embroideries, they yet give content to the actions of individuals and institutions who have decided to adopt a conservative attitude towards them. *Taken by themselves* the standards are incapable of forbidding the most outrageous behaviour.[47] *Taken in conjunction* with the kind of conservatism just described they have a subtle but firm influence on the scientist. And this is precisely how Lakatos wants them to be used. Considering a degenerating programme he suggests that 'editors of scientific journals should refuse to publish...papers [by scientists pursuing the programme]...Research foundations, too, should refuse money'.[48] The suggestion is not in conflict with the standards, as we have seen. Nor can it be used to raise the charge of irrationality against alternative suggestions: measured by the standards of the methodology of research programmes the conservative attitude expressed by the suggestion is neither rational nor irrational. It is an interesting sociological fact – nothing more. *But it is eminently rational* according to other standards, for example, according to the standards of common sense.[49] This wealth of meanings of the word

[47] For a minor exception which by now seems to have become the only point of resistance of the methodology of research programmes cf. this volume, p. 323.

[48] This volume, p. 16.

[49] 'In such decisions', says Lakatos, referring to decisions such as those leading to a conservative use of standards, 'one has to use one's *common sense*' – this volume, p. 16, footnote 58 – as long as we recognise that in doing so we *leave* the domain of rationality as

'rational' is used by Lakatos to maximum effect. In his arguments against naive falsificationism he emphasises the new 'rationalism' of his standards which permits science to survive. In his arguments against Kuhn and against anarchism he emphasises the entirely different 'rationality' of common sense but without informing his audience of the switch and so he can have his cake (have more liberal standards) and eat it too (have them used conservatively) and he can even expect to be regarded as a rationalist in both cases. Indeed, there is a great similarity between Lakatos and the early Church Fathers who introduced conservative doctrines in the guise of familiar prayers (which formed the common sense of the time) and who thereby gradually transformed common sense itself.[50]

4 *Using the methodology of research programmes as our theory of rationality we must therefore regard the enclosed case studies as sociological studies and we must disregard the frequent judgements of rationality or irrationality that occur in them (on the other hand, these judgements may be retained if we adopt a different theory of rationality, for example Hegel's). But though the aim of rationalising history is never reached, the attempt to reach it has produced a history that is richer in content and more conceptual than its predecessors*

Taking all these things into consideration it is clear that Lakatos has not succeeded in showing 'rational change' where 'Kuhn and Feyerabend see irrational change'.[51] A revolution occurs when a new research programme has accumulated a sufficient number of successes and the orthodox programme suffered a sufficient number of failures for both to be regarded as serious rivals, and when the protagonists of the new programme proclaim the demise of the orthodox scheme. Seen from the point of view of the methodology of research programmes they do this not just because of their standards – the standards are not strong enough for making such a judgement – but because they have adopted a conservative attitude towards the standards (all this assumes, of course, that both sides use the

defined by the standards and move to an 'external' medium, or to other standards. Lakatos does not always make the change clear. Quite the contrary. In his attack upon opponents he makes full use of our inclination to regard common sense as inherently rational and to use the word 'rational' in accordance with *its* standards. He accuses his opponents of 'irrationality'. We instinctively agree with him, quite forgetting that the methodology under debate does not support the judgement and does not provide any reasons for making it.

[50] Using the *psychological* hold which the baptismal confession had over the members of the early Christian Church and taking the non-Gnostic interpretation 'as its self-evident content' (Harnack [1961], vol. 2, p. 26), Irenaeus succeeded in defeating the Gnostic heresy. Using the psychological hold which common sense has over philosophers of science and taking the conservative interpretation of his standards as *its* self-evident content, Lakatos has almost succeeded in convincing us of the reality of his own law-and-order philosophy and the non-ornamental character of his standards: now as before the best propagandists are found in the Church, and in conservative politics.

[51] This volume, pp. 31–32.

methodology of research programmes in their deliberations – a matter that is open to considerable doubt[52]). Their orthodox opponents have what one might call a 'liberal' attitude; they are prepared to tolerate a lot more degeneration than the conservatives. The standards permit both attitudes. They have nothing to say about the 'rationality' or 'irrationality' of these attitudes and of the developments initiated by them. It follows that the fight between the conservatives and the liberals and the final victory of the conservatives is not a 'rational change'[53] but a 'power struggle' pure and simple, full of 'sordid personal controversy'.[54] It is a topic neither for methodology, nor for the theory of rationality, but for 'mob psychology',[55] or, to use a more traditional term, *it is a topic for the sociology of knowledge.*

Exactly the same is true of the essays that have been collected in this volume. Each of these essays describes a battle between alternative research programmes and the victory of one of them. Each essay reconstructs the battle in terms of the methodology of research programmes, using the standards of this methodology as a basis for the proper application of its evaluative terms. The standards are not strong enough to guide such application. Basing our judgements on them we can only say that one programme was *accepted* while the other *receded into the background*; we cannot add that the acceptance was *rational*, or that a rational development took place. Thus, the essays are interesting contributions to the *history*, or the *sociology* of science no matter how hard the authors try to present them as something different. The reader must not be misled by the frequent and rather assured use of terms such as 'rational' or 'irrational' which suggests that the authors have some deeper insight into the historical process. Using the methodology of research programmes they cannot have such insight, as I have tried to show. Of course, they have their *preferences*, they know what kinds of things they 'would like to have' and they defend these things with religious fervour. But neither they nor the authorities on whom they rely have succeeded in turning this fervour into a rational procedure.

Let us therefore from now on regard the case studies as historical studies and let us evaluate them on that basis. We see at once that they are superior to earlier studies of the same kind. The procedure is always the same. First, a certain historical episode is identified: the Copernican Revolution (not in this volume); the Einsteinian Revolution; the Chemical Revolution of Lavoisier and its altercation with the phlogiston theory; the rejection of Young's version of the wave theory of light; the battle between phenomenological thermodynamics and the kinetic theory of heat. Then follow explanations of the episodes that have played a role in the literature. Some of these explanations are mere narratives, with con-

[52] See below, §5.
[53] This volume, p. 32.
[54] This volume, p. 34.
[55] Lakatos [1970], p. 178 – italics in the original.

ceptual connections, others are psychological, still others are methodological; they try to show how the events arose, in a 'rational' manner, as a result of the determined use of methodological rules. Among the methodologies surveyed are inductivism, naive falsificationism, conventionalism, and the views of Kuhn. Next comes the demolition of the traditional explanations. This demolition is almost always historical: the explanations omit important facts, they conflict with others (inductivism alone is removed by logical arguments). Finally, we have an account in terms of the methodology of research programmes. Guided by a complex methodology this account is richer and more sophisticated than the alternative accounts. It is history of ideas in the best sense of the word. It is *history* because it deals with facts. It is history of *ideas* because it shows conceptual connections between these facts. It is *sophisticated* history of ideas because it uses a rich inventory of conceptual tools (hard core, protective belt, heuristic, progressiveness, degeneration, monster adjustment, recovery of hidden lemmas, and so on) rather than relying on intuition in all cases. The researcher is equipped with instruments that aid him on his way and are open to inspection so that he can criticise them and replace them by better instruments. It is true – the wish of the writers to arrive at some 'objective' judgement has made these instruments overly intellectual: a researcher who wants to inquire into motivation, or sociological causes hardly receives any help. Even worse, he is discouraged from putting too much weight on causes of this kind. This explains the rather primitive sociology and psychology of some of the papers and the complete absence of any inquiry into when, how and why certain standards were accepted. But there are definite advantages when one compares the history with what other methodologists have to offer. An inductivist, for example, will consider a theory (or, rather, a 'logical reconstruction' of it), the 'evidence' (which again is a reconstruction of the complex experimental results that govern science), and the relation between the two. He has two abstract elements and he examines them irrespective of the historical surroundings in which they arose. Writing history he is interested in the 'rational' parts of science only which are again the elements and their relation. This is why inductivist history is so arid or, if richer, so lacking in conceptual penetration. A naive falsificationist is not much better, for all *he* wants is *some* evidence (no matter in what historical surroundings) that contradicts the theory. The methodology of research programmes, on the other hand, does not examine theories; it examines sequences of theories connected by hard cores, heuristics and intuitive attitudes not all of which need to be formulated explicitly. Already at this point it goes much deeper into the structure of a theory than do the rivals. Secondly, the methodology of research programmes does not examine research programmes by themselves: it examines them in comparison with other research programmes. So the investigation spreads and must ultimately reach every research programme at the period in question. The whole intellectual scene must be taken into account. In the sixteenth and seventeenth centuries this includes theology, Aristotelian physics and metaphysics, magic (neoplatonism), the philosophy of Paracelsus (which centred around medicine and chemistry), alchemy – all subjects which were studied with care by the great Newton himself. Then we must examine whether the predictions made were novel predictions, or whether they were repetitions of things already known. This means we must examine the way in which the research programme was originally introduced,[56] the expectations of the age, the 'accepted facts' and the relation of these facts to current theory. We must know a great amount of material that belongs to the history of ideas and that is often missing even from rather detailed *historical* accounts. Nor is this material just *aufgerafft.* It is collected with an aim in mind, described in terms adapted to this aim and thus essentially ideational. It is true that the basic scheme lends a certain uniformity to the procedure which in the hands of less gifted writers can introduce an element of boredom. It is also ironical to realise that the aim – to give a 'rational' account of developments – is never reached and that we are left with a historical narrative only.[57] But on the way to the aim history has been transformed to such an extent that a slight change in our standards, say from research programme standards to Hegelian standards, enables us to read it as a history of reason itself.[58]

(5) *Remaining lacunae can be explained by the rationalising tendency of the authors and their blind acceptance of modern science*

Still, the rationalising tendencies of the authors and their assumption of the excellence of science make themselves felt and are responsible for some lacunae even in the historical account. Let me mention two of them.

[56] Zahar writes: 'My redefinition of novelty amounts to the claim that *in order to assess the relation between theories and empirical data, one has to take into account the way in which a theory is built and the problems it was designed to solve.* This new criterion of novelty of facts also implies that the traditional methods of historical research are even more vital for evaluating experimental support than Lakatos had already suggested. The historian has to read the private correspondence of the scientist whose ideas he is studying; his purpose will not be to delve into the psyche of the scientist, but to disentangle the heuristic reasoning which the latter used in order to arrive at the new theory.' (This volume, p. 219.)

[57] Environmentalists therefore need not be intimidated by Peter Urbach's [1974]. What is shown in these papers is that the relation of environmentalism to the evidence is different from the relation of some versions of geneticism to the evidence. *This is all that is shown* though both terminology ('degenerating', 'progressive') and philosophical insinuations *create the impression* that the one type of relation is better, more 'scientific' than the other. My remarks in the text above make it clear that this is not so. Value judgements of this kind are completely arbitrary and subjective and nobody needs to be intimidated by them. But alas! The propagandistic genius of Lakatos has concocted a mixture of propaganda and sham argument that is only too difficult to unravel and so he will have his way, because of the power of his rhetoric. One might call this the *List der Unvernunft.*

[58] Cf. the introduction to Hegel's *Geschichte der Philosophie.*

All the authors assume that the defenders of rival research programmes (and perhaps even the onlookers) are influenced by, and proceeding in accordance with, the methodology of research programmes. They use this methodology as a measure of good science and they behave in accordance with the advice given in the standards. It has emerged that the standards are incapable of giving any advice – but let us now disregard this drawback. Let us assume that the standards do indeed favour progressive programmes (they don't just *state* their progressiveness) and condemn degenerating programmes. It is then assumed by the authors that all good scientists distribute praise and blame in accordance with the methodology of research programmes. They may not be able to give an account of the *principles* of this methodology, but they still proceed as if they held such principles to be the basis of research.

Now this assumption may have some plausibility in periods of peace and uniformity which are most likely ruled by a single methodological framework. It becomes improbable when we are dealing with developments such as the Copernican Revolution, or the rise of twentieth-century science, especially quantum theory. Lakatos has emphasised that his standards are based on outstanding science of 'the last two centuries'[59] and he has thereby conceded the possibility of different standards *before* this period. But in his joint paper with Zahar on the Copernican Revolution he contends that 'Copernicus' and Kepler's and Galileo's adoption of the heliocentric theory' are 'rationally explainable'.[60] 'The Copernican Revolution', the authors write,[61] 'became a great scientific revolution... simply because it was scientifically superior' where 'scientifically superior' means, of course, superior in terms of the methodology of research programmes. Even Aristarchus's claim should have been taken seriously because 'the geocentric programme had already heuristically degenerated'.[62] Similarly Peter Clark in summarising the result of his paper says 'It was the degeneration of these attempts to provide a foundation for thermodynamics in terms of some deeper theory, compared with the empirical progress of the phenomenological programme, which led to the elevation of thermodynamics into a "paradigm" of great science.'[63] This again assumes that the moves of contemporary scientists were motivated by an implicit understanding of the methodology of scientific research programmes. Is this assumption correct?

I do not think it is.

In the case of the Copernican Revolution we know that the methodology of research programmes, if it was used at all, was not the only methodology in existence. One influential methodology was connected with the slogan of *saving the phenomena*. The slogan presupposes a distinction between *basic*

[59] This volume, p. 23.

[60] Lakatos and Zahar [1976], p. 380.

[61] *Op. cit.*, p. 31.

[62] *Ibid.*, p. 372, footnote 53.

[63] This volume, p. 44.

physics and *auxiliary assumptions*. Basic physics describes the processes that are expected to occur in this world, auxiliary assumptions link the processes to the phenomena. In Aristotle basic physics is decidedly empirical, much more so than modern science could ever aspire to be. It starts from 'phenomena' which are either observed facts or assumptions of common sense,[64] develops terminology for their description and principles for their explanation[65] and incorporates 'new facts' in a 'degenerating' manner. This is regarded as an advantage: the fact that theory and phenomena are related in a manner described as 'degenerating' by Lakatos shows the truth of the principles and the adequacy of the terminology used. The auxiliary assumptions are eventually separated from basic theory and assembled into various disciplines (astronomy, optics, mechanics, and so on). The task of these disciplines is to save the phenomena, not to give a physical account of the processes (motions) that created them.[66] Attempts to *re-absorb* them into basic theory in a 'degenerating' way continue throughout the Middle Ages.[67] They continue with Copernicus who tries to find an arrangement of circular movements that agrees with the basic Aristotelian philosophy[68] to such an extent that it can again be regarded as an account of real motions leading to a coherent *system* of the world rather than as a set of independent *hypotheses* for the calculations of planetary phenomena. Neither the tradition of saving the phenomena nor the attempt to absorb the mathematical devices of this tradition into basic theory look for novel facts or novel explanations of existing facts in the sense of the methodology of research programmes. The former cannot be found for all facts are related to the same basic principles and formulated in their terms, the latter cannot be admitted for that would deny the existence of a stable basic theory. To a certain extent this is true even of Copernicus himself (though not of Rheticus, and certainly not of Kepler[69]). But if that is the case then the revolution that was started

[64] Owen [1961].

[65] For the role of principles in Aristotle cf. Wieland [1970]. Wieland makes it clear how principles in Aristotle are designed with the explicit purpose of achieving 'degenerating' adaptations of facts (or course, he does not use this terminology).

[66] Originally, in Aristotle and his immediate successors the task is to *give an account of phenomena* in terms of basic physics which in turn must be constructed in such a way that the phenomena can be accounted for. Later on basic physics is taken for granted and phenomena must be explained in its terms. This is how the idea of *saving* the phenomena (rather than giving an account of them) arises. Cf. Krafft [1965]. An account of phenomena deals with the nature of things. The auxiliary assumptions that are used to save the phenomena have no such pretensions. The distinction is prepared by *Physics* B 2.

[67] The attempts start with Ptolemy's *Planetary Hypotheses*, are continued by Arab astronomers of the eleventh and twelfth centuries who demand a realistic account of planetary motions and last until the sixteenth century when the system of Alpetragius is taken up by Purbach. Cf. Duhem [1915], vol. 2, pp. 130ff.

[68] Cf. *De Revolutionibus*, i, 5–8, with Birkenmaijer's comments in footnote 82ff of Klaus [1959]. Cf. also Zahar, this volume p. 254.

[69] 'Copernicus was not aware of his own riches' writes Kepler in the *Mysterium Cosmographicum*, chapter 1, footnote 4, by which he means that Copernicus was not aware

by Copernicus's hypothesis was much more dramatic than a transition, *within the methodology of research programmes*, from one research programme to another. It brought in new standards and thus constituted a true paradigm change in the sense of Kuhn.[70]

A second historical lacuna is created by the authors' insistence on the 'objective' character of their evaluations: 'objectively' Copernicus (Einstein, the phenomenologists, etc.) progressed, hence it was 'rational' to follow their lead. (I must again repeat that the standards of the methodology of research programmes are not strong enough to permit us to make such a judgement of rationality; let us forget this drawback now that we are dealing with another difficulty.) But we have seen above[71] that 'correct' actions ('correct' in the sense of the methodology of research programmes) are only rarely carried out for the 'correct' reasons. It is therefore quite possible that the Copernicans, wanting to do one thing, ended up doing another. (Newton wanted to achieve certainty, what he did achieve was unceasing progress.) In this case they acted rationally (in the sense of the methodology of research programmes) for irrational (again in the sense of the methodology of research programmes) reasons and their rationality was a matter of luck, of accident, of the propitious collaboration of irrational causes (Hegel called this the *List der Vernunft*). The authors, concentrating on the internal features of a perhaps quite fortuitous result, leave such possibilities unexamined and thereby add to the false impression of the overwhelming 'rationality' of modern science.

6 *Judgements of progress and of degeneration are often arbitrary, for they depend on arbitrary selection of the research programmes to be compared. Altogether the appraisals made by the authors are arbitrary in at least five different ways. Once more it becomes clear that science needs and uses a plurality of standards and that scientists work best without any authority, the authority of 'reason' included*

In §3 I argued that the standards of the methodology of research programmes are not strong enough to recommend particular actions or to characterise such actions as being rational, or irrational. We may describe the heuristic of a research programme, we may say that the programme progresses, or degenerates, we cannot infer that it may be retained, or that

(*a*) that he was dealing with a research programme rather than with a single theory and (*b*) that this research programme was capable of producing novel predictions. One novel prediction, mentioned by Kepler, is that the synodic anomaly of the planets depends on the *true* motion of the sun not on its mean motion as had been assumed by Ptolemy.

[70] The Aristotelian methodology of finding a point of view that could accommodate facts by degenerating absorption was not the only alternative in existence. Neoplatonists and reformers of magic such as Agrippa von Nettesheim emphasised the *hidden virtues* of objects that were not accessible to normal observation and had to be brought forth by special methods. In our period they may well have been the only thinkers to come close to the methodology of scientific research programmes. Cf. Kocher [1953].

[71] Cf. text to footnotes 15ff.

it must be abandoned, or that continued work on it is rational, or irrational. It was then assumed that at least the judgements of progress and degeneration themselves are made in an unambiguous and objective manner. This is not always the case.

To see this, consider two rival research programmes, R, and R'. According to the methodology of research programmes both R and R' are plagued by anomalies. Let r and r' be corresponding sub-programmes of R and R' and let the anomalies of R and R' be so distributed that r progresses while r' degenerates. A follower of the methodology of research programmes will then support r over r'. It seems that the authors of some of the papers are in this situation prepared to support R over R' as well. But in doing so they may go against their own principles for it is quite possible that R' progresses while r' degenerates and r progresses.

As an example, let R' be the Aristotelian cosmology, r' Ptolemaic astronomy, r the astronomy of Copernicus and R a dynamical research programme consisting of r and suitable dynamical principles. Lakatos and Zahar have shown that r progresses while r' degenerates. Now they are either instrumentalists or realists. In the first case they will remain content with what they have shown. In the second case they will regard the motions of r as real motions and infer that R' must be abandoned. But R' continued to progress long after Copernicus as can be seen from the work of Harvey.[72]

Similar remarks apply to Zahar's comparison of the research programmes of Lorentz and Einstein. We have to distinguish L and E, the research programmes of Lorentz and Einstein, L' and E', the programmes that are compared by Zahar, L^t and E^t, the *theories* that are usually regarded as the decisive rivals in 1905, and L'' and E'', those parts of the programmes that deal with inertial spacetime only. L^t and E^t are often asserted to be equivalent, and Zahar repeats that judgement.[73] He is mistaken,[74] but the mistake does not matter. Not theories, but research programmes are decisive. Lorentz's programme is said to consist of Maxwell's equations, Newton's laws of motion (with Galilean transformations) and the Lorentz force.[75] Zahar never mentions the hard core of Einstein's programme though he mentions a programme, E'', that contains the relativity principle together with the principle of the constancy of

[72] Pagel [1967], [1969]. Cf. also Schmitt [1974]. Schmitt's problem is: what is the reason for the 'dogged persistence of the Aristotelian tradition in the sixteenth and seventeenth centuries'? (p. 171). His answer: the ability of the system 'to adapt itself and to absorb within itself many novel elements' (p. 178). The success of such absorption in the eyes of the contemporaries indicates that the methodology of research programmes was not universally taken as a basis of evaluation. In addition there were progressive developments.

[73] This volume, p. 250.

[74] The Lorentz contraction involves real forces and should therefore lead to oscillations. No oscillations are to be expected on Einstein's account. No oscillations were found in the experiment of Wood, Tomlinson and Essen, *Proc. Roy. Soc.*, **158** (1937), p. 606.

[75] This volume, p. 215.

c.[76] In 1905 this programme started degenerating while Lorentz's programme L' was advancing,[77] and had been advancing for quite some time. Now Zahar wants to explain, in a rational way, 'why...brilliant mathematicians and physicists like Minkowski and Planck abandon[ed] the classical programme in order to work on Special Relativity'[78] and he also wants to show that the success of general relativity in the explanation of the perihelion of Mercury was 'a success for the *whole* relativistic programme'.[79] To achieve both aims he replaces E'' by E' in the following manner:[80] c is used not for 'empirical' reasons (it is not used in a degenerating way), but because of Einstein's belief in (*a*) the basic nature of Maxwell's equations[81] and (*b*) their limited validity.[82] The principle of the constancy of light velocity is all that can be salvaged of (*a*) in view of (*b*), which means that it is a fundamental principle not only because of its position in the theory, but because of the nature of things. Thus E'' no longer degenerates, but it does not seem to advance either. To explain its acceptance by Planck, Minkowski and others in an 'internal' way, i.e. without recourse to psychosociology, Zahar turns to heuristics. To preserve continuity with the general theory of relativity and to refute Whittacker's conjecture that the ether programme was 'developed, into the Relativity Programme,[83] he concentrates on this heuristic to the exclusion of any hard core.[84] This brings him very close to E, the always unknown research programme underlying all of Einstein's activities and thus he seems to move in the right direction. But E is still compared with L', the *truncated and frozen* programme of Lorentz, and L, which contains atomism as well as the possibility of a more fundamental explanation of electromagnetic phenomena is never considered. But L can take care of all the facts which E'' obtains from the relativistic formulation of the electrodynamics of media.[85] It yields the constancy of c as a contingent fact and is in this respect closer to the general theory of relativity than E'' and E' where the constancy of c is a basic law.[86] And its heuristics are at least as

[76] This volume, p. 245.

[77] The light principle 'is thrown out with no justification whatever' (*ibid.*), while Lorentz 'explained Michelson's result in a non-*ad hoc* way...and he explained the invariance of c' (this volume, p. 236). Also 'there was no build up of unsolved anomalies which Einstein's theory dissolved better than Lorentz's' (this volume, p. 251).

[78] This volume, p. 251.

[79] This volume, p. 262.

[80] This volume, p. 248, footnote 94.

[81] This volume, p. 247.

[82] This volume, p. 248, footnote 94.

[83] This volume, p. 252.

[84] This volume, p. 265.

[85] $E = mc^2$ as well as the one-sided nature of electromagnetic emission was obtained by Poincaré in 1900 without invoking the relativistic point of view. Cf. *Archives Néerland*, **5** (1900), p. 252. Hasenöhrl arrived at a more restricted result four years later. It is quite true that Lorentz himself gives no indication 'that the rest mass is a variable quantity' (this volume, p. 262), but we are not talking about Lorentz; we are talking about his research programme.

[86] Cf. Einstein's comparison between 'constructive theories' such as the theory of Lorentz and 'theories of principle' such as the special theory of relativity in his autobiographical notes. (Einstein [1951], p. 53.)

powerful as the heuristics of E'', for every law produced by a research programme can, of course, be used in the heuristics of that research programme.[87] We see: the choice of research programmes and rivals is fairly arbitrary, and so are the judgements that rest on them. But these judgements are the basis for Zahar's 'objective' or 'internal' appraisal of the actions of Planck, Minkowski and others.

Such appraisals (and the corresponding appraisals in the other papers of this volume) therefore turn out to be arbitrary, or subjective, or 'irrational' in at least four different ways. They are arbitrary, because they proceed from an arbitrarily chosen authority: science of 'the last two centuries' (see above, §2). Science is chosen not because its excellence has been shown by argument, but because everybody is impressed by it.[88] They are arbitrary because it is not really science that decides the issue – science is much too chaotic for that – but a streamlined image of it, and there are no independent arguments for the principles of streamlining chosen. Thirdly, the appraisals are arbitrary because the standards that are obtained via steps one and two are not strong enough to support judgements of rationality, or irrationality. Any such judgement is independent of the standards, it receives no authority from the standards, there are no other arguments in its favour, and so it is arbitrary, or subjective in a very strong sense of the word. Fourthly, the appraisals are arbitrary because they rest on an arbitrary choice of rival research programmes. In the case of Zahar we find still another source of arbitrariness, and it is rather amusing: Planck and Minkowski start working on relativity, Planck and Minkowski are great scientists, hence their actions must be explained in an 'objective' manner. But there were many great scientists who either rejected the theory, or did not pay any attention to it. As a matter of fact, 'It was only in Germany that the theory was elaborated upon.'[89] How are the actions of these dissenters to be explained? Are they o be explained in the same way as the acceptance of the theory by Planck and Minkowski? Not likely. Are they to be explained 'internally'? This would mean that a theory has advantages as well as disadvantages, that different people look at it in different ways and come to different results though they use the same set of standards. It does not seem that Zahar would accept such an explanation. But then the only way out is to admit that the dissenters acted 'irrationally', and for external reasons. But if *they* can act irrationally, then why not Planck and Minkowski? This is the fifth type of arbitrariness found in the case studies. It is surprising to see

[87] Thus the derivation on pp. 259ff. of Elie Zahar's paper might have been carried out by a Lorentzian, though he would have given a very different interpretation to its result. Planck himself, who carried out the derivation, always spoke of the 'Lorentz–Einstein theory'.

[88] 'Is it not *hubris* to try to impose some *a priori* philosophy of science on the most advanced sciences?. . . I think it is.' See above, p. 318.

[89] Goldberg [1970], vol. 2, p. 97.

that a philosophy that makes such a fuss about 'rationality' and 'objectivity' should possess so abjectly little of either.

Zahar tries to show that Planck and Minkowski acted rationally when deciding to work on Einstein's programme and is thus faced with the problem of the rationality of those who stayed either with Lorentz, or with ether models. He neither states the problem, nor does he indicate how he would solve it. Peter Clark, in his essay on the kinetic theory, perceives an analogous problem. Towards the turn of the century the kinetic theory 'was subject to severe attacks from some of the leading scientists of the day'.[90] This is not supposed to be due to purely philosophical preferences, such as a preference for positivism. All these are 'external explanations'.[91] The correct explanation, according to Clark, is that 'in the last decades of the nineteenth century [the kinetic research programme] was *degenerating* [while] from the two laws of thermodynamics a number of startling novel facts were deduced'.[92] Was it therefore irrational to work on the kinetic theory? Not at all! The kinetic theory had a heuristic, there are means 'of systematically *improving* the theory'[93] while the phenomenological theory 'lacks a heuristic'. It is therefore also 'rational'[94] to try to make the latter supersede the former.

Now it would seem that such an attempt is rational only if a heuristic is *needed*. The kinetic theory which ran into one difficulty after another definitely needed a heuristic. The phenomenological theory which was applied to an ever increasing domain of problems without ever failing to live up to its promise (this is Clark's description of the situation, not mine!) does not. It is too good to be in need of becoming a research programme. Besides, it was this very universality of the theory, this absence of models that made it so attractive to Einstein. Einstein[95] distinguished between 'constructive theories' which move through various stages and gradually conquer one problem after another and 'theories of principle' which remain valid no matter how far they are extended, and he preferred the latter, mentioning thermodynamics as an outstanding example. He viewed his own theory of relativity as a theory of principle, not as a constructive theory. We see: Clark does not succeed in explaining the rationality *both* of those who objected to the kinetic theory and of those who continued working on it. *At least one party must be criticised as being irrational.* Given the methodology of research programmes it is arbitrary which party we choose.[96]

This result can be generalised. Research programmes are superseded

[90] This volume, p. 42.

[91] This volume, p. 43. This definition of 'external' makes, of course, many of Einstein's reasons external.

[92] This volume, p. 91.

[93] This volume, p. 75.

[94] This volume, p. 90.

[95] Cf. footnote 86.

[96] Peter Clark no longer believes the phenomenological theory to be without any heuristic. But, he says 'that heuristic was a weak one in a very specific sense, namely that it was *fact dependent* in much the same way that the Ptolemaic heuristic was.' (Letter of March 26, 1975.)

by other research programmes only because people are not invariably impressed by success and progressiveness. They are not impressed by success either because they have standards different from those accepted by the proponents of the 'successful' programme, or because they are moved by external considerations. In the first case we have a plurality of *standards*, in the second case a plurality of *motives*. *Science as described in the papers in this volume* cannot exist without such a plurality and where it seems to succeed it does so only because certain problems have been overlooked (the problem of the opponents of relativity in the case of Zahar) or because of the biased terminology (talk of 'rationality' or 'objectivity' when no rational grounds for such talk have been given). The methodology of research programmes most certainly had led to some interesting historical discoveries. This is not surprising. Any hypothesis, however implausible, can widen our horizon. It has not led to a better understanding of science and it is even a hindrance to such a better understanding because of its habit of beclouding facts with sermons and moralising phrases.

References

Benacerraf, P. and Putnam, H. (eds.) [1964]: *The Philosophy of Mathematics.*

Duhem, P. [1915]: *Le Systeme du Monde.*

Einstein, A. [1951]: Autobiographical section of P. A. Schilpp (ed.): *Albert Einstein: Philosopher-Scientist.*

Feyerabend, P. K. [1975]: *Against Method.*

Feyerabend, P. K. [1976]: *Einführung in die Naturphilosophie.*

Forsdyke, J. [1964]: *Greece before Homer.*

Goldberg, S. [1970]: 'In Defense of Ether', R. McCormmach (ed.): *Historical Studies in the Physical Sciences*, **2.**

Harnack, A. [1961]: *History of Dogma.*

Klaus, G. (ed.) [1959]: *Copernicus ueber Kreisbewegung.*

Kocher, P. H. [1953]: *Science and Religion in Elizabethan England.*

Körner, S. (ed.) [1957]: *Observation and Interpretation.*

Krafft, F. R. [1965]: 'Der Mathematikos und der Physikos – Bemerkungen zu der angeblichen Platonischen Aufgabe, die Phaenomene zu retten', in *Alte Probleme – Neue Ansaetze. Drei Vortraege von Fritz Krafft.*

Lakatos, I. [1970]: 'Falsification and the Methodology of Scientific Research Programmes', in I. Lakatos and A. Musgrave (eds.): *Criticism and the Growth of Knowledge.*

Lakatos, I. and Zahar, E. [1976]: 'Why did Copernicus's research programme supersede Ptolemy's?' in R. Westman (ed.): *The Copernican Achievement.*

Lenin [1968]: 'Backward Europe and Advanced Asia', *Collected Works*, **19.**

Medvedev, Z. A. [1969]: *The Rise and Fall of T. D. Lysenko.*

Mitchell, A. D. (ed.) [1974]: *Psychic Explanation, A Challenge for Science.*

Musgrave, A. [1976]: 'Method or Madness? – Can the Methodology of Research Programmes be rescued from epistemological Anarchism?', to appear in the Imre Lakatos Memorial Volume of *Boston Studies in the Philosophy of Science* (page references refer to the circulated manuscript).

Ostrander, S. and Schroeder, L. [1970]: *Psychic Discoveries behind the Iron Curtain*

Owen, G. E. L. [1961]: 'Tithenai ta phainomena in *Aristote et les problèmes de méthode.*

Pagel, W. [1967]: *William Harvey's Biological Ideas.*
Pagel, W. [1969]: 'William Harvey Revisited', *History of Science*, **8**, pp. 1–31.
Popper, K. R. [1947]: *The Open Society and its Enemies*, 2 volumes.
Sagan, C. [1975]: *The Cosmic Connection.*
Schmitt, C. B. [1973]: 'Towards a Reassessment of Renaissance Aristotelianism', *History of Science*, **11**, pp. 159–93.
Urbach, P. [1974]: 'Progress and Degeneration in the "I.Q. Debate"', *British Journal for the Philosophy of Science*, **25**, pp. 99–135, 235–59.
Waters, F. [1963]: *Book of the Hopi.*
Watkins, J. [1975]: Position Paper: 'Criteria of Scientific Progress, a Critical Rationalist View' (mimeographed).
Wieland, W. [1970]: *Die Aristotelische Physik.*
Worrall, J. [1975]: Position Paper: 'Criteria of Scientific Progress, A critical Rationalist View' (mimeographed.)

Index of Names